CONTRIBUTORS

John B. Adams
James R. Aronson
J. A. Bastin
Peter M. Bell
Roger G. Burns
Alfred G. Emslie
J. E. Geake
A. A. Green
W. P. Griffith
Arthur L. Hiser
Graham R. Hunt
Clarence Karr, Jr.
John J. Kovach
Homer C. Liese
Lloyd M. Logan
Robert P. Lowndes
R. J. P. Lyon
H. K. Mao
Clive H. Perry
M. J. Pugh
George R. Rossman
John W. Salisbury
David J. Vaughan
G. Walker
William B. White

Infrared and Raman Spectroscopy of Lunar and Terrestrial Minerals

---

Edited by CLARENCE KARR, Jr.

U.S. Department of the Interior
Bureau of Mines
Morgantown Energy Research Center
Morgantown, West Virginia

ACADEMIC PRESS New York San Francisco London 1975

A Subsidiary of Harcourt Brace Jovanovich, Publishers

ACADEMIC PRESS, INC.
111 Fifth Avenue, New York, New York 10003

United Kingdom Edition published by
ACADEMIC PRESS, INC. (LONDON) LTD.
24/28 Oval Road, London NW1

Library of Congress Cataloging in Publication Data

Karr, Clarence.
Infrared and Raman spectroscopy of lunar and terrestrial minerals.

Includes bibliographies and index.
1. Mineralogy, Determinative. 2. Spectrum analysis.
3. Lunar petrology. I. Title.
QE369.S65K37 549'.133 74-17975
ISBN 0-12-399950-2

PRINTED IN THE UNITED STATES OF AMERICA

CONTENTS

7 Reflectance and Emittance of Terrain in the Mid-Infrared (6–25 μm) Region

R. J. P. Lyon and A. A. Green

8 Selected Terrestrial Minerals and Their Infrared Absorption Spectral Data (4000–300 cm^{-1})

Homer C. Liese

9 Far-Infrared Spectroscopy of Minerals

John J. Kovach, Arthur L. Hiser, and Clarence Karr, Jr.

10 Far-Infrared Observation of Lunar Rock

M. J. Pugh and J. A. Bastin

LIST OF CONTRIBUTORS

Numbers in parentheses indicate the pages on which the authors' contributions begin.

JOHN B. ADAMS* (91), Planetary Astronomy Laboratory, Department of Earth and Planetary Sciences, Massachusetts Institute of Technology, Cambridge, Massachusetts

JAMES R. ARONSON (143), Arthur D. Little, Inc., Cambridge, Massachusetts

J. A. BASTIN (255), Physics Department, University of London, Queen Mary College, London, England

PETER M. BELL (1), Geophysical Laboratory, Carnegie Institution of Washington, Washington, D.C.

ROGER G. BURNS (39), Department of Earth and Planetary Sciences, Massachusetts Institute of Technology, Cambridge, Massachusetts

ALFRED G. EMSLIE (143), Arthur D. Little, Inc., 15 Acorn Park, Cambridge, Massachusetts

J. E. GEAKE (73), Physics Department, University of Manchester, Institute of Science and Technology, Manchester, England

A. A. GREEN (165), Mineral Physics Division, Commonwealth Scientific and Industrial Research Organization, North Ryde, Australia

W. P. GRIFFITH (299), Department of Chemistry, Imperial College of Science and Technology, London, England

ARTHUR L. HISER (231), U.S. Department of the Interior, Bureau of Mines, Morgantown Energy Research Center, Morgantown, West Virginia

GRAHAM R. HUNT (117), Optical Physics Laboratory, Air Force Cambridge Research Laboratories, Hanscom Air Force Base, Bedford, Massachusetts

CLARENCE KARR, JR. (231), U.S. Department of the Interior, Bureau of Mines, Morgantown Energy Research Center, Morgantown, West Virginia

JOHN J. KOVACH (231), U.S. Department of the Interior, Bureau of Mines, Morgantown Energy Research Center, Morgantown, West Virginia

* Mailing address: West Indies Laboratory, Fairleigh Dickenson University, St. Croix, U.S. Virgin Islands.

Homer C. Liese (197), Department of Geology, University of Connecticut, Storrs, Connecticut

Lloyd M. Logan (117), Optical Physics Laboratory, Air Force Cambridge Research Laboratories, Hanscom Air Force Base, Bedford, Massachusetts

Robert P. Lowndes (273), Department of Physics, Northeastern University, Boston, Massachusetts

R. J. P. Lyon (165), Department of Applied Earth Sciences, Stanford University, Stanford, California

H. K. Mao (1), Geophysical Laboratory, Carnegie Institution of Washington, Washington, D.C.

Clive H. Perry (273), Department of Physics, Northeastern University, Boston, Massachusetts

M. J. Pugh (255), Physics Department, University of London, Queen Mary College, London, England

George R. Rossman (1), Division of Geological and Planetary Sciences, California Institute of Technology, Pasadena, California

John W. Salisbury (117), Optical Physics Laboratory, Air Force Cambridge Research Laboratories, Hanscom Air Force Base, Bedford, Massachusetts

David J. Vaughan (39), Department of Geological Sciences, University of Aston, Birmingham, England

G. Walker (73), Physics Department, University of Manchester, Institute of Science and Technology, Manchester, England

William B. White (325), Materials Research Laboratory and Department of Geosciences, The Pennsylvania State University, University Park, Pennsylvania

PREFACE

As one of the Principal Investigators in NASA's Lunar Sample Program, I was aware that reports of research on infrared and Raman spectroscopy of lunar minerals and their terrestrial counterparts were buried in a very large number of papers from other scientific disciplines and of additional unreported spectroscopic research on terrestrial minerals that developed from the impetus and inspiration provided by investigation of lunar samples. The purpose of this volume is to make available in a single reference work original descriptions and summaries of the research on infrared and Raman spectroscopy of lunar and terrestrial minerals so that this information will be readily available not only to those researchers in the continuing programs on lunar samples from the completed Apollo series and on the remote sensing of solar system objects, but, in particular, to that much larger group of researchers in government, industry, and universities involved in the many programs on terrestrial minerals and earth sciences by remote sensing.

Sometimes a multiauthor book can be merely a hodgepodge of unrelated chapters. In this volume, however, not only has an attempt been made to arrange the chapters in logical progression, but the authors have written their contributions with the contents of all of the chapters in mind. As complete a description of the present status of this field of research as is practical is presented. To establish a broadly useful work, maximum use of cross references is made (over eighty altogether) and a comprehensive Subject Index included.

The chapters are arranged according to spectroscopic technique and/or frequency range rather than application, although Chapters 4–7, dealing largely with remote sensing, are grouped together. Thus there are chapters on visible and near-infrared, followed by those on mid-infrared, far-infrared, and Raman spectroscopy. Applications are roughly divided between lunar and terrestrial, although the broad range of interchangeability of applications is obvious in many instances.

This theme of applicability to either the theoretical or practical aspects of the study of minerals runs through all the chapters, sometimes subtly, sometimes boldly, but always as a practical result evolved from the infrared or Raman analysis of lunar samples or from parallel developments in the analysis of terrestrial samples. Thus there are chapters on remote sensing of space targets and earth sciences, chapters on lunar mineralogy and terrestrial mineralogy and geology, and chapters on structures of lunar minerals and structures of terrestrial minerals, with these various topics and related subjects listed in the Subject Index.

My heartfelt thanks go to all the contributors for their unfailing interest, enthusiasm, and cooperation in the preparation of this volume without which it would not have been possible to make this reference work available to the scientific community. I also wish to thank Beverly R. Goudy for her assistance with correspondence and other matters involved in the preparation of this book.

Clarence Karr, Jr.

INFRARED AND RAMAN SPECTROSCOPY OF LUNAR AND TERRESTRIAL MINERALS

1

ABSORPTION SPECTROSCOPY OF IONIC AND MOLECULAR UNITS IN CRYSTALS AND GLASSES

*Peter M. Bell, H. K. Mao, and George R. Rossman**

I. Introduction

All minerals have regions of strong absorption of electromagnetic radiation in the ultraviolet, visible, or infrared portions of the spectrum,

* Contribution #2400, California Institute of Technology.

which arise from electronic and vibrational transitions of the various ions in the minerals. The study of the ultraviolet spectra of minerals is a comparatively neglected subject because the absorption bands in the ultraviolet are so intense that it is experimentally difficult to prepare samples thin enough to transmit the amount of light necessary for measurements of the spectra. The study of the infrared spectra of minerals is an important and expanding field, and is the topic of several of the following chapters. The region in between, the visible and near-infrared portions of the spectrum, extending from 400 to 2500 nm (0.4–2.5 μm), is the spectral region discussed in this chapter.

The absorption of light in the visible region extending from 400 to 700 nm (25,000–14,300 cm^{-1}) is responsible for the attractive colors of the minerals. The study of processes from which the colors originate is a challenging and complex aspect of mineralogy. The quantitative study of the color of minerals by observing spectroscopic events is directed both to the understanding of the basic physical phenomena that govern the behavior of atoms of the mineral and to the application of the data to problems of geochemical and geophysical interest. Although color remains a useful indicator of processes taking place when light interacts with the atoms in a mineral, the spectral studies must extend into the near-infrared spectral region because the same electronic processes that give rise to color as perceived by the human sensory system also give rise to "color" in the near-infrared.

Several factors influence the spectra of minerals. Some of these relate to the chemical composition and atomic structure of the mineral, whereas others relate to the external conditions to which the mineral is subjected. In this chapter several important factors that influence the spectra of minerals will be examined.

The most important factor necessary for color in silicate minerals is the presence of transition metal ions such as iron, manganese, chromium, and titanium. Iron is the most abundant and widespread transition element in terrestrial minerals and in the returned lunar samples. It is also one of the most important sources of color in these materials. Accordingly, much of the discussion in this chapter will be directed to the understanding of the role of these elements in mineral spectroscopy. Important rock-forming minerals and several less common silicates that illustrate important principles of mineral spectroscopy will be considered. The opaque oxides and sulfides such as magnetite and pyrite will not be included in this discussion of the spectra of ions in minerals because the spectra of these minerals do not contain features characteristic of the individual constituent ions. The optical properties of these substances, reflecting the extensive delocalization of their electrons, are better treated with regard to band gaps in semiconducting crystals.

Metal ions such as those of iron, chromium, and manganese can exist in multiple oxidation states. Each oxidation state of a metal ion has a spectrum different from those of the other oxidation states. Because of this factor, optical spectra can be used to identify the oxidation state of ions in minerals. In terrestrial minerals iron commonly occurs as both Fe^{3+} and Fe^{2+}, whereas in the lunar samples it occurs predominantly as Fe^{2+} and metallic iron, with little spectroscopically significant Fe^{3+}. The occurrence in one lunar sample of a small quantity of the Fe^{3+} mineral akaganeite, β-FeO(OH), is thought to have resulted from oxidation of a reactive Fe^{2+} chloride in the sample after it was brought to earth (Taylor *et al.*, 1973).

A third important factor is the coordination number of the metal ion in the mineral. Fe^{2+} is found in coordination numbers from four to eight, six-coordination being most commonly encountered. Another important factor, related to the coordination number, is the coordination geometry. Iron in four-coordination can occur in both tetrahedral and square-planar geometries. Geometries of low symmetry also commonly occur in minerals. Such differences in coordination geometry have profound effects upon the spectra of the mineral.

Nearest neighbor and next-nearest neighbor chemistry are also important. The discussion in this chapter is limited to metal ions bound to oxygen ions, although there is ample evidence to show that where metal ions are bound to a different ion, such as chloride or sulfide, this difference in local chemistry will produce significant changes in the position of absorption bands in the spectrum. It is important to consider whether the metal ion is bonded to an oxygen ion derived from a silicate group, a sulfate, or a hydroxide group. Such differences will change the covalency of the metal–oxygen bond, which in turn will be reflected in the spectroscopy. Next-nearest neighbor chemistry can be especially important in the case of two metal ions. An iron ion sharing an octahedral edge with another iron ion can behave quite differently from the same ion sharing an octahedral edge with a titanium ion. In practice, many of these factors will simultaneously enter into the spectroscopy of any given mineral to subtly influence its color.

II. Theoretical Background: Ti^{3+}

The absorption bands in the visible and near-infrared portions of the spectrum occur because of transitions of the valence electrons among the various 3d orbitals in the transition metal ions in the crystals. The concepts are illustrated with Ti^{3+} in an octahedral crystallographic site. The 3d orbitals are energetically degenerate in the absence of a crystalline environment (Fig. 1a) but when placed in the crystal are raised in their overall energy by the electrostatic field of the neighboring atoms (Fig. 1b) and also split by the octahedral symmetry of that environment into a set of two orbitals at somewhat higher energy and a set of three orbitals at somewhat lower energy (Fig. 1c). The energy difference between these two sets of orbitals is symbolized by Δ_0, the subscript 0 denoting that the orbitals were split by an octahedral electrostatic field. Ti^{3+} has one d electron, which would reside in one of the three degenerate lower energy orbitals. For convenience, the two sets of orbitals in the octahedral electrostatic field are designated e_g and t_{2g}. The symbols t, e, and g describe the degeneracy and symmetry properties of the orbitals in three-dimensional space. The symbol t refers to a triply degenerate orbital, and e to a doubly degenerate orbital. Additionally, the symbols a and b are used for low-symmetry environments to denote nondegenerate orbitals. Other symbols, such g and 2g, denote the presence of combinations of symmetry elements, such as mirror planes and inversion centers, in the distribution of the electrons in space. These symbols are derived from the formal procedures of group theory described in reference texts (Cotton, 1971).

When the lone electron of Ti^{3+} resides in the lower t_{2g} orbital, the system comprised of the titanium ion and its immediate environment has a

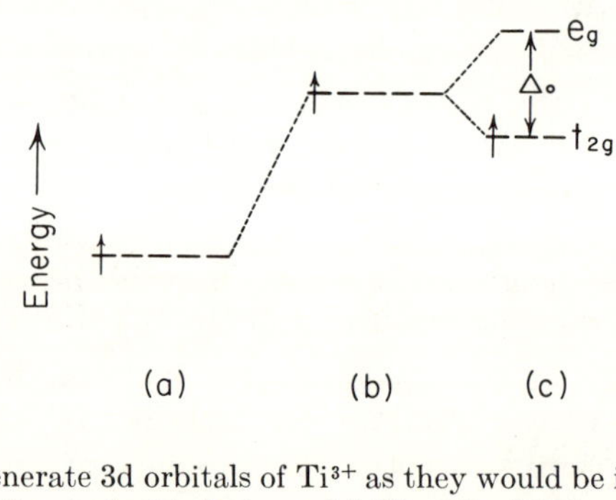

FIG. 1. (a) Five degenerate 3d orbitals of Ti^{3+} as they would be in the case of Ti^{3+} gas. The arrow symbolizes the single 3d electron. (b) The 3d orbitals raised in energy by the electrostatic field of a crystalline environment. (c) The degeneracy of the 3d orbitals partially removed by the octahedral symmetry of the electrostatic field at the crystal site of the Ti^{3+} ion. (d) Energy states of Ti^{3+} in an octahedral site. The T_{2g} state corresponds to the single 3d electron residing in the t_{2g} orbitals and the energetically higher E_g state to the electron in the e_g orbitals. The arrow symbolizes the electronic transition between the two states.

particular energy. Under the influence of incident light of the appropriate wavelength the titanium electron will be excited into the e_g orbital with concurrent absorption of the light. This is the process that gives rise to absorption bands and color. The system, titanium and immediate environment, will then have more energy than the initial system. The energy difference corresponds to the energy of the incident photon that was absorbed. On the average, this energy difference will be equal to Δ_0. In order to understand mineral spectra, one must ultimately be concerned with the various energy states of the system rather than the energy of the orbitals.

A diagram of the energy states of Ti^{3+} in an octahedral environment is shown in Fig. 1d, where the ground state of the system is designated $^2T_{2g}$ and the excited state is designated ^2E_g. The symbols that designate states are obtained from the same group theory rules that are used to designate orbitals (Cotton, 1971). To distinguish states from orbitals, the states are designated in upper-case letters and the orbitals are designated in lower-case. Conveniently, for Ti^{3+}, although not in general, the names of the states are the same as the orbitals occupied in each state. To classify the states further, the spin multiplicity of the state is added as a superscript. The transition from the $^2T_{2g}$ state to the ^2E_g state is indicated by the arrow in Fig. 1d. The absorption bands that arise from the transitions of electrons within the d orbitals are referred to as d–d bands or ligand-field bands, the latter name derived from the name of the theory used to analyze them.

Ligand-field theory predicts that octahedral Ti^{3+} would have one electronic transition and, consequently, just one absorption band. The wavelength at which this transition is centered is not accurately predicted by this theory but must instead be determined empirically. Studies of Ti^{3+} compounds indicate that this absorption is usually found near the center of the visible spectrum. Once this position is determined empirically, parameters can be evaluated that can be used with the theory to predict the change in this position in response to changes in the chemical and structural environment.

Claims of the occurrence of Ti^{3+} in minerals are few and often speculative in nature. Silicate glasses containing Ti^{3+} can be prepared in the laboratory under carefully controlled oxygen fugacities (Mao and Bell, 1973). The optical spectrum of one of these glasses (Fig. 2) shows the single absorption band from Ti^{3+} centered at about 550 nm (18,200 cm^{-1}). Another useful calibration point is provided by a synthetic Ti^{3+} pyroxene, $NaTiSi_2O_6$ (Prewitt *et al.*, 1972), for which a single absorption band at about 640 nm (15,600 cm^{-1}) has been reported. The occurrence of Ti^{3+} in a pyroxene

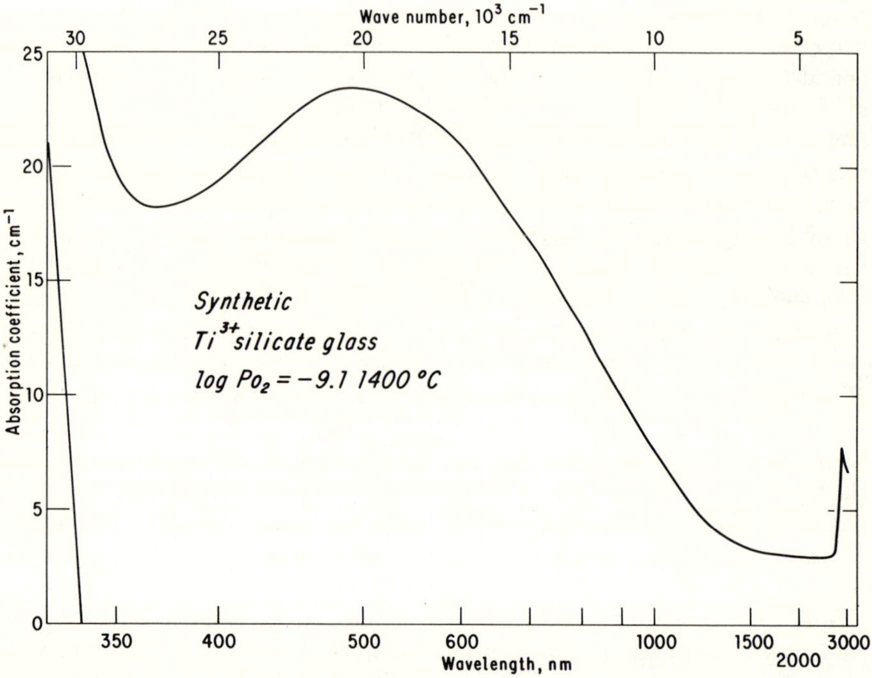

FIG. 2. Absorption spectrum of Ti^{3+} in a silicate glass of the composition (weight %) SiO_2, 50.85; Al_2O_3, 8.19; TiO_2, 11.66; CaO, 9.99; MgO, 19.31; where the ratio $Ti^{3+}/(Ti^{3+} + Ti^{4+}) = 0.46$. The absorption band at 530 nm corresponds to the $T_{2g} \rightarrow E_g$ electronic transition of the Ti^{3+} ion.

crystal obtained from the Allende meteorite has been reported (Dowty and Clark, 1973; Chapter 2, Fig. 1), and also from terrestrial minerals such as titanaugite (Burns and Fyfe, 1967; Manning and Nickel, 1969). The spectra of these minerals are more complicated than would be expected from isolated Ti^{3+} in an undistorted octahedral environment (see Chapter 2, Section III, A, 1, b). To examine the effects of distortions, the discussion now turns to iron, an element better characterized spectroscopically than titanium.

III. Iron

A. Fe^{2+}: Near-Octahedral

1. Background

Fe^{2+} has six valence electrons to be distributed among the five d orbitals. When the iron is located in a perfectly octahedral crystallographic site the d orbitals will be split into a t_{2g} set and an e_g set as in the titanium example previously considered. In the ground state of the system the electrons will be distributed one per orbital with one of the t_{2g} orbitals containing the sixth electron in a spin-paired configuration (Fig. 3a). In the light absorption process the sixth, spin-paired electron is excited into the upper set of e_g orbitals so as to preserve the total spin alignment (Fig. 3b). Electronic transitions in which the total number of spin-up and spin-down electrons are unchanged are called spin-allowed transitions.

Again the primary concern is the states of the system rather than the transitions of electrons between orbitals. The ground state of the octahedral Fe^{2+} system, like the Ti^{3+}, is a T_{2g} state and the excited state is an E_g state. When the spin multiplicity is included, the ground and excited states become $^5T_{2g}$ and ^5E_g, respectively (see Chapter 2, Fig. 6). One spin-allowed transition occurs between the ground and excited states, namely, the $^5T_{2g} \rightarrow {}^5E_g$ transition.

2. Lunar Glass

The coordination polyhedra of Fe^{2+} are invariably distorted from regular octahedral geometry. If the distortions are small enough,* the spectrum will have an absorption band in the 1000 nm (10,000 cm^{-1}) region, arising from the $^5T_{2g} \rightarrow {}^5E_g$ transition, which is a characteristic feature of near-octahedral Fe^{2+} in oxygen coordination. The Fe^{2+} ions in the green lunar glass are in sites that are reasonably close to octahedral. The salient spectral features shown in Fig. 4 consist of a relatively featureless absorption band in the near-infrared centered at 1077 nm (9300 cm^{-1}), a region of maximum transmission in the visible centered at 575 nm (17,400

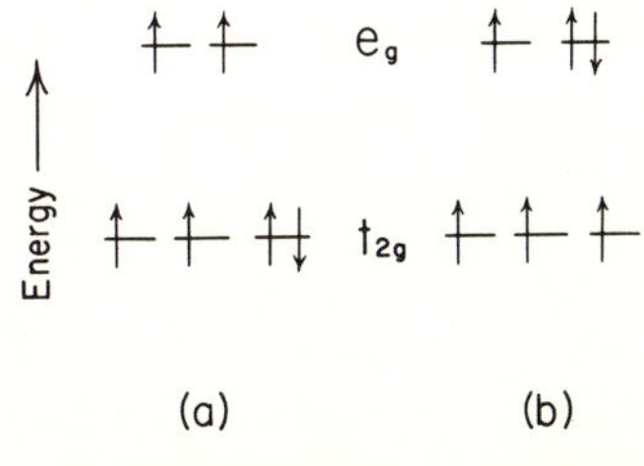

FIG. 3. Electron occupancy of the d orbitals in octahedral Fe^{2+} before (a) and after (b) excitation to an excited state by absorption of light. The electrons are symbolized by arrows whose up-down orientation represents the alignment of the electron spin. Because the excitation from the ground state to the excited state preserves the total spin alignment it is a spin-allowed transition.

*Generally, this means bond distances within 0.1 Å and bond angles within 5° of 90°.

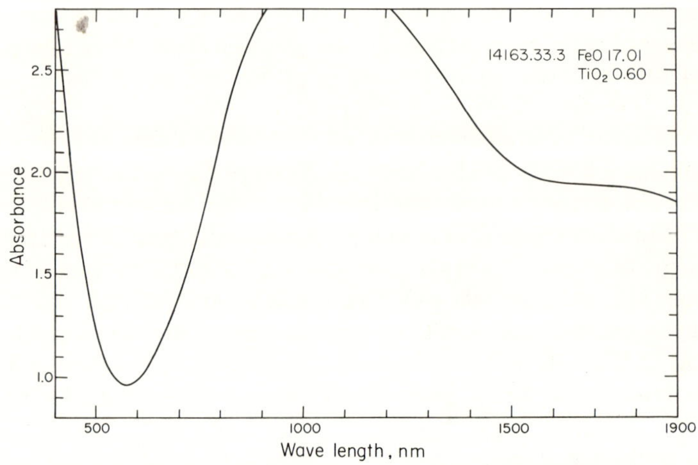

FIG. 4. Optical spectrum of green lunar glass showing the absorption band centered at 1077 nm arising from the $^5T_{2g} \rightarrow {}^5E_g$ transition of octahedral Fe^{2+}.

cm^{-1}), and an absorption feature beginning in the visible and rising to great intensity in the ultraviolet. The color of this glass is the result of the transmission of the blue-green, green, and yellow in the 500–600 nm region.

3. *Crystalline Minerals*

The spectra of green, crystalline, ferromagnesian silicates are characterized by dominant absorption bands in the near infrared. The spectrum of olivine is an example (Fig. 5a). The prominent absorption in the 800–1400 nm (12,500–7,100 cm^{-1}) range is again evident, but owing to the crystalline environment in which the iron is located, significant structure develops in the 1000 nm (10,000 cm^{-1}) feature. A hint of such structure was already apparent in the spectrum of the green lunar glass appearing in the vicinity of 1820 nm (12,550 cm^{-1}).

Because olivine is an important mineral in terrestrial mineralogy, considerable effort has been devoted to understanding the details of its spectroscopy (Burns, 1970a,b; Runciman *et al.*, 1973b). The analysis of the spectrum is complicated by the fact that the iron can occupy two sites with different crystallographic environments. The two sites, M1 and M2, have different iron–oxygen bond lengths, bond angles, and distortions from octahedral symmetry. The models used generally account for the splitting of the 1000 nm feature into resolvable components and for the observed polarization of the spectra of the 1000 nm features.

Other weak absorption features that were not observed in the spectrum of the lunar glass appear as comparatively narrow absorption bands in the olivine spectrum between 300 and 700 nm. These low-intensity absorption bands are called spin-forbidden bands. Their origin is discussed in the following section. As will be presented in subsequent sections, these features are important because they respond to changes in both the pressure and temperature of the system.

All silicate minerals that contain Fe^{2+} in a reasonably octahedral environment of oxygens have absorptions in the near-infrared similar to those of olivine. The exact wavelength at which the absorption maximum occurs depends in part upon the iron–oxygen bond distance and the nature of the other atoms (such as aluminum, silicon, or beryllium) bound to the oxygen. One of the more extreme shifts from the 1000 nm region is found

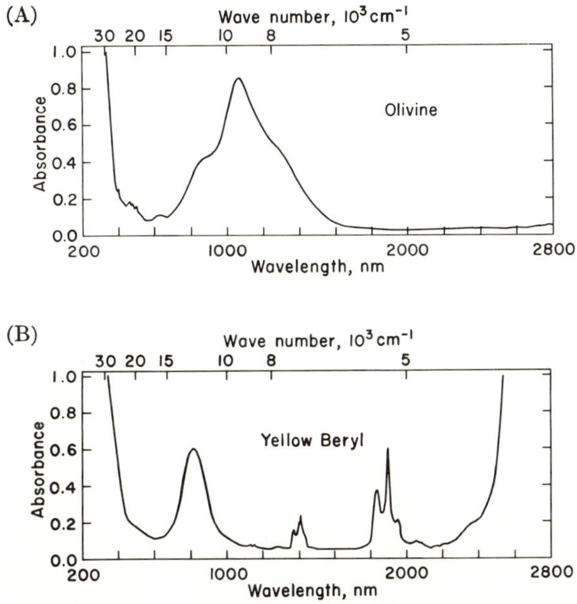

FIG. 5. A. Spectrum of an olivine crystal, $(Mg,Fe)_2SiO_4$, showing the absorption due to Fe^{2+} in the near-infrared region around 1000 nm. Unpolarized light incident upon the ac plane, 0.85 mm thick. B. Spectrum of a yellow beryl, $Be_3Al_2Si_6O_{18}$, showing the Fe^{2+} absorption at 820 nm. The features near 1400 nm are the first overtones of the symmetric and antisymmetric stretching vibrations of molecular water in the channels of the beryl. The features near 1900 nm are combination modes of the stretching and bending motions of the water. Unpolarized light looking down the c axis, 0.9 mm thick. C. Spectrum of the orthopyroxene bronzite, $(Mg_{0.84}Fe_{0.16})SiO_3$, from Balme, Norway, showing the two prominent bands at 901 and 1829 nm in the near infrared due to Fe^{2+} in the C_{2v} low-symmetry site. Unpolarized light incident upon the $\alpha\beta$ plane, 0.71 mm thick. D. Absorption spectrum of plagioclase feldspar from Lake County, Oregon, containing both Fe^{3+} and Fe^{2+}.

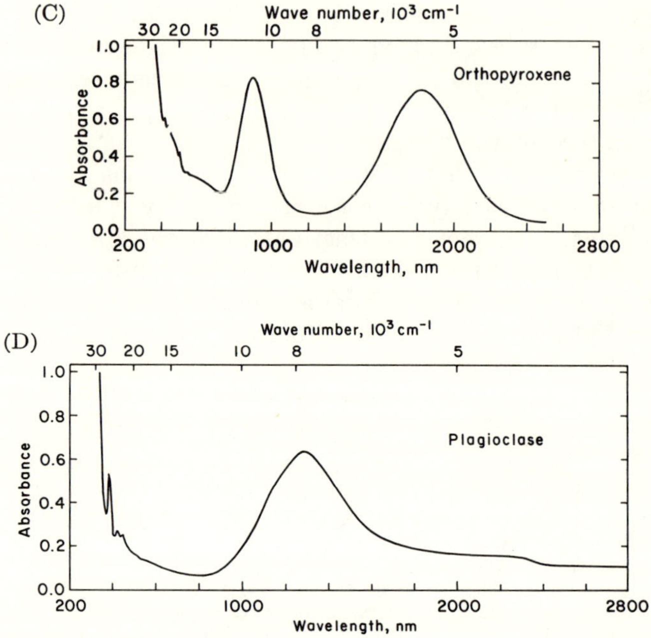

FIG. 5. C and D.

in the spectrum of yellow beryl (Fig. 5B), which can have the Fe^{2+} absorption maximum as low as 810 nm (12,300 cm^{-1}) (Wood and Nassau, 1968).

B. Spin-Forbidden Bands

The spin-forbidden bands are low intensity absorption bands resulting from electronic transitions from the ground state to excited states that have a spin multiplicity different from that of the ground state. They can be considered to arise from transitions in which an electron changes its spin-up/spin-down orientation. Spin-forbidden bands are generally about two orders of magnitude less intense than spin-allowed bands. Nevertheless, they are often observed as low-intensity features, frequently at higher energies than the spin-allowed bands. Unfortunately, it is not possible to portray both simply and accurately the distribution of electrons in the system after most spin-forbidden transitions. Instead, one must be content to speak of them somewhat abstractly in the language of spectroscopic states.

C. Low-Symmetry, Six-Coordinate Fe^{2+}

The coordination polyhedra of common ferromagnesian silicates can deviate significantly from octahedral symmetry. A low-symmetry site invariably influences the appearance of the absorption spectrum by removing the degeneracy of the electronic states of the system, resulting in a greater number of transitions. Simply stated, the complexity and structure of absorption spectra tend to increase as the symmetry of the local environment is lowered. The symmetry can be lowered by geometric structural factors that change bond angles and bond lengths from their idealized octahedral values, or the symmetry can be lowered by chemical factors, such as differences in next-nearest neighbor populations, which alter the electrostatic environment and the covalency of bonds. The geometric factors, which are generally more important, are illustrated in the case of orthopyroxene. In low-iron orthopyroxenes, $(Mg,Fe)SiO_3$, the majority of the iron is situated in a crystallographic site close to C_{2v} symmetry, which can be thought of as an octahedron with one edge stretched out from the center (Ghose, 1965; Morimoto and Koto, 1969). The two e_g orbitals, which are energetically degenerate in octahedral symmetry, have different energies in the C_{2v} symmetry. Similarly, the three t_{2g} orbitals which were degenerate in octahedral symmetry, also lose their degeneracy. In Fe^{2+} with six d electrons there is a one-to-one correlation between the removal of the degeneracy of orbitals and the establishment of new energy states of the metal ion system. It is conceptually useful to think of the electronic states of this system as being derived from the states of an octahedral system but having been split by the lower symmetry. Figure 6 indicates that the E_g state, like the e_g orbitals, is a degenerate state that will split into components, A_1 and B_1, when the symmetry is reduced to C_{2v}, as it is in the pyroxenes. Likewise, the T_{2g} state can be split into three states in the C_{2v} symmetry. The three electronic transitions that occur in C_{2v} symmetry are indicated by the arrows. The optical spectrum of the low iron orthopyroxenes has been analyzed by Bancroft and Burns (1967), White and Keester (1967), and more recently by Runciman *et al.* (1973a). The spectrum (Fig. 5C) is dominated by two intense absorption bands at 901 nm (11,100 cm^{-1}) and at 1829 nm (5530 cm^{-1}). With reference to the energy level scheme in Fig. 6, the absorption bands would correspond to the $^5A_1 \rightarrow {}^5A_1$ and $^5A_1 \rightarrow {}^5B_1$ transitions, respectively. Burns (1965) has shown that the wavelength of the 900 nm absorption band varies smoothly as the composition of the orthopyroxene varies from $MgSiO_3$ to $FeSiO_3$. Other weaker absorption bands that occur in the 400–600 nm region are spin-forbidden bands.

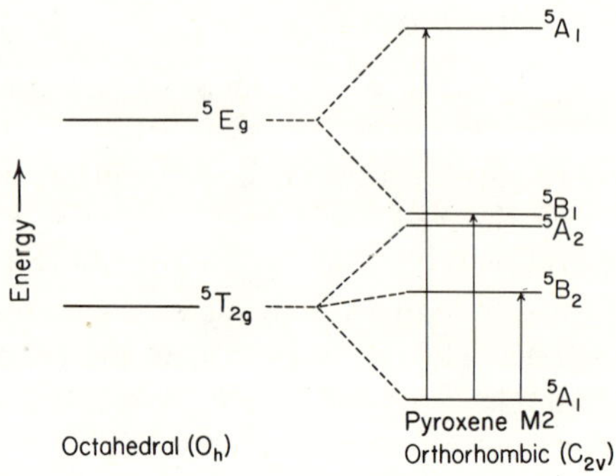

FIG. 6. Electronic energy states of Fe^{2+} in the C_{2v} crystallographic site in low-iron orthopyroxene minerals according to the analysis of Runciman *et al.* (1973a). Arrows indicate the electronic transitions allowed by the selection rules under C_{2v} symmetry. The two absorption bands at 901 and 1829 nm in Fig. 5C correspond to the two highest energy transitions, $A_1 \rightarrow A_1$ and $A_1 \rightarrow B_1$, respectively.

D. *Tetrahedral* Fe^{2+}

Another coordination geometry that occurs about Fe^{2+} is that in which the iron is situated in the center of a tetrahedron of four oxygens. The tetrahedral coordination is designated T_d. If the effects of changes in iron–oxygen bond distance and the covalency of the bond are ignored, the energy of the peak of the absorption band for T_d Fe^{2+} is predicted by ligand-field theory to be four-ninths that of octahedral Fe^{2+}. When the bands at approximately 1000 nm (10,000 cm^{-1}) of the lunar glass and olivine are used as typical examples of octahedral Fe^{2+} spectra, the tetrahedral bands are expected to occur in the vicinity of 2000 nm (5000 cm^{-1}) and to have no effect on the visible color of the minerals that contain Fe^{2+} in this geometry.

The spinels containing low concentrations of Fe^{2+} [the (Mg,Fe)Al_2O_4 spinels, in particular] provide examples of tetrahedral Fe^{2+} spectroscopy. These spinels, which occur in both terrestrial and lunar rocks, are rich in Fe^{2+} but also contain Fe^{3+}, Cr^{3+}, Cr^{2+}, and Ti^{4+} as transition metal cations. Figure 7 shows absorption spectra of Fe^{2+} and Cr^{3+} in lunar and terrestrial spinels. Strong absorption by T_d Fe^{2+} dominates the spectra, as would be predicted, because the transition is allowed from both spin and symmetry considerations. A theoretical analysis of these spectra, based in part on spectra obtained at liquid helium temperatures, has been presented (Slack *et al.*, 1966).

E. Isolated Octahedral Fe^{3+}

The ligand-field bands of Fe^{3+} arise entirely from spin-forbidden transitions. Consequently, Fe^{3+} minerals are pale colored as long as other types of absorption processes are not occurring simultaneously. A nearly ideal example of the Fe^{3+} spectrum is observed in andradite garnet. This spectrum of the gem variety, demantoid, is shown in Fig. 8A. The pattern of two broad, weak bands at lower energy followed by a comparatively sharp band at 440 nm (22,700 cm^{-1}) is the characteristic spectrum of an isolated Fe^{3+} ion in regular octahedral coordination. The spectrum of octahedral Fe^{3+}, free of the interfering interactions that will be discussed later in this chapter, is seldom encountered.

F. Tetrahedral Fe^{3+}

Fe^{3+} occurs in four-coordination but only in tetrahedral environments. Tetrahedral Fe^{3+} has been best characterized in the feldspars, where it substitutes for aluminum into the aluminosilicate framework, imparting a yellow color whose intensity depends on the extent of substitution.

(A)

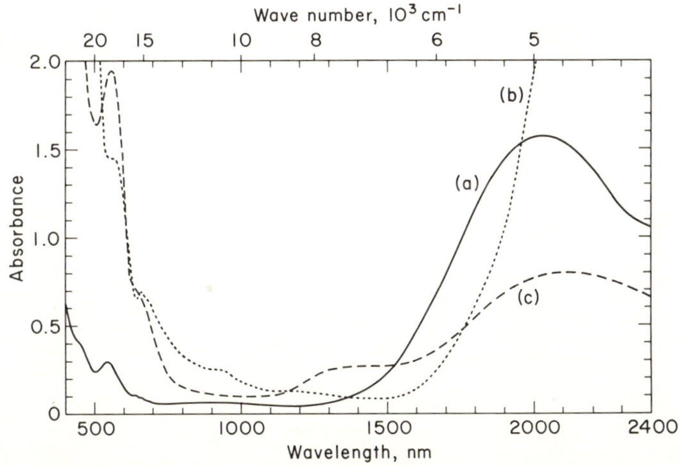

FIG. 7 A. Crystal-field spectra of iron, chromium, titanium spinels: (a) Apollo 17, (b) Philippines, (c) South Africa. The main feature is the absorption band at 2000 nm caused by tetrahedral Fe^{2+}. Other bands are tetrahedral Cr^{2+} at 1300 nm, octahedral Fe^{2+} at 950 nm, octahedral Cr^{3+} at 550 nm, and the strong absorption tail caused by Fe^{3+}. B. Minor bands in the spectra shown in Fig. 7A. The bands in (a) are caused by tetrahedral Fe^{2+}; those in (b) are caused by tetrahedral Fe^{2+} at 685 nm and tetrahedral Fe^{3+} at 655 nm; the broad band in (c) is caused by octahedral Cr^{3+}.

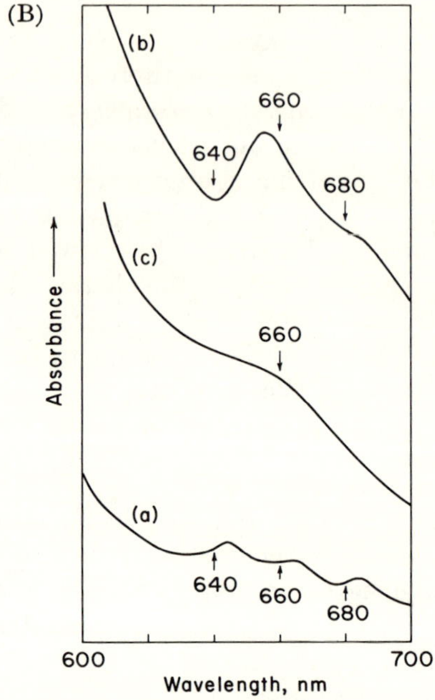

FIG. 7. B.

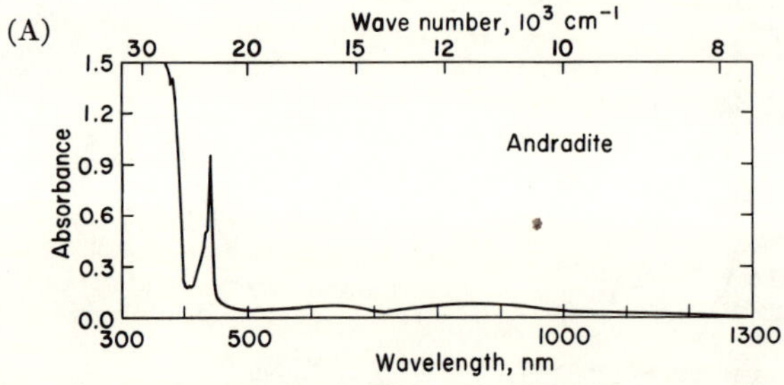

FIG. 8 A. Absorption spectrum of yellow-green andradite garnet, $Ca_3Fe_2(SiO_4)_3$, from Val Malenco, Italy, showing the spin-forbidden bands of Fe^{3+}. The sharp band at 441 nm arising from the $^6A_{1g} \rightarrow (^4A_{1g}, {}^4E_g)$ transition is useful for identifying small amounts of ferric iron in a variety of minerals; 0.45 mm thick. B. Absorption spectrum of orthoclase feldspar, $KAlSi_3O_8$, from Itrongahy, Madagascar, containing tetrahedral Fe^{3+}. Unpolarized light incident upon the ab face; 17.5 mm thick. C. Absorption spectrum in the xz plane of goethite, α-FeO(OH), showing the charge-transfer tail extending less into the visible wavelengths than that of the hematite shown in Fig. 8D; 0.015 mm thick. D. Absorption spectrum of red hematite, α-Fe_2O_3, showing the charge-transfer tail extendtending well into the visible portion of the spectrum; 0.015 mm thick, (0001) cleavage fragment.

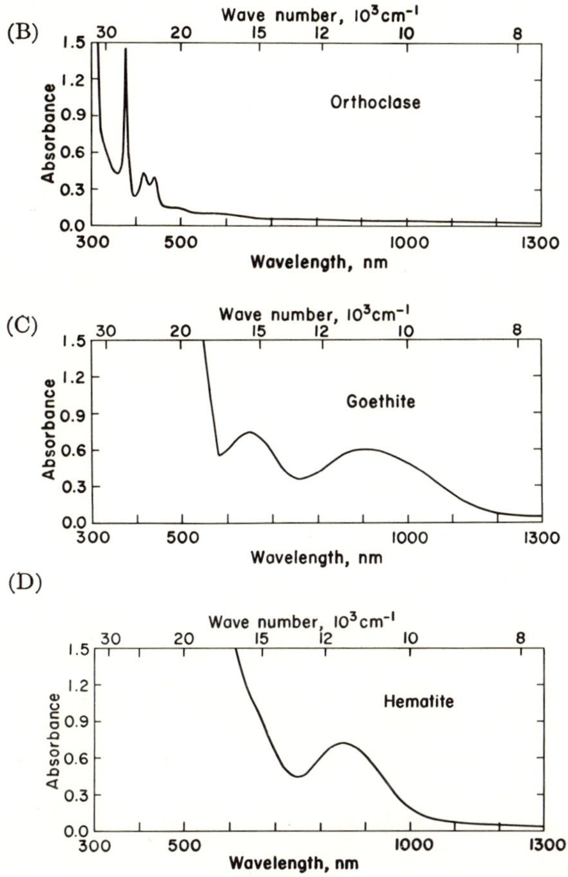

FIG. 8. B, C, and D.

Although tetrahedral iron is commonly found in both the alkali and plagioclase feldspars, the yellow color is seldom observed in crystals other than those of optical quality because of scattering from exsolved phases or grain boundaries. Orthoclase feldspar contains predominantly tetrahedral Fe^{3+}, whereas the plagioclase feldspars contain comparable amounts of Fe^{2+} in the calcium sites, which absorbs in the near-infrared (Hafner *et al.*, 1971).

The spectrum of tetrahedral iron in orthoclase is shown in Fig. 8B. This spectrum has been discussed by Faye (1969) and Manning (1970), who consider alternative, more detailed interpretations of the origin of these bands. The sharpness of the 376 nm (26,600 cm^{-1}) feature makes it useful for determining low concentrations of ferric iron in feldspars in spite of its relatively low intensity. Figure 5D illustrates the spectrum of a plagioclase feldspar containing both tetrahedral Fe^{3+} and Fe^{2+} in the irregular calcium site. The sharp 376 nm feature in orthoclase appears at 383 nm (26,100 cm^{-1}) in plagioclase but is still easily recognized. For example, 5 ppm of Fe^{3+} can be reliably determined in a 1-cm thick plagioclase crystal of suitable clarity. Studies of lunar plagioclase have failed to yield any evidence for ferric iron by this technique, placing the upper limit of Fe^{3+} in the samples studied at 0.001 weight % (for a further discussion of Fe^{3+} in lunar plagioclase, see Chapter 3, Section III, C). The absorption features between 1000 and 1800 nm in Fig. 5D are due to the Fe^{2+}.

IV. Charge-Transfer Bands

A. General Considerations

The origin of the intense absorption in the ultraviolet should be considered at this time in conjunction with the spectra of minerals. Electrons localized predominantly on the oxygen ions that surround the iron can shift their distributions toward the iron under the influence of incident radiation to enter an excited state of the iron–oxygen system wherein part of the electron density around the oxygens has been transferred to the iron. This process is referred to as a charge-transfer transition, and the absorption bands produced by this process are called charge-transfer bands. In the case where the electron density was transferred from the oxygen to the iron, the corresponding absorption band would be called an oxygen-to-metal charge-transfer band.

A charge-transfer band can be seen in Fig. 4, beginning at 550 nm (18,200 cm^{-1}) and rising off scale toward shorter wavelengths. It is important in determining the color of the green glass because it removes the violet and much of the blue. If the charge-transfer band were not present, this glass would have had a color close to blue. Charge-transfer bands are generally thousands of times more intense than the ligand-field bands that have previously been considered. However, because they are generally caused by high-energy electronic transitions, they reach maximum intensity in the ultraviolet portion of the spectrum. They are important in the discussion of mineral spectra because even though the absorption maximum is located well into the ultraviolet, the side wings of the band extend into the visible, commonly causing absorption of the shorter wavelengths and obscuring weaker features that may fall beneath the tail of the band.

Charge-transfer transitions can occur between any anion–cation combination, although many combinations require ultraviolet photons. All other factors being comparable, it is the trend that cations with higher charge favor oxygen-to-cation charge transfer. A high charge on the central ion is not a sufficient factor by itself to cause low-energy charge transfer; witness the fact that oxides of cations of high oxidation states such as TiO_2, SiO_2, and Al_2O_3 are colorless. Still, it is the case that charge-transfer phenomena are important in Fe^{3+} spectroscopy.

B. Next-Nearest Neighbors

Tails of charge-transfer bands that extend into the visible have a pronounced effect upon the spectra of iron-containing minerals. The effects of the charge-transfer tail are commonly observed in systems where an iron has another iron as its next-nearest neighbor. Two Fe^{3+} ions can be next-nearest neighbors by sharing corners, edges, or faces of the coordination polyhedra. The ions common to the two irons are usually oxide, hydroxide, or combinations of the two.

Goethite, α-FeO(OH), when chemically pure, is yellow-orange. The absorption spectrum of goethite (Fig. 8C) shows two ligand-field bands at 650 nm (15,400 cm^{-1}) and 900 nm (11,100 cm^{-1}). The dominant spectral feature, however, is the charge-transfer band that extends into the visible spectral region to about 600 nm (16,700 cm^{-1}). The goethite structure contains Fe^{3+} bridged by both oxide and hydroxide.

The charge-transfer tail is commonly observed to extend farther toward the red when ions bridging the shared polyhedra are oxide rather than hydroxide groups. Hematite, α-Fe_2O_3, is the classic example of this type of system. Hematite structurally consists of octahedra of ferric iron sharing edges and faces (Blake *et al.*, 1966) bridged by oxide ions. It is the common, widely distributed red pigment in soils and weathered rocks. The absorption spectrum of hematite (Fig. 8D) displays bands and shoulders arising from ligand-field transitions at 850 and 660 nm (11,800 and 15,200 cm^{-1}). The dominant spectral feature is the charge-transfer tail that extends into the edge of the near infrared. The small amount of light being transmitted in the 750 nm region is responsible for the deep red color of the mineral.

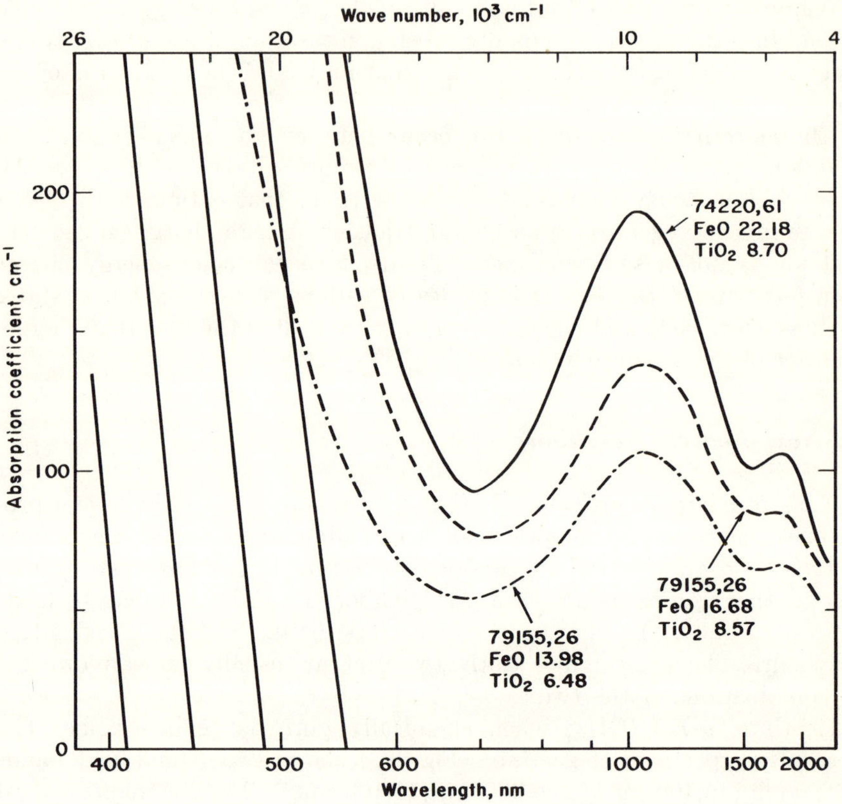

FIG. 9. Absorption spectrum of the orange lunar soil. The material consists of orange glass with a high titanium content. The color is the result of the region of maximum transmission at 700 nm bounded by the Fe^{2+} ligand-field band at 1030 nm and the charge-transfer tail extending into the visible wavelengths.

A special case of adjacent metal ions sharing a common corner, edge, or face of a coordination polyhedron is two neighboring iron ions in different oxidation states. A type of charge transfer known as intervalence charge transfer occurs in this case at energies much lower than the charge-transfer processes that have been discussed so far. The intervalence charge-transfer process, which is extremely important in terrestrial minerals, is considered in Chapter 2, Section III, B, 1.

The final type of next-nearest neighbor interaction that will be considered is the case of dissimilar transition metal neighbors. One of the most interesting combinations is the iron–titanium pair. In several systems the addition of titanium to an iron-containing silicate results in dramatic intensification of color as a result of the extension of a charge-transfer tail well into the visible or even the near-infrared spectral regions. The melanite variety of andradite garnet is one example. Amounts of titanium as small as 0.05 weight % cause the charge-transfer tail to move from the ultraviolet into the visible with an accompanying change in color from green to yellow-brown. When the concentration of titanium is raised to 2%, the mineral becomes black as the charge-transfer tail shifts into the near-infrared. These black, titanium-rich andradites are called melanite. The melanite spectra have been discussed by several authors (Dowty, 1971; Moore and White, 1971; Manning and Harris, 1970), who consider possible origins of the black color, ranging from trivalent titanium to displacement of ferric iron to the tetrahedral sites.

The spectrum of lunar glass is in many ways similar to that of garnet. Titanium-free glass is light green and has the characteristic spectrum of

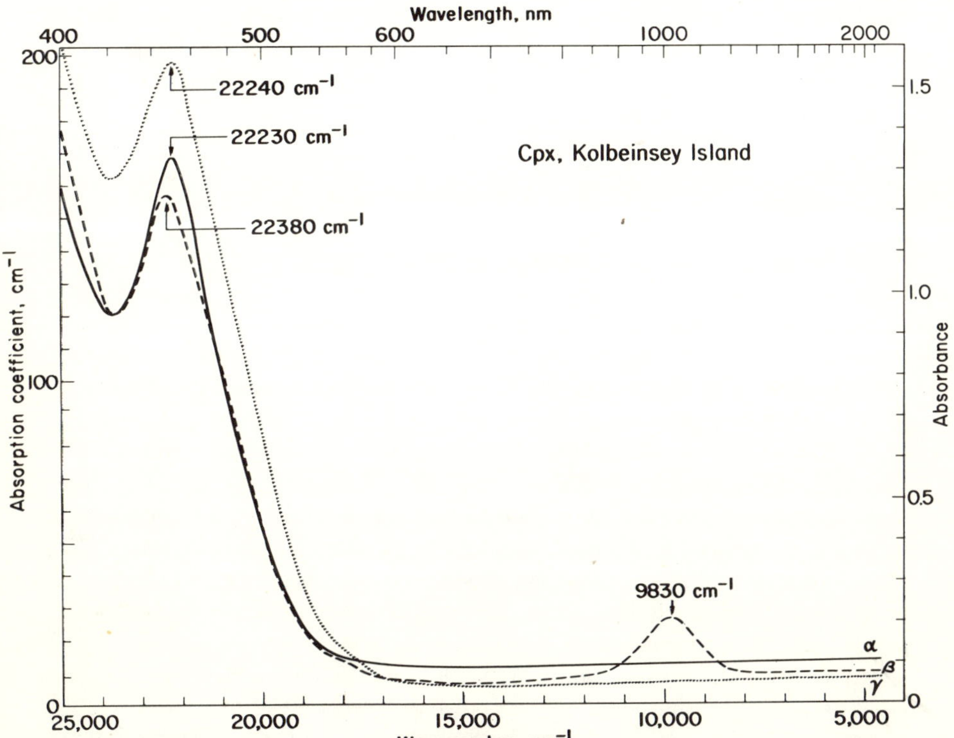

FIG. 10. Spectrum of clinopyroxene from Kolbeinsey Island showing octahedral Fe^{2+} band at 9830 cm^{-1} and charge-transfer band of Fe^{3+} near 22,300 cm^{-1}. Spectra are polarized in α, β, and γ directions.

Fe^{2+} shown in Fig. 4. As the titanium concentration is increased, the charge-transfer tail moves across the visible into the infrared (Fig. 9) and the glass becomes black. The orange soil of the Apollo 17 flight is actually an example of orange glass that, in addition to the features of Fe^{2+}, has its charge-transfer tail shifted by incorporation of titanium into the glass (Mao *et al.*, 1974).

The spectrum of jadeite, $NaAl(SiO_3)_2$, containing a few tenths of a percent of Fe^{3+} substituting for the Al^{3+} resembles the spectrum of andradite in Fig. 8A. The spectrum of a clinopyroxene containing tens of a percent of Fe^{3+} (Bell and Mao, 1972) is unlike the low-Fe^{3+} pyroxenes (Fig. 10). The strong absorption band at 450 nm (22,200 cm^{-1}) is probably caused by nearest neighbor interactions or may be caused by several overlapping bands. Many of the various factors are operating simultaneously to influence the color and spectra of minerals. The complex spectra of common minerals like the amphiboles and micas are the result of the simultaneous effects of multiple site occupancy for the metal ions, intervalence processes, and the charge-transfer process modified by titanium. So far, attempts to analyze these spectra are rendered ambiguous by the possibility of multiple causes for a particular observable feature.

V. Water and Hydroxide

Water and hydroxide ions are commonly constituents of terrestrial minerals. Neither of these species has electronic transitions in the visible or near-infrared, but the overtones and combinations of the vibrational motions of these two species are frequently observed in the near-infrared. Beryl provides a good example. The fundamental symmetric and asymmetric stretching motions of the water present as a nonessential impurity trapped in the open channels of the beryl structure occur at frequencies in the range 2700–2800 nm (3555 to 3694 cm^{-1}). The first overtones of these motions, which represent the water molecule vibrating twice as fast, occur in the near-infrared at twice the energy in the vicinity of 1400 nm (7150 cm^{-1}) (Fig. 5b). The combination of simultaneous bending and stretching of water molecules can also occur. The resulting combination band absorbs at the sum of the energy of the two component motions. In the beryl spectrum, the absorption bands near 1900 nm (5250 cm^{-1}) represent this combination, and the weak absorption bands near 1150 nm (8700 cm^{-1}) correspond to the combination of the first overtone of the stretching motion with the fundamental bending mode. Similar spectra of a variety of beryls have been analyzed by Wickersheim and Buchanan (1959, 1968) and Wood and Nassau (1967, 1968). Water and hydroxide ions commonly occur as essential constituents of minerals and are found to be ubiquitous trace impurities in minerals. Whereas study of trace amounts of these substances is best done with infrared spectroscopic techniques (Wilkins and Sabine, 1973,) overtones and combination modes will be commonly encountered in the near-infrared spectra of minerals such as amphiboles and micas, which contain significant amounts of these species. Some normally anhydrous minerals that are examined in relatively thick sections show significant hydroxide overtones. Many garnets, in particular the grossularite and spessartite varieties, show hydroxide overtones near 1400 nm.

VI. Spectroscopy at Low Temperatures

An important experimental method employed in the basic study of mineral spectra is cryogenic temperature spectroscopy. Samples cooled to liquid nitrogen (78°K) or liquid helium (4.2°K) temperatures commonly show significant improvements in resolution of spectral features because of bandwidth decreases accompanied by increases in peak intensities. Overlapping bands are experimentally resolved, weak features, such as spin-

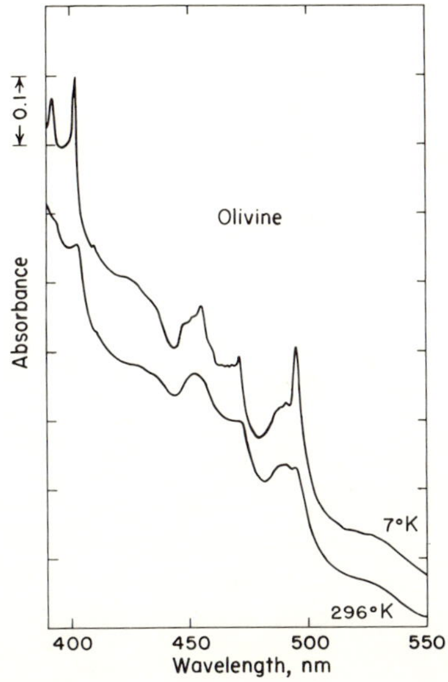

FIG. 11. Absorption spectrum of an olivine crystal from Arizona at room temperature, 296°K, and near liquid helium temperature, 5.5°K, illustrating the improved resolution of the overlapping spin-forbidden bands at low temperature; 1.90 mm thick sample, β-spectrum in polarized light.

forbidden bands, are more clearly seen, and bands buried under intense charge-transfer bands are revealed as the tail of the charge-transfer band recesses toward the ultraviolet.

Figure 11 shows a portion of the olivine spectrum at room temperature and liquid helium temperature. The numerous overlapping bands are more clearly resolved in the low-temperature spectrum. A complete analysis of the low-temperature olivine spectrum has been presented by Runciman *et al.* (1973b). The 5°K spectrum of gillespite, $BaFeSi_4O_{10}$, a mineral that contains square-planar Fe^{2+}, shows the improved resolution of the spin-forbidden bands superimposed upon the spin-allowed band near 500 nm (20,000 cm^{-1}) (Fig. 12). The shift of the 500 nm band maximum toward higher energies at low temperatures is in response to the contraction of the Fe–O bonds, which will tend to increase the electrostatic field at the iron owing to the surrounding oxygen ions. It is reasonable to hope that such temperature shifts in ligand-field bands would occur in the same direction as pressure-induced shifts, which also occur in response to a forced contraction of the crystal. For example, pressure-spectral studies

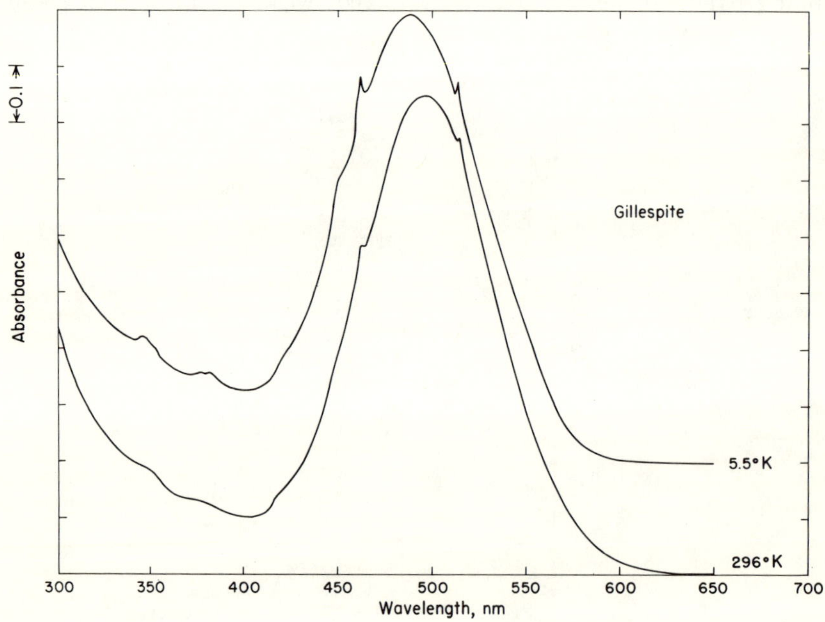

FIG. 12. Expanded scale absorption spectrum of gillespite from Incline, California, at 296° and 5.5°K, showing the improved resolution of the spin-forbidden bands. Unpolarized light incident upon a cleavage flake 0.67 mm thick.

of gillespite have shown that the 500 nm band also shifts toward higher energies at pressures up to 25 kbar (Abu–Eid *et al.*, 1973), as will be discussed in a later section.

A more dramatic change occurs in the spectrum of andradite garnet at 78°K. A portion of the spectrum of andradite garnet that was shown in Fig. 8 is given in expanded scale in Fig. 13. The $^6A_{1g} \rightarrow (^4A_{1g}, {}^4E_g)$ transition near 440 nm (22,700 cm^{-1}) and especially the $^6A_{1g} \rightarrow {}^4T_{2g}$ transition near 380 nm (26,400 cm^{-1}) undergo major improvement in resolution at low temperatures. Some of the features develop because structurally there are actually small deviations from perfect octahedral symmetry (Novak and Gibbs, 1971) that remove the degeneracy of the ^4T and ^4E states. When the distortions are small, the energy states derived from the ^4T and ^4E states are close enough in energy that thermal motion of the local environment will smear the components together. At cryogenic temperatures the thermal broadening effects are reduced to the point that the individual components are resolved.

(a)

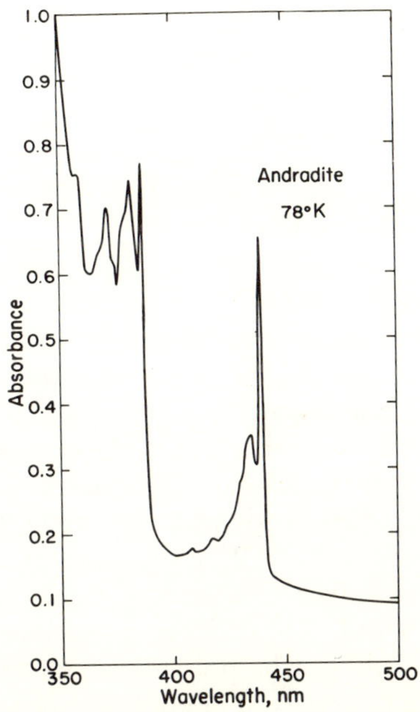

FIG. 13. Expanded scale absorption spectrum of the Val Malenco andradite garnet at (a) 78°K and (b) 296°K, showing the features that develop in the spin-forbidden Fe^{3+} bands at low temperature; 0.227 mm thick.

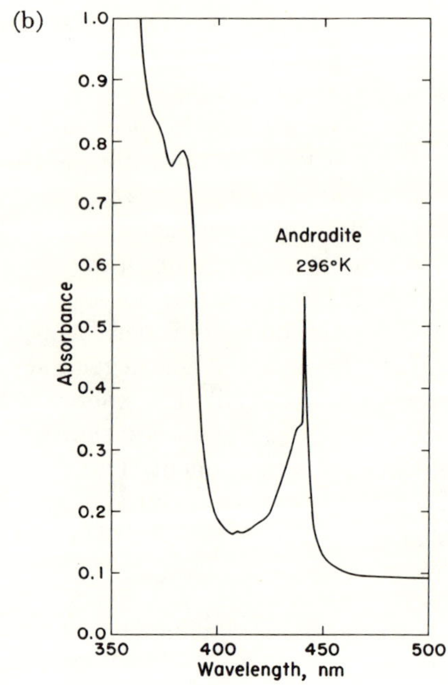

FIG. 13 b.

Spectroscopic study of minerals at high temperatures is not commonly practiced. By extrapolation of cryogenic and room-temperature data one would expect that absorption bands would widen out at high temperatures and that charge-transfer tails moving toward the infrared and near-infrared ligand-field (Fe^{2+}) bands would have greater absorption in the lower energy regions than at room temperature. Such changes in absorption of minerals at high temperatures would be important in the consideration of the overall radiative heat transport in the interior of the earth.

Not all minerals show dramatic changes upon cooling to cryogenic temperatures. For example, the 4.2°K spectrum of enstatite (Runciman *et al.*, 1973a) does not differ significantly from the room-temperature spectrum. Amphiboles also fail to undergo noticeable improvement at low temperatures. No firm rule can be presented for predicting the extent to which the spectrum of a mineral will improve at low temperature, although the following trends are evident. Minerals with high degrees of covalency in the bonding to the transition metal ions tend to show greater changes at low temperatures than the more ionic materials, such as orthoclase and olivine, which show small to moderate amounts of improved resolution at low temperatures.

VII. Spectroscopy at High Pressures

A. General Considerations

The same spectral features of the visible and near-infrared that produce colorful minerals on the earth's surface are immensely important in minerals that occur deep in the earth and other planets. In the earth's mantle, absorption of near-infrared energy affects the transfer of heat and the earth's temperature distribution. Temperature and pressure strongly affect electronic processes that cause spectral absorption in minerals and therefore also the processes that affect the thermodynamic behavior of minerals. As a result, most chemical processes and most physical properties such as thermal, electrical, and magnetic properties, and those involving the strength of the earth are interrelated with the spectral properties of minerals in the earth's interior. There are new methods of obtaining spectral measurements under the intense conditions of the earth's interior. Studies of mineral spectra at high temperatures and pressures are made to understand the nature of electronic processes in minerals and their application to the physical and chemical properties of the earth.

The approach to the study of the influences of pressure and temperature on the optical absorption of minerals traditionally has not been systematic because spectra are difficult to obtain and because the complete spectral response has not been successfully calculated from first principles. Nevertheless, theoretical models and approximations are employed, and many successful applications of spectral measurements to the study of the earth's interior have been made.

In general, high temperature causes an apparent diffuseness and broadening of spin-allowed absorption bands resulting from vibrational effects (vibronic states) at lower energies. Spin-forbidden absorption bands broaden at high temperatures but gain intensity because of interaction between electronic and vibrational processes. The broadening obscures the bands and makes their spectroscopy a difficult and often impossible task. In contrast, high pressure usually does not broaden absorption bands, and one can successfully measure the spectral response during the process of compressing atoms of a mineral together.

Compression of a mineral has the effect of reducing interatomic distance, and this effect is greater than thermal expansion in the earth at depths greater than 35–40 km. In theory, the crystal fields that cause most of the spectral features in minerals increase in strength exponentially with the decrease in interatomic distance.* Observations of the shifts in energy and intensity and of the appearance and disappearance of spectral features as pressure increases provide a useful, if not unique, method of testing the validity of the assignment of absorption bands to electronic processes, as will be described.

B. Pressure and the Absorption Spectrum

Fundamentally, when pressure on a mineral causes a decrease in its volume and a shortening of its interatomic distances, the forces involved are those that oppose the repulsive forces of the individual atoms. In a typical symmetrical, but not spherical, electrical field that surrounds a transition metal cation in a mineral, compression causes a strong effect on the d electrons as the atoms move closer together. This is the same effect that would arise if the bands of a cation were shorter in one mineral than in another of different crystal structure. The increased energy of the d electrons of the cation is proportional to the repulsion caused by the electrons of the coordinating anions. Figure 14 shows a useful plot of octahedral d^6 energy versus Δ_0, the energy difference of e_g and t_{2g}. Starting at $\Delta_0 = 0$, the energies of some of the spectroscopic "states" (spectroscopic lines of the free atom) are shown for all the d electrons in a d^6 ion. As the field changes from spherical to octahedral and as Δ_0 increases linearly (and interatomic distance nonlinearly), the states are split. Within the limits of the point-charge model from which the diagram was constructed, it is possible to observe directly how the energy of electronic transitions varies with Δ_0 and, therefore, how optical absorption bands of these energies shift with pressure.

The diagram is artificial in that the "ground state" ($^5T_{2g}$) does not vary with Δ_0 and both axes are normalized (to B, a value related to the repulsion between d electrons). The state ^5E_g plots with a slope of 45° because its value of E/B is, by definition, exactly equal to Δ_0/B. If interatomic distances were plotted instead of Δ_0, ^5E_g would plot as an exponential to the fifth power.

Spin-forbidden transitions such as $^5T_{2g} \rightarrow {}^3T_{1g}$ (change from spin multiplicity 5 to 3) are of higher energy at low Δ_0 and, as was stated earlier, are low in intensity because of their lower probability. This particular transition has the property that its energy decreases with pressure.

*The increase in Δ_0, the crystal-field energy, is inversely proportional to the fifth power of the interatomic distance in models that are based on the approximation of point charges.

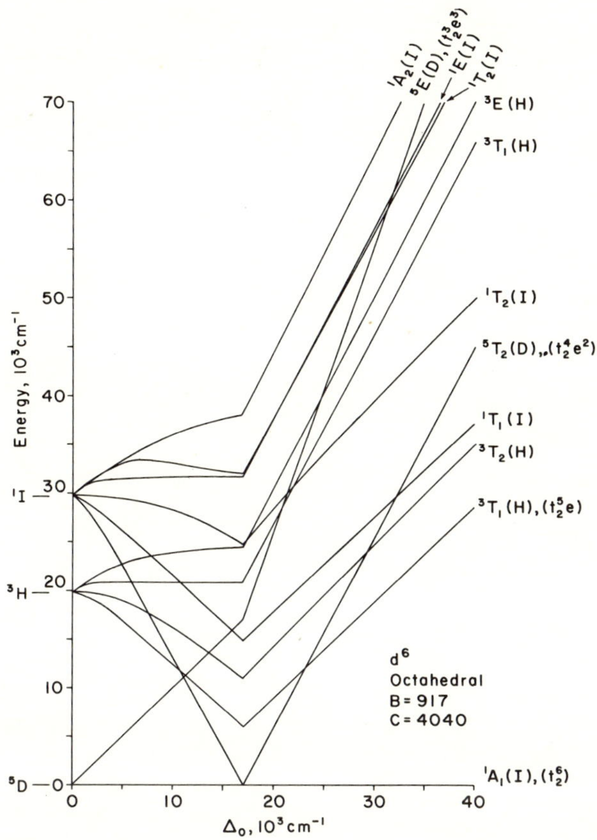

FIG. 14. Energy-level diagram of transitions in octahedrally coordinated transition ions that contain d^6 electrons. (From Berkes, 1968.)

Other spin-forbidden transitions such as $^5T_{2g} \rightarrow {^3T_{2g}}$ remain nearly constant with pressure, and $^5T_{2g} \rightarrow {^3E_g}$ increases with pressure. Qualitatively, the energies of many spin-forbidden transitions are relatively independent of Δ_0 and, therefore, of pressure, resulting in a series of sharp absorption bands of high energy and of low intensity.

C. Octahedral Fe^{2+} d^6 and Cr^{3+} d^3 under Pressure

Octahedral Fe^{2+} in olivine and pyroxene responds to pressure differently, depending on the range of conditions. At moderate depths in the earth (300 km), electronic processes such as crystal-field effects from which the absorption spectra arise are localized to the vicinity of the iron cations. In the olivine spectrum, for example (Fig. 15a), one can observe intense

(A)

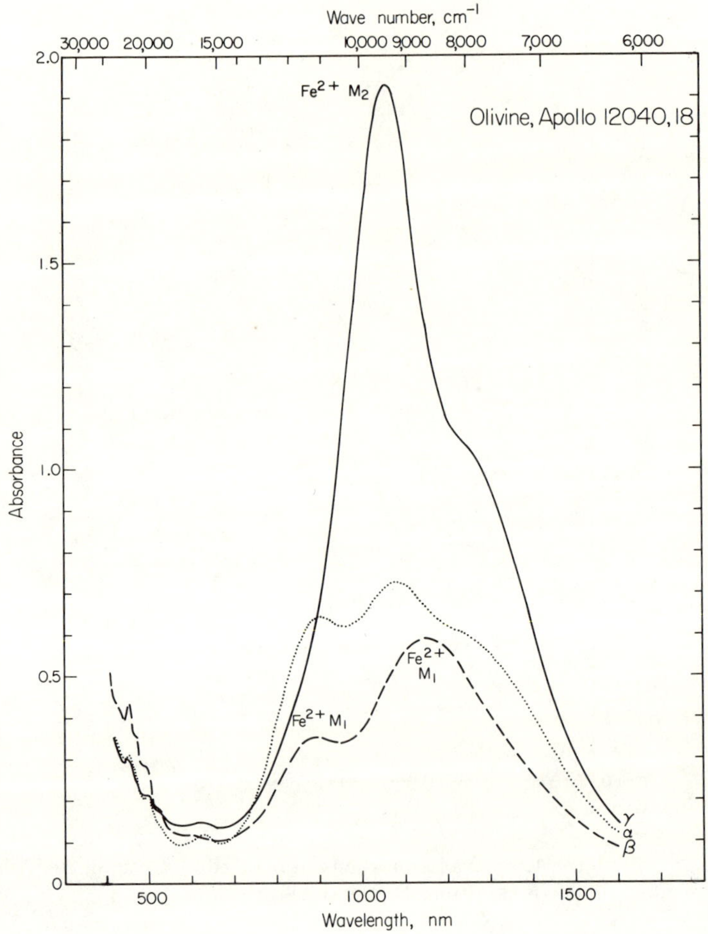

FIG. 15 A. Polarized spectra (α, β, γ) of a lunar olivine crystal from the Apollo 12 mission, showing Fe^{2+} absorption bands in two (M1, M2) octahedral sites of different distortion. The minor bands near 500 nm are spin-forbidden bands of Fe^{2+}. B. β-Polarization of Fe^{2+} bands in fayalitic olivine at 1 bar, 20 kbar, and 25 kbar, showing the energy shift of absorption. C. β-Polarization of spin-forbidden Fe^{2+} bands at 1 bar, and 25 kbar. D. α-Polarization of spin-forbidden Fe^{2+} bands at 1 bar, 20 kbar, and 25 kbar.

spin-allowed bands in the wavelength region 900–1300 nm caused by octahedral Fe^{2+}. Figure 15B shows plots of two of these octahedral bands (β-polarization) of fayalite at 1 bar and 20 and 25 kbar. Referring to Fig. 14 one can see that these two octahedral Fe^{2+} absorption bands move to higher energy with increasing pressure.

The fayalite absorption spectra plotted in Fig. 15A require further comment in regard to the existence of two bands for the spin-allowed transition ($^5T_{2g} \rightarrow {}^5E_g$) in octahedral Fe^{2+}. The two bands result from additional splitting of energy caused by distortion or departure from a purely octahedral field in this particular crystalline site (M1). One interpretation of the fact that one band (near 8000 wave numbers) shifts

(B)

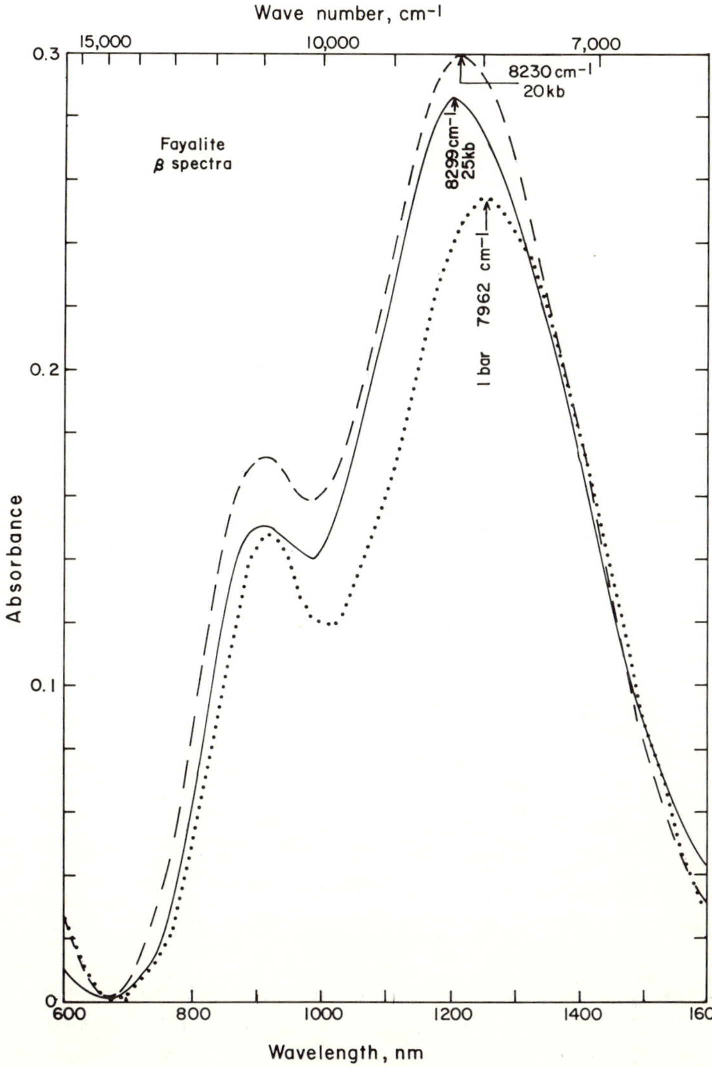

FIG. 15 B.

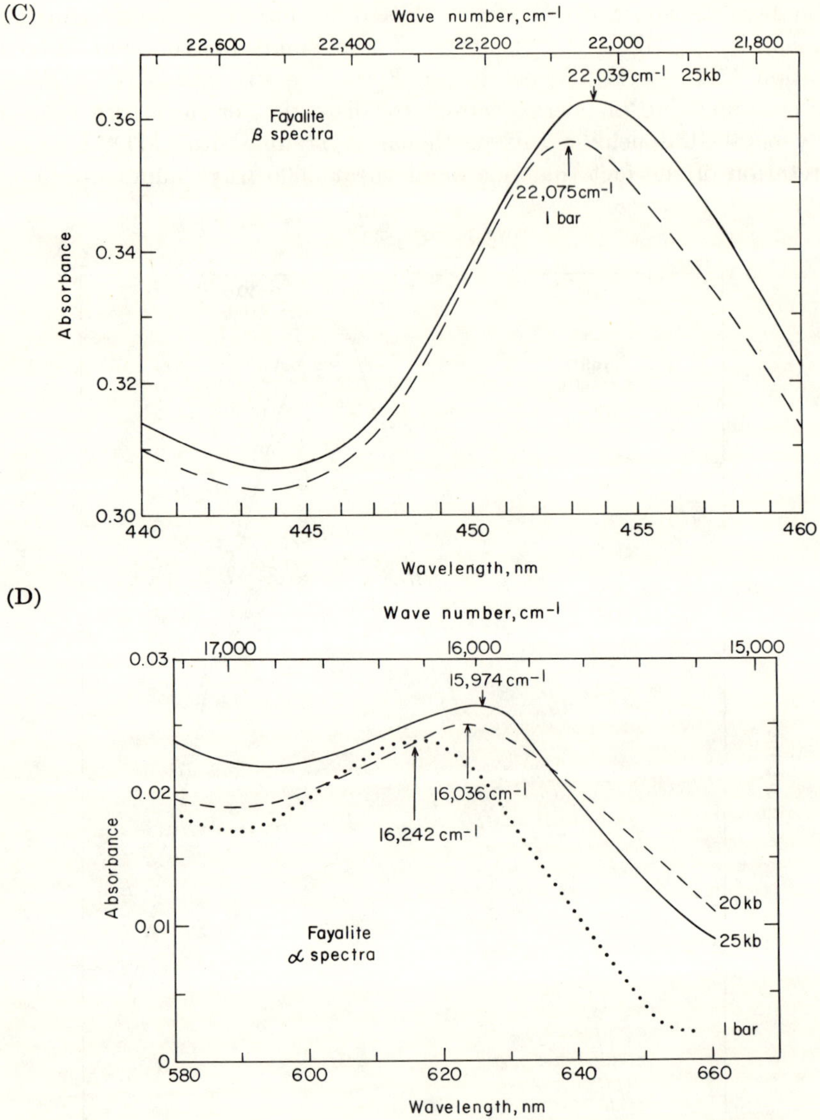

FIG. 15 C. and D.

more rapidly with high pressure is that the distortion is being reduced. Figure 16 shows similar results for Fe^{2+} absorption in two octahedral sites (M1, M2) of different distortion in the pyroxene orthoferrosilite. Figure 16A is a plot of the change in energy (wave numbers, cm^{-1}) of two bands of each site from 1 bar to approximately 25 kbar. Figure 16B shows energy-level diagrams for the two sites at 1 bar and 25 kbar.

In both pyroxene and olivine spectra, one can observe small bands as "shoulders" on a steep absorption tail from the ultraviolet. The steep tail is probably caused by a charge-transfer process in Fe^{3+}, and it is never certain how many, if any, of the smaller bands are actually caused by the charge-transfer process. The small bands are sharp, however, and their response to shift with pressure makes a convincing case for their assignment as spin-forbidden bands in octahedral Fe^{2+}. Comparison of the intensities of the minor bands between wavelengths of 400 and 700 nm in lunar olivine (see Fig. 15A, and Chapter 2, Fig. 11); and pyroxene (see Chapter 2, Fig. 9), which contain relatively small amounts of Fe^{2+} and only trace amounts of Fe^{3+}, with the intensities of spin-forbidden octahedral Fe^{3+} bands in andradite (Fig. 8A), which contains all its iron in the trivalent state, makes it clear that if the minor bands in iron-bearing olivine and pyroxene are spin-forbidden bands of octahedrally coordinated iron, they are caused by Fe^{2+}. Using the energy-level diagram for the d^6 configuration symmetry (Fig. 14), one can deduce that the spin-forbidden

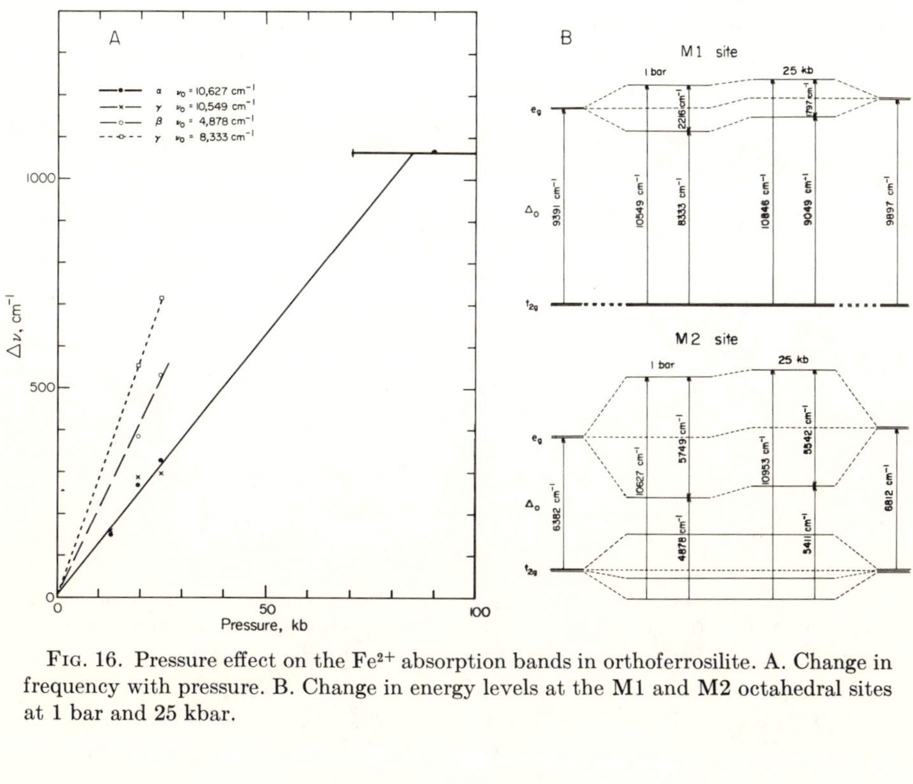

FIG. 16. Pressure effect on the Fe^{2+} absorption bands in orthoferrosilite. A. Change in frequency with pressure. B. Change in energy levels at the M1 and M2 octahedral sites at 1 bar and 25 kbar.

bands of Fe^{2+} are probably caused by transitions between split terms of the "quintet" (^5D) and "triplet" (^3H) states. In olivine and pyroxene there will be more transitions than those shown in Fig. 14 (i.e., more than $^5T_{2g} \rightarrow {}^3T_{1g}$, $^5T_{2g} \rightarrow {}^3T_{2g}$, $^5T_{2g} \rightarrow {}^3E_g$, all of which involve a change in spin state from 5 to 3). Fe^{2+} in olivine and pyroxene exists in more than one type of octahedral site, and distortions of the sites cause further splitting.

Figure 15D shows the pressure shift in energy of the small band at about 615 nm in olivine (see also Fig. 15A). The shift is significant toward lower energy, consistent with what one would predict for the spin-forbidden transition $^5T_{2g} \rightarrow {}^3T_{1g}$ from Fig. 14. At higher energy in the olivine spectrum there is another small band at approximately 453 nm, whose pressure shift is shown in Fig. 15C. This band scarcely shifts to lower energy (approximately (1 nm/25 kbar) and behaves like the spin-forbidden transition $^5T_{2g} \rightarrow {}^3T_{2g}$ shown in Fig. 14.

The d^3 ion, Cr^{3+}, has similarly predictable properties under pressure. The spin-allowed transition $^4A_2 \rightarrow {}^4T_2$ in octahedral Cr^{3+} is equivalent in energy to Δ_0 and shifts to higher energy with increasing pressure. Figure 17 shows a spin-allowed band for octahedrally coordinated Cr^{3+} in spinel at 1 bar, 10 kbar, and 20 kbar. There would be little chance of confusing the assignment of this band with a spin-forbidden band of Fe^{3+}* because virtually all the Fe^{3+} bands shift to lower energy with pressure.

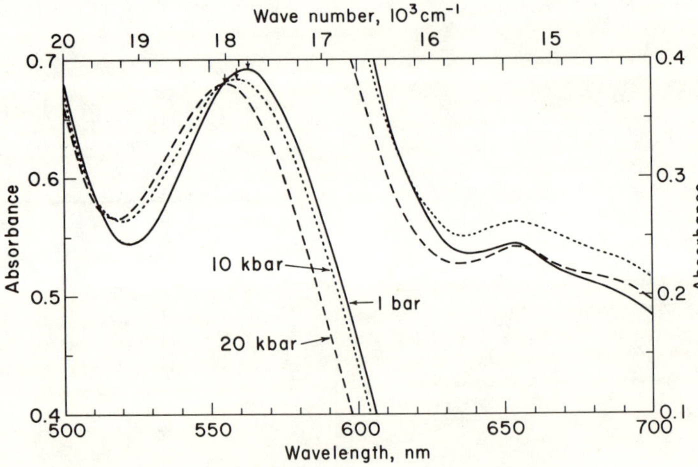

FIG. 17. Pressure effect on the crystal-field band of octahedral Cr^{3+} in spinel (near 560 nm) at 1 bar, 10 kbar, and 20 kbar. Section of spectra between 500 and 600 nm is displaced downward with an appropriate change in absorbance scale.

*Fe^{3+} d^5 has no spin-allowed transitions.

D. Change of Spin State with Pressure: Fe^{2+} d^6

In Fig. 3 a spin-allowed transition was depicted in terms of removing a spin-paired electron from one level (t_{2g}) to another, preserving the gross number of paired and unpaired electrons. Electronic transitions that change this number were defined as "spin-forbidden." This discussion considers a type of configuration in which a transition involves a new ground state with maximum spin-pairing of d electrons, rather than a configuration of maximum spin multiplicity. The former configuration of this type is called the "low-spin" state, the latter the "high-spin" state. A high–low spin effect in minerals in the earth is most important because many physical properties accompany the change from one spin state to another. The density and magnetic, electrical, and thermal properties are affected. The colors of minerals are often drastically changed, because the absorption spectrum shifts during a high–low spin transition.

The case for octahedral d^6 can be demonstrated from the diagram in Fig. 3. The high-spin state is shown. If electrons in the e_g level were moved to t_{2g} with all the spins paired, the low-spin state would obtain. This transition occurs at high pressure as the electrons are crowded into a smaller volume or ionic radius in the low-spin state. One can see the spectral responses in the transition energies shown in Fig. 14. When Δ_0 reaches the inflection point, the crystal field is strong enough to counteract the pairing energy, and the high–low transition occurs. At Δ_0 greater than the value at the inflection, the ground state changes, and now the first spin-allowed transition is ${}^1A_{1g} \rightarrow {}^1T_{1g}$, which is accompanied by a sudden drop to lower energy and then begins to shift to higher energy as pressure and Δ_0 increase. Spin-forbidden transitions are different in the low-spin state, and they cease their shift to lower energy and begin an apparent shift to higher energy with pressure.

In perfect octahedral and tetrahedral coordination symmetry, the high–low spin transition can occur only at sufficient Δ_0 and only in octahedral d^{4-7} and tetrahedral d^{3-6} configurations, because in all other d configurations the high-spin states are identical with the corresponding low-spin states. As was previously discussed, however, distortions of coordination symmetry cause additional terms and therefore a greater complexity of spin states. The nature of intermediate spin states has yet to be explored in minerals, although theoretically their stability could be lower than either high- or low-spin configuration. High-spin, low-spin transitions in an octahedral field have not been studied extensively either, probably because of the high pressures required.

The Δ_0 of gillespite (Fig. 12) is close to the critical value of spin-pairing energy at 1 atm, and attempts have been made to observe spin-pairing at high pressure in gillespite's square-planar Fe^{2+}. Gillespite is bright red in color because of the strong absorption band at 500 nm. Early pressure experiments (Strens, 1966) showed a change to colorless at approximately 26 kbar. More recently, polarized spectra at the same pressures (Abu-Eid *et al.*, 1973) showed that gillespite changed from bright red to bright blue as the band at 500 nm suddenly shifted to approximately 600 nm. The 1200 nm band simultaneously shifted to 1400 nm. The sudden shift in absorption bands to lower energy is consistent with a high–low spin transition in iron, but further experiments on the spectra at higher pressure to determine changes in interatomic distance suggest a structural change (Hazen and Abu-Eid, 1974).

The principle of spin-pairing of Fe^{2+} in minerals has also been studied in synthetic analogs of pyrite and its solid solution toward the manganese end-member hauerite. Fe^{2+} is low-spin in pyrite and high-spin in hauerite, and judging from the cell parameters (pyrite, 5.5 Å; hauerite, 6.1 Å), it was deduced that the spin configurations are caused by iron's different bond lengths in the two minerals. Confirmation was obtained by raising synthetic hauerite (containing 2 weight % Fe) to pressures above 40 kbar, where an onset of the high–low spin transition was observed (Bargeron *et al.*, 1971).

Spin-pairing transitions in Fe^{2+} will not occur in most minerals below pressures of several hundred kilobars because of the strong repulsion between spin-paired d electrons. However, interelectronic (d) repulsion is known to decrease at high pressure (Drickamer and Frank, 1973), and possibly the low-spin state of Fe^{2+} does not require as strong crystal fields deep in the earth as might otherwise be suspected. In any respect, if d electrons behave predictably in minerals at great depths in the earth, pressures are sufficiently high (3.3 megabars at the core) to stabilize the low-spin state of iron.

E. *Major Pressure Effects: Fe^{2+} in Olivine, Spinel, and Magnesiowüstite*

The attempt in this chapter at a heuristic discussion of optical absorption in minerals has, up to this point, followed the system of relating spectral features to the nature of the coordinating field and its cation. However, it is known that at sufficiently high pressure (estimated at about a megabar in most minerals) and with shortening of interatomic distance, minerals that are insulators at low pressure change to semiconductor or metallic state, and furthermore, there are numerous known electronic processes that can cause the change. Recently, other processes have been discovered in iron that occur at lower pressure and cause minerals to assume metal character and to greatly absorb light.

As the electrons of a cation become delocalized from their atomic orbitals, they become essentially involved in processes (such as electrical conduction) that include most of the atoms of a mineral. In addition, competing processes that simultaneously include charge transfer, spin-pairing, and chemical reduction are known. The spectral effects in minerals at very high pressures are major, and they are significant in the earth because physical and chemical properties are greatly changed.

Figure 18A shows the absorption spectra of fayalite at eight pressures in the range 45–270 kbar. Because powdered samples were employed in

(A)
(B)

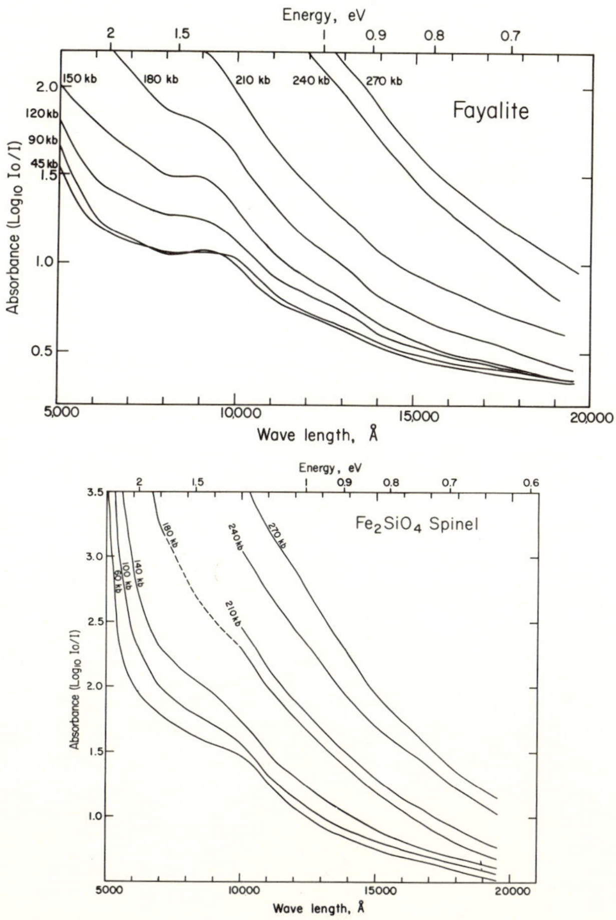

FIG. 18. Nonpolarized spectra of polycrystalline (A) fayalite (Fe_2SiO_4), (B) spinel (Fe_2SiO_4), and (C) magnesiowüstite (wüstite$_{22}$periclase$_{78}$) in the pressure range 45–310 kbar.

these experiments, the spectral features are diffuse, but one can observe the octahedral Fe^{2+} band near 10,000 Å and the steep absorption tail at higher energy, thought to be caused by charge transfer in Fe^{3+}. At 1 atm the fayalite spectrum is identical with the 45 kbar spectrum, and there is no change as the pressure is doubled to 90 kbar. As pressure exceeds 120 kbar, the absorption edge or tail moves sharply to lower energy, and the spectral features are lost to total absorption in the visible and near infrared regions. Similar pressure effects occur in spinel (Fig. 18B) and magnesiowüstite (Fig. 18C). This behavior is also generally observed in other iron-bearing minerals. During the experiments whose results are plotted in Fig. 19, electrical resistance was also measured and found to decrease rapidly in the same pressure range. The resistance of fayalite dropped by a factor of nearly one million.

Studies of olivine, spinel, and magnesiowüstite were made because these minerals are thought to occur in the earth's interior. The implications for the earth of the observed changes in electrical and near-infrared (thermal radiation) properties are major. Pressure effects of this kind were previously unknown and were omitted in constructing physical

(C)

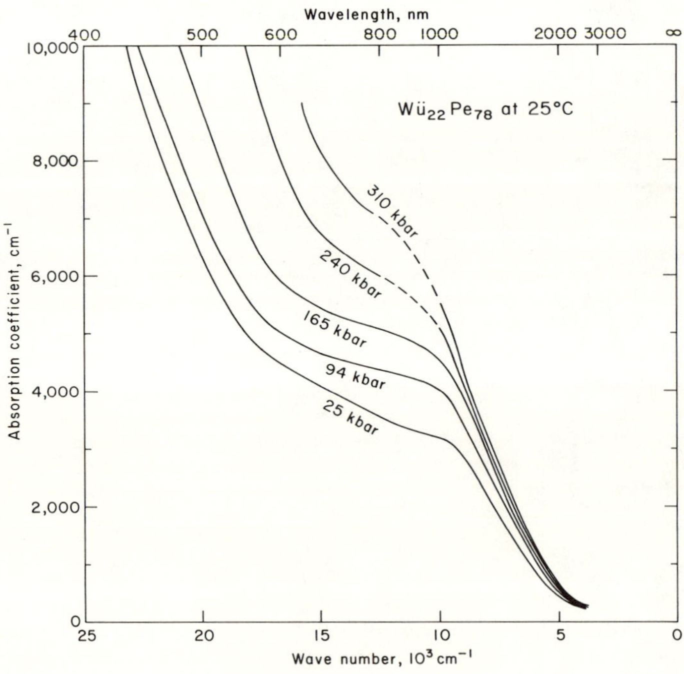

FIG. 18 C.

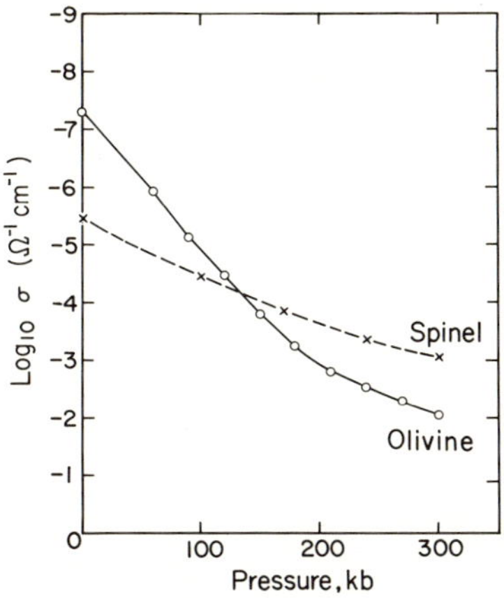

FIG. 19. Electrical resistance (Ω, ohm; cm, centimeter) of olivine and spinel (Fe_2SiO_4) in the pressure range 1 bar to 300 kbar.

models of the earth. Revised models should provide a more accurate understanding of electrical conductivity and the magnetic field and of heat transfer and the earth's temperature distribution, if olivine, spinel, and magnesiowüstite actually exist at depth. Possibly effects such as charge transfer and spin-state changes will be important in stabilizing other phases. At present an experimental approach to the problem of predicting the composition and properties of the earth's interior appears to be justified. The study of electronic properties of minerals at high pressure provides data needed in establishing the theoretical framework as well.

References

Abu-Eid, R. M., Burns, R. G., and Mao, H. K. (1973). *Eos, Trans. Amer. Geophys. Union* **54**, 498.

Bancroft, G. M., and Burns, R. G. (1967). *Amer. Mineral.* **52**, 1278–1287.

Bargeron, C. B., Aviror, M., and Drickamer, H. G. (1971). *Inorg. Chem.* **10**, 1338–1339.

Bell, P. M., and Mao, H. K. (1972). *Carnegie Inst. Wash., Yearb.* **71**, 531–534.

Berkes, J. S. (1968). "MRL Monograph No. 2." Mater. Res. Lab., Pennsylvania State University, University Park, Pennsylvania.

Blake, R. L., Hessevick, R. E., Zoltai, T., and Finger, L. W. (1966). *Amer. Mineral.* **51**, 123–129.

Burns, R. G. (1965). Ph.D. Thesis, University of California, Berkeley.

Burns, R. G. (1970a). *Amer. Mineral.* **55**, 1608–1632.

Burns, R. G. (1970b). "Mineralogical Applications of Crystal Field Theory." Cambridge Univ. Press, London and New York.

Burns, R. G., and Fyfe, W. S. (1967). *Res. Geochem.* **2**, 259–285.

Cotton, F. A. (1971). "Chemical Applications of Group Theory." Wiley (Interscience), New York.

Dowty, E. (1971). *Amer. Mineral.* **56,** 1983–2009.

Dowty, E., and Clark, J. R. (1973). *Amer. Mineral.* **58,** 230–242.

Drickamer, H. R., and Frank, C. W. (1973). "Electronic Transitions and the High Pressure Chemistry and Physics of Solids." Chapman & Hall, London.

Faye, G. H. (1969). *Can. Mineral.* **10,** 112–118.

Ghose, S. (1965). *Z. Kristallogr., Kristallgeometrie, Kristallphys., Kristallchem.* **122,** 81–99.

Hafner, S. S., Virgo, D., and Warburton, D. (1971). *Earth Planet. Sci. Lett.* **12,** 159–166.

Hazen, R. M., and Abu-Eid, R. M. (1974). *Eos, Trans. Amer. Geophys. Union* **55,** 463.

Manning, P. G. (1970). *Can. Mineral.* **10,** 677–688.

Manning, P. G., and Harris, D. C. (1970). *Can. Mineral.* **10,** 260–271.

Manning, P. G., and Nickel, E. H. (1969). *Can. Mineral.* **10,** 71–83.

Mao, H. K., and Bell, P. M. (1973). *Carnegie Inst. Wash., Yearb.* **72,** 629–631.

Mao, H. K., El Goresy, A., and Bell, P. M. (1974). *Proc. Lunar Sci. Conf., Geochim. Cosmochim. Acta, 5th, 1974,* Vol. 1, Suppl. 5, pp. 673–683.

Moore, R. K., and White, W. B. (1971). *Amer. Mineral.* **56,** 826–840.

Morimoto, N., and Koto, K. (1969). *Z. Kristallogr., Kristallgeometrie, Kristallphys., Kristallchem.* **129,** 65–83.

Novak, G. A., and Gibbs, G. V. (1971). *Amer. Mineral.* **56,** 791–825.

Prewitt, C. T., Shannon, R. D., and White, W. B. (1972). *Contrib. Mineral. Petrol.* **35,** 77–82.

Runciman, W. A., Sengupta, D., and Marshall, M. (1973a). *Amer. Mineral.* **58,** 444–450.

Runciman, W. A., Sengupta, D., and Gourley, J. T. (1973b). *Amer. Mineral.* **58,** 451–456.

Slack, G. A., Ham, F. S., and Chrento, R. M. (1966). *Phys. Rev.* **152,** 376–402.

Strens, R. G. J. (1966). *Chem. Commun.* **21,** 777–778.

Taylor, L. A., Mao, H. K., and Bell, P. M. (1973). *Proc. Lunar Sci. Conf., Geochim. Cosmochim. Acta, 4th, 1973,* Vol. 1, Suppl. 4, pp. 829–839.

White, W. B., and Keester, K. L. (1967). *Amer. Mineral.* **52,** 1508–1514.

Wickersheim, K. A., and Buchanan, R. A. (1959). *Amer. Mineral.* **44,** 440–445.

Wickersheim, K. A., and Buchanan, R. A. (1968). *Amer. Mineral.* **53,** 347.

Wilkins, R. W. T., and Sabine, W. (1973). *Amer. Mineral.* **58,** 508–515.

Wood, D. L., and Naussau, K. (1967). *J. Chem. Phys.* **47,** 2220–2228.

Wood, D. L., and Nassau, K. (1968). *Amer. Mineral.* **53,** 777–800.

2

POLARIZED ELECTRONIC SPECTRA

Roger G. Burns and David J. Vaughan

I. Introduction

The first series transition metals (Sc, Ti, V, Cr, Mn, Fe, Co, Ni, Cu) constitute an important group of elements in the earth and the moon. These metals occur not only in the economically important opaque oxide and sulfide ore minerals, but also as major and minor components in transparent silicate and oxide phases, to which they impart distinctive colors and magnetic properties. In the past decade, the electronic structures and stabilities of the transition elements have been explained by several theories of the chemical bond, including crystal field theory, ligand field theory, and molecular orbital theory. Many of these interpretations are based on data from a variety of spectral techniques, including electronic absorption and reflectance, x-ray photoelectron (ESCA), Mössbauer, and x-ray absorption and emission spectroscopies (Burns, 1970a, 1973a; Bancroft, 1973; Tossell *et al.*, 1974; Vaughan *et al.*, 1973).

Prior to the Apollo and Luna missions, a large body of spectral data had been acquired for a variety of minerals and synthetic phases containing ions of the transition elements. Many of the geochemical and crystal chemical properties of transition metal ions were elucidated by crystal field theory (Bethe, 1929; Orgel, 1966; Burns *et al.*, 1964; Burns and Fyfe, 1967) using absorption spectral measurements in the visible and nearby-infrared regions. These results and applications were reviewed by Burns (1970a). With the return of lunar samples, there has been a renewed interest in the measurement and interpretation of the electronic spectra of silicate minerals containing transition elements. This is due partly to the discovery of new minerals and of elements in unusual oxidation states in the lunar rocks. It is timely, therefore, to review the spectral data published since 1970 on lunar silicate minerals, together with recent relevant information on terrestrial extraterrestrial minerals and synthetic phases. This chapter summarizes the results of such electronic spectral measurements of transition metal-bearing minerals in the near-ultraviolet, visible, and near-infrared regions (approximately 25,000–3,000 Å or 4,000–33,300 cm^{-1}). The emphasis is on crystal field or intraelectronic transitions and intervalence charge transfer spectra which are relevant to the lunar minerals. Most of the data are derived from polarized absorption spectra of single crystals. However, important information from unpolarized absorption and diffuse relectance spectral measurements is also included.

Following a review of the spectra of individual transition metal ions in relevant host silicate and oxide structures, chiefly with the metal ion in 6-fold coordination, the absorption spectra of each transparent silicate mineral found in the lunar samples are discussed in the light of available data for terrestrial and synthetic analogues. The chapter concludes with an outline of the relevance of the spectral data to lunar science.

II. Experimentation and Theory

A. Boundaries Imposed by Lunar Petrology and Geochemistry

Numerous chemical and microprobe analyses of bulk rock samples and mineral separates from the moon rocks have revealed that iron, titanium, chromium, and manganese predominate over the remaining transition elements in silicate and oxide minerals. Thus, the concentrations of Sc, V, Co, Ni, and Cu are well below 0.1 weight % in the transparent silicate phases. As a result, spectral contributions from ions of these elements are extremely unlikely to be resolved amidst the numerous absorption bands and peaks arising from Fe, Ti, Cr, and Mn. Furthermore, the relatively low abundance of manganese in the moon rocks, together with the weak (spin-forbidden) spectral features of Mn^{2+} ions, make this element an unlikely contributor to the electronic spectra of lunar minerals. Therefore, only the spectra of the Ti^{4+}, Ti^{3+}, Cr^{3+}, Cr^{2+}, Fe^{3+}, and Fe^{2+} ions in host silicate and oxide minerals will be discussed in this review.

The principal transparent silicate minerals occurring in lunar rocks belong to the pyroxene (pigeonite, augite, orthopyroxene), pyroxenoid (pyroxferroite), olivine (forsterite–fayalite), and feldspar (plagioclase series) groups. Of these, the principal transition metal-bearing phases are pigeonite, augite, pyroxferroite, and olivine, although significant amounts of iron occur in lunar plagioclase feldspars. Therefore, the emphasis of this review is on the spectral profiles of these minerals only.

The petrography of the lunar rocks shows that the crystallites rarely exceed 5 mm in dimensions, although 1–3 cm long acicular pyroxene phenocrysts occur in several rocks collected in the Apollo 12 and 15 missions. Thus, no gem variety, large mineral crystals of the type known on earth have yet been found on the moon. Although it is a simple procedure to extract pure mineral separates from crushed lunar rocks for spectral measurements on powdered samples, somewhat more sophisticated techniques are required for measuring the polarized absorption spectra of single crystals occurring in the moon rocks.

B. Techniques for Measuring Spectra of Lunar Minerals

The transparent minerals in lunar rocks are typically 1 mm or less in diameter. This means that microscope techniques must be employed to study spectra of individual crystals in the moon rocks. Such techniques are not new to the absorption spectroscopy of chemical compounds (Tsuchida and Kobayashi, 1938; Melankholin, 1956; Vishnevshii and Klimovskaya, 1958). However, the methods have been refined in mineral spectroscopy so as to incorporate essential features of a petrologic microscope into modern, high-resolution spectrophotometers. For example, Burns (1966) described a system employing two petrologic microscopes, to which three-axis universal stages could be attached, which were mounted horizontally in a Cary model 14 or 17 recording spectrophotometer. The calcite Nicol polarizers of the microscopes produce polarized light over the spectral range of interest for transition metal ions in minerals, and the universal stage enables single crystals of minerals in rock thin sections to be orientated three-dimensionally with respect to the plane of polarization. Spectral features are accentuated by using thick, mounted, single crystals or rock sections (50–100 μm). This system has been used to measure the spectra of a large variety of terrestrial minerals (Burns, 1965, 1970a; Faye, 1971a; Faye and Nickel, 1970) and the minerals in Apollo rock samples (Burns *et al.*, 1972a,c, 1973) with dimensions exceeding 0.75 mm (750 μm) in diameter.

Recently, the microscope system of Burns (1966) has been redesigned so as to employ petrographic microscopes in a vertical alignment (Mao and Bell, 1973). Modifications to the light path from the spectrophotometer through the microscope system to the detectors has enabled spectral measurements to be made on crystals as small as 50 μm in diameter and 5 μm thick. Such a system is particularly useful for measuring polarized spectra of single crystals extracted from moon rocks (Bell and Mao, 1972a, b; Mao and Bell, 1971, 1973) and minute crystals in diamond cells at high pressures (Bell and Mao, 1969; Mao and Bell, 1971, 1972, 1973; Abu-Eid *et al.*, 1973).

C. Theory of Electronic Spectra

The theory of transition metal spectra is described in several authoritative texts and reviews (Ballhausen, 1962; Cotton, 1963; Dunn *et al.*, 1965; Figgis, 1967; Lever, 1968; Schläfer and Gliemann, 1968; Orchin and Jaffé, 1971). In this section, only a brief outline will be presented so as to provide a background to the spectral data described in Sections III and IV. For further details, the reader is referred to the texts cited above.

The spectra of transition metal ions in the visible and nearby regions originate principally from excitations of electrons between unfilled 3d orbitals. These excitations may either be confined to individual ions giving rise to crystal field or intraelectronic transitions, or take place between neighboring ions in a crystal structure producing charge-transfer transitions. Electron transfer between adjacent elements may involve cation → cation, cation → anion, and anion → cation transitions. Charge transfer between cations and neighboring oxygens in silicates leads to absorption in the ultraviolet region. In this review of transition metal-bearing silicates and oxides, only cation → cation charge transfer (also called intervalence transfer) processes occurring predominantly in the visible region will be discussed.

When a transition metal ion is octahedrally coordinated by oxygens, crystal field transitions in 3d orbitals involve electron excitations between more stable t_{2g} orbitals (which project between the surrounding oxygens) and the higher energy e_g orbitals (which are directed toward the negatively charged oxygens).* Such transitions lead to absorption bands in the visible and near-infrared regions (20,000–8,000 cm^{-1}), depending on the type and valence of the central cation. When the symmetry of oxygens surrounding a transition metal ion is lower than octahedral (i.e., the cation occurs in a distorted six-coordinate site), the five 3d orbitals are split into several energy levels so that additional electronic transitions are possible, leading to absorption bands at different energies. Many of these transitions within cations in low symmetry coordination sites are polarization dependent and lead to absorption bands of different intensity when light is polarized in three mutually perpendicular directions through an anisotropic crystal. Spin-allowed transitions, in which the electronic configurations of the ground and excited states have identical numbers of unpaired electrons, are generally one or two orders of magnitude more intense than spin-forbidden transitions (Figgis, 1966; Lever, 1968; Schläfer and Gliemann, 1968) in which electron pairing occurs in excited states. As a result, the spectra of Mn^{2+}- or Fe^{3+}-bearing silicates are less intense than minerals containing Ti^{3+}, Cr^{3+}, Cr^{2+}, and Fe^{2+} ions. Similarly, if the transition metal ion occurs in a coordination site which is centrosymmetric (e.g., regular or tetragonally distorted octahedron), the intensities of absorption bands are generally 1–2 orders of magnitude smaller than those originating from cations located in noncentrosymmetric sites (e.g., regular tetrahedron, trigonally distorted octahedron). In a given coordination symmetry, the electronic configurations of the ground state and each excited state are denoted by group theory representations such as $^2T_{2g}$ or ^2E_g for Ti^{3+} or $^4A_{2g}$, $^4T_{1g}$, and $^4T_{2g}$ for Cr^{3+}. These symbols summarize the symmetry, degeneracy, and spin-multiplicity of the electronic configurations of each crystal field state. Energy level diagrams provide a convienient method for showing the relative energies of the crystal field states, and depicting the transitions portrayed in electronic absorption spectra.

Light absorption in the visible region also results from electron transfer between two different cations in adjacent coordination sites sharing a common edge or face. The t_{2g} orbitals of neighboring transition metal ions in octahedral coordination in silicates and oxides may overlap across the shared O–O edge or triangular oxygen face of the coordination polyhedra, with the result that metal–metal charge transfer processes are strongly polarization dependent (Burns, 1973b). Allen and Hush (1967) have proposed that light absorption resulting in transfer of an electron from

* See Chapter 1, Section II, for definitions of these symbols.

a lower to a higher oxidation state of the same element be termed homonuclear intervalence–transfer absorption (e.g., $Fe^{2+} + Fe^{3+} \rightleftharpoons Fe^{3+} + Fe^{2+}$). Transfer is also possible between ions of different elements, resulting in heteronuclear intervalence-transfer absorption (e.g., $Fe^{2+} + Ti^{4+} \rightleftharpoons Fe^{3+} + Ti^{3+}$). Such reversible excitations are therefore photochemical oxidation–reduction processes.

Theories of intervalence transfer have been reviewed by Hush (1967), who points out the important connection between optical intervalence transfer and the corresponding thermally activated process. Since homonuclear and heteronuclear intervalence transfer processes take place between cations in adjacent sites in a crystal structure, the degree of overlap of the cation orbitals, both with the neighboring cation and its own coordinated anions, are important factors. However, if the degree of overlap is so great that the valence electrons are no longer localized on the cation, the material will have metallic properties. In Fe^{2+} (a) $\rightarrow$ Fe^{3+} (b) charge transfer, an electron is excited from a t_{2g} orbital on cation (a) to the corresponding orbital on cation (b). Transitions at higher energy of the type t_{2g} (a) $\rightarrow$ e_g (b) have been observed (Hush, 1967) but have much weaker intensity. Intervalence transitions exhibit a wide range of intensities, since both concentration of the relevant species and the degree of overlap are important factors.

D. Calculation of Electronic Spectra

The crystal field, ligand field, and molecular orbital theories are three theories of the chemical bond which can be used both to interpret and to predict the experimental electronic absorption spectra of transition metal ions. Crystal field theory (Bethe, 1929; Orgel, 1952) uses the methods of atomic quantum mechanics, and is confined to describing the relative energies of the d orbitals of the transition metal in different crystalline environments. The anions are regarded as point charges and any covalent interaction or mixing between metal and anion is ignored. Thus, crystal field theory has been extremely successful in explaining features of the electronic absorption spectra which are due to intraelectronic (d $\rightarrow$ d) transitions, but is inappropriate in discussions of cation $\rightleftharpoons$ anion or cation $\rightleftharpoons$ cation charge transfer transitions. The same is true of ligand field theory, which is essentially a parameterized modification of crystal field theory in which covalency effects between cation and anion are treated (Orgel, 1966).

In contrast, molecular orbital theory provides a complete description of the electronic structure of a compound, including all the orbitals of cation and anion. Until recently, it has been little used in mineralogy both because of the success of crystal field theory in interpreting spectra due to d $\rightarrow$ d transitions, and because of the poor agreement between calculation and experiment in quantitative treatments. However, with the development of new spectroscopic techniques yielding information on electrons other than those of the metal d shell and with much improved calculational methods, the molecular orbital theory is certain to be more widely used. As this chapter is principally concerned with the metal d $\rightarrow$ d transitions, the formalisms and terminology of crystal field theory will largely be used. However, reference will be made to the results of recent molecular orbital calculations, particularly when cation $\rightarrow$ cation charge transfer transitions are discussed. Results obtained using a new first-principles molecular orbital method, the SCF-X_α scattered wave cluster method (Slater and Johnson, 1972; Johnson, 1973), will be discussed in Section III, B.

III. Spectra of Individual Transition Metal Ions

A. Crystal Field Transitions

1. Titanium

a. Ti^{4+}. Elemental titanium has the electronic configuration $1s^2\ 2s^2\ 2p^6\ 3s^2\ 3p^6\ 4s^2\ 3d^2$. In forming Ti(IV) compounds, the outermost $4s^2$ and $3d^2$ electrons are removed leaving a species with the electronic configuration of argon ([A] = $1s^2\ 2s^2\ 2p^6\ 3s^2\ 3p^6$). Therefore, there are no electrons in 3d orbitals of Ti^{4+} to be excited between t_{2g} and e_g orbital levels in the visible region. As a result, most pure Ti(IV) compounds are colorless or white, including pure rutile TiO_2. However, impurities, defects, and the presence of Ti^{3+} ions in the rutile structure lead to color and absorption bands over a wide range of energies (Robin and Day, 1967, p. 274; Table IV). Examples are discussed in Section III, B, 2.

b. Ti^{3+}. The Ti^{3+} ion, with one 3d electron, might be expected to give rise to a single-peak crystal field spectrum. In octahedral coordination, one absorption band is predicted, corresponding to electron transfer between the t_{2g}-and e_g-orbital energy levels. This transition is designated as $^2T_{2g} \rightarrow {}^2E_g$. However, the excited ^2E_g crystal field state is split into two levels as a result of the dynamic Jahn-Teller effect (Cotton and Meyers, 1960), so that transitions to these levels produce asymmetric or two-peaked spectra in the visible region for Ti^{3+} ions in aqueous solution and chemical compounds (Hartmann and Schläfer, 1951; Hartmann *et al.*, 1957; Schläfer, 1964; Hoff and Brubaker, 1971; Clark, 1968, p. 153). The crystal field states are also split when the environment about the Ti^{3+} ion is distorted from octahedral symmetry, such as in the crystal structures of most Ti (III) compounds and minerals. For example, the absorption spectrum of a single crystal of cesium titanium alum [containing octahedral $Ti(H_2O)_6^{3+}$ groups], contains two bands separated by about 1900 cm^{-1} (Schläfer, 1964). Similar energy separations of up to 8000 cm^{-1} have been reported in a variety of Ti(III) compounds and minerals (Clark, 1968, p. 153–156). Examples are listed in Burns and Huggins (1973). The polarized spectra of a unique titanium pyroxene from the Allende meteorite, having the formula $Ca(Mg, Ti^{4+}, Ti^{3+})(Si, Al)_2O_6$, is shown in Fig. 1. It also contains two well-resolved absorption bands of comparable intensity in all three spectra centered at about 21,000 cm^{-1} and 16,500 cm^{-1}. The positions, separation, and intensities of these two bands are consistent with crystal field transitions within Ti^{3+} ions which are located in the very distorted octahedral M1 coordination sites of the calcic clinopyroxene structure (Clark et al., 1969; Burns and Huggins, 1973). An alternative interpretation of the 16,500 cm^{-1} band of this pyroxene as representing a $Ti^{3+} \rightarrow Ti^{4+}$ intervalence transition was proposed (Dowty and Clark, 1973a), which has led to debate (Burns and Huggins, 1973; Dowty and Clark, 1973b). The problem is discussed further in Section III, B, 2.

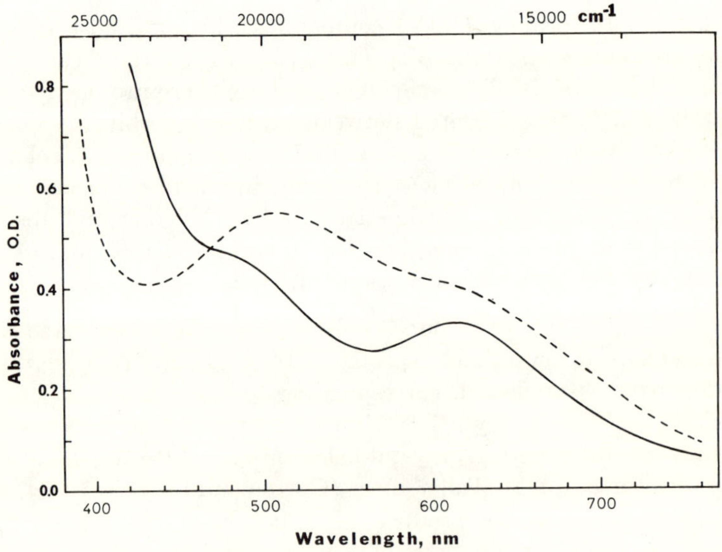

FIG. 1. Polarized spectra of the titaniferous fassaite in the Allende meteorite. The two bands represent crystal field transitions in Ti^{3+} ions in the pyroxene structure $[Ca(Mg, Ti^{4+}, Ti^{3+})(Si, Al)_2O_6]$ —— α spectrum; - - - - γ spectrum.

2. Chromium

a. Cr^{3+}. The ground state for the Cr^{3+} ion, [A]$3d^3$, is denoted by $^4A_{2g}$ in octahedral coordination, corresponding to the electronic configuration $(t_{2g})^3(e_g)^0$. At least three spin-allowed transitions are expected, corresponding to $^4A_{2g} \rightarrow {}^4T_{2g}$, $^4A_{2g} \rightarrow {}^4T_{1g}$ (F), and $^4A_{2g} \rightarrow {}^4T_{1g}$ (P), as one ($^4T_{2g}$) and two ($^4T_{1g}$) electrons are excited from the t_{2g} orbitals into the e_g orbitals. In many Cr(III) compounds and most minerals containing Cr^{3+} ions, only the two lower energy spin-allowed bands may be resolved because the third high energy feature is obscured by intense oxygen–metal charge-transfer bands. The energies of the crystal field absorption bands of Cr^{3+}, like most transition metal ions, depend on the distances between the metal (chromium) and the coordinated oxygens. The positions of the absorption bands also determine the colors of the compounds. Thus, bands for red ruby (average Al-O distance = 1.91 Å) occur at 17,990 cm^{-1} and 24,390 cm^{-1}, while corresponding bands for green eskolaite (Cr–O = 2.00 Å) occur at 21,750 cm^{-1} and 16,670 cm^{-1}. Other data for Cr^{3+}-bearing minerals are summarized in Table I.

b. Cr^{2+}. The chromous ion, [A]$3d^4$, occurs in a high-spin state in oxide structures at atmospheric pressures. In octahedral coordination, the ground state electronic configuration $(t_{2g})^3(e_g)^1$ corresponds to the ^5E_g crystal field state, while in tetrahedral coordination the $(e)^2(t_2)^2$ configuration gives rise to the $^5T_2$ ground state. In Cr^{2+} ions in octahedral environments, one spin-allowed absorption band is expected. This corresponds to electron excitation to the $(t_{2g})^2(e_g)^2$ configuration (i.e., a $^5E_g \rightarrow {}^5T_{2g}$ transition). Thus, aqueous solutions containing $Cr(H_2O)_6{}^{2+}$ give rise to a broad, asymmetric band centered around 13,900 cm^{-1} (Holmes and McClure, 1957).

The d^4 configuration is particularly susceptible to the Jahn–Teller effect, so that Cr(II) compounds are distorted from type structures (Burns, 1970a, p. 109). On the other hand, the Cr^{2+} ion is stabilized in a distorted octahedral site. The olivine structure contains two distorted six-coordinate sites, suggesting that significant amounts of Cr^{2+} ions might occur in natural and synthetic olivines (Burns, 1969, 1970a, p. 124). Initially, synthetic Cr_2SiO_4 was suggested to be isostructural with forsterite (Healy and Schottmiller, 1964), but this was disproved by Scheetz and White (1972) on the basis of crystal field spectral measurements. Scheetz and White succeeded in synthesizing a Cr^{2+}-doped forsterite containing traces of Cr^{3+} ions. The latter give rise to characteristic peaks for Cr(III) compounds at 23,500 and 16,900 cm^{-1} (Fig. 2). The additional broad, asymmetric bands at 11,800 cm^{-1} and 6,400-6,700 cm^{-1} shown in Fig. 2 are due

TABLE I

POSITIONS OF CRYSTAL FIELD BANDS IN THE SPECTRA OF Cr^{3+}-BEARING MINERALS

MINERAL	AVERAGE M–O DISTANCE (Å)	ABSORPTION BANDS (cm^{-1})		COLOR	REFERENCE
Ruby (Al_2O_3/Cr^{3+})	1.91	17,990	24,390	Red	Neuhaus (1960)
Eskolaite (Cr_2O_3)	2.00	16,670	21,750	Green	Neuhaus (1960)
Spinel ($MgAl_2O_4/Cr^{3+}$)	1.91	18,000	25,000	Pink	Poole (1964)
Olivine (Mg_2SiO_4/Cr^{3+})	2.12	16,900	23,500	Green	Scheetz and White (1972)
Pyrope ($Mg_3Al_2(SiO_4)_3/Cr^{3+}$)	1.905	17,606	24,272	Red	Moore and White (1972)
Uvarovite $Ca_3Cr_2(SiO_4)_3$	1.985	16,529	24,814	Green	Neuhaus (1960)
		16,191	22,676	Green	Moore and White (1972)
Chrome diopside ($CaMgSi_2O_6/Cr^{3+}$)	2.08	16,129	22,989	Green	Neuhaus (1960)
Ureyite ($NaCrSi_2O_6$)	1.998	15,600	22,000	Green	White *et al.* (1971)
Cr-tremolite	2.07	16,310	23,530	Green	R. G. Burns (unpublished); Neuhaus (1960)
Cr-epidote	1.93 or 2.05	≃16,300	≃24,000	Green	Burns and Strens (1967)
Cr-chlorite	—	18,450	25,000	Red-violet	R. G. Burns (unpublished); Neuhaus (1960)
Cr-mica	—	15,820	23,580	Green	R. G. Burns (unpublished); Neuhaus (1960)
Cr-tourmaline	1.93 or 2.05	17,000	24,000	Green	Manning (1969a)

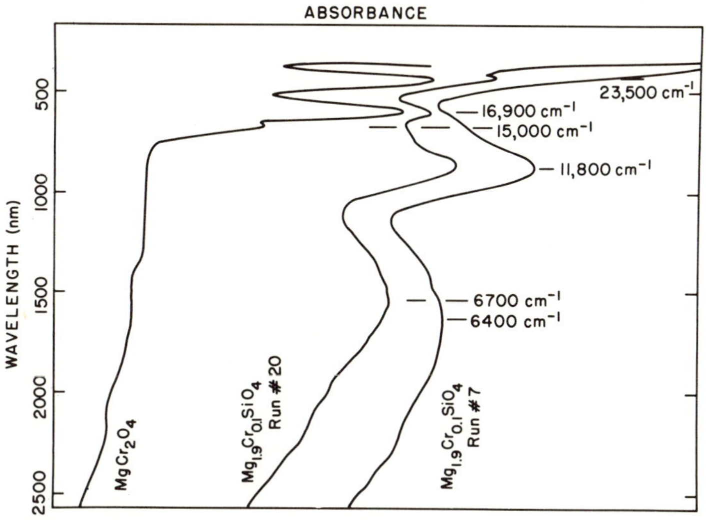

FIG. 2. Diffuse reflectance spectra of Cr-bearing forsterite and picrochromite. The peaks at 23,500 cm^{-1} and 16,900 cm^{-1} arise from Cr^{3+} ions, the other broad bands are due to Cr^{2+} ions. (From Scheetz and White, 1972.)

to crystal field transitions in Cr^{2+} ions. Evidence for Cr^{2+} ions in the pyroxene structure has been demonstrated by spectral measurements of a synthetic chrome diopside reproduced in Fig. 3 (Mao *et al.*, 1972). Absorption around 22,000 and 15,000 cm^{-1} is due to Cr^{3+} ions. However, the additional bands centered around 17,000, 13,200, and 9,500 cm^{-1} may be attributed to crystal field transitions in Cr^{2+} ions. Spectral measurements have also demonstrated the existence of tetrahedrally coordinated Cr^{2+} ions. Thus, Ulmer and White (1966) and Greskovich and Stubican (1966) synthesized magnesium chromium spinels and attributed absorption features at 6250 and 6670 cm^{-1} to crystal field transitions in Cr^{2+} ions in the tetrahedral sites.

3. *Iron*

a. *Fe^{3+}*. The Fe^{3+} ion, [A]$3d^5$, with one electron in each of the five 3d orbitals, gives rise to the nondegenerate $^6A_{1g}$ ground state. Excited states involve pairing of electrons in different orbitals, giving rise to a variety of crystal field states of lower spin multiplicities (quartets and doublets, compared to the sextet ground state). As a result, all electronic transitions in Fe^{3+} ions are spin-forbidden and lead to weak absorption bands. The energy level diagram for a d^5 ion, including Mn^{2+} and Fe^{3+}, is illustrated

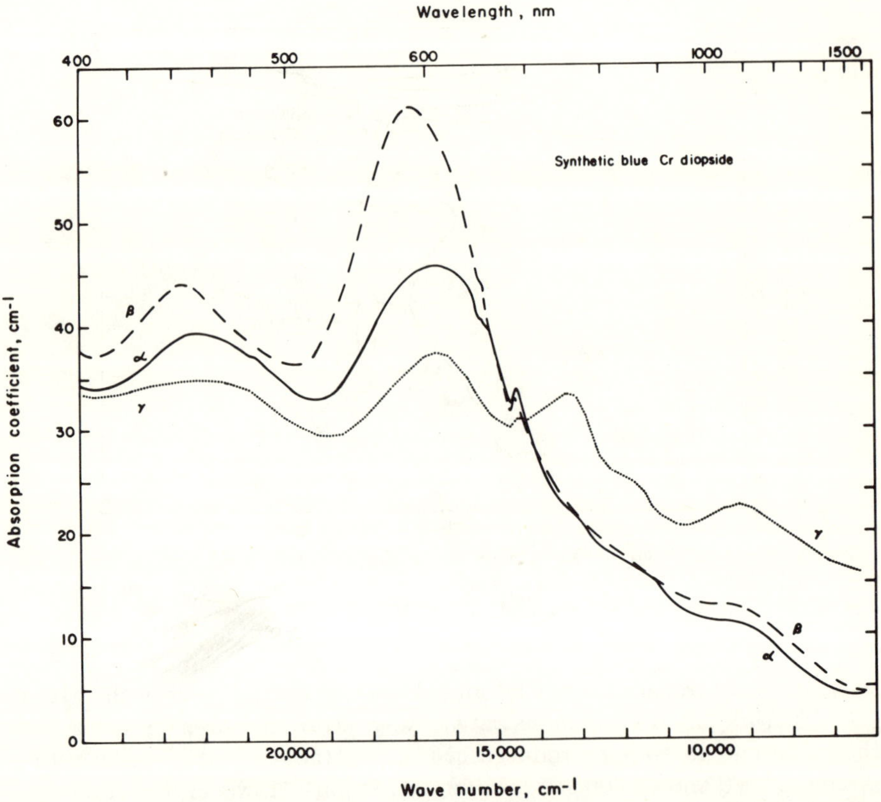

FIG. 3. Polarized absorption spectra of blue synthetic chrome diopside. The spectra contain crystal field bands originating from Cr^{2+} and Cr^{3+} ions. (From Mao *et al.*, 1972.)

in Fig. 4. Significant features of this diagram are first, the accidental degeneracy of the $^4A_{1g}$, ^4E_g (G) states; and second, the field independence of the $^4A_{1g}$, ^4E_g (G), and ^4E(D) states, which are parallel to the $^6A_{1g}$ ground state. As a result, absorption bands that characterize the electronic transitions $^6A_{1g} \rightarrow {}^4A_{1g}$, ^4E_g (G), and $^6A_{1g} \rightarrow {}^4E_g$ (D) are particularly sharp relative to other transitions such as $^6A_{1g} \rightarrow {}^4T_{1g}$ (G) and $^6A_{1g} \rightarrow {}^4T_{2g}$ (G). This is illustrated by the spectrum of andradite shown in Fig. 5. The sharp peak around 22,000–23,000 cm^{-1} is diagnostic of Fe^{3+} ions coordinated by oxygen. Other spectral data for Fe^{3+} ions in silicates are summarized in Table II and Manning (1970).

b. Fe^{2+}. The electronic configuration of Fe^{2+}, [A]$3d^6$, is such that, at normal pressures, ferrous ions have four unpaired electrons and occur in high-spin states in silicates and oxides. In octahedral coordination, the ground state is $^5T_{2g}$, corresponding to the electron distribution $(t_{2g})^4(e_g)^2$. Excited states include ^5E_g, corresponding to the high-spin configuration $(t_{2g})^3(e_g)^3$ and a variety of spin-paired configurations with two unpaired electrons (triplet states) and zero unpaired electrons (singlet states) Fig. 6). The spectra of octahedral Fe(II) oxide compounds contain one broad, asymmetric band centered around 10,000 cm^{-1} ($^5T_{2g} \rightarrow {}^5E_g$) and several weaker spin-forbidden peaks in the visible region representing transitions to triplet states. The asymmetry of the spin-allowed band results from the dynamic Jahn-Teller effect (Cotton and Meyers, 1960), whereby the excited ^5E_g state is split into two energy levels. The degeneracies of the ^5E_g and $^5T_{2g}$ states are removed when Fe^{2+} ions occur in distorted octahedral environments, such as the six-coordinated sites in most silicate structures (Burns and Huggins, 1973). Thus, a variety of spin-allowed and spin-forbidden peaks occur in the crystal field spectra of most rock-forming Fe^{2+}-bearing silicates.

The polarized spectra of ferromagnesian silicate minerals have been studied extensively, and data prior to 1970 are summarized in Burns (1970a, Chapter 5). Noteworthy trends are

1. The wide range of energies for the spin-allowed transitions, which depend on the point symmetry of oxygen ions around the Fe^{2+} ions. In six-coordinate sites, the positions of the absorption bands range from about

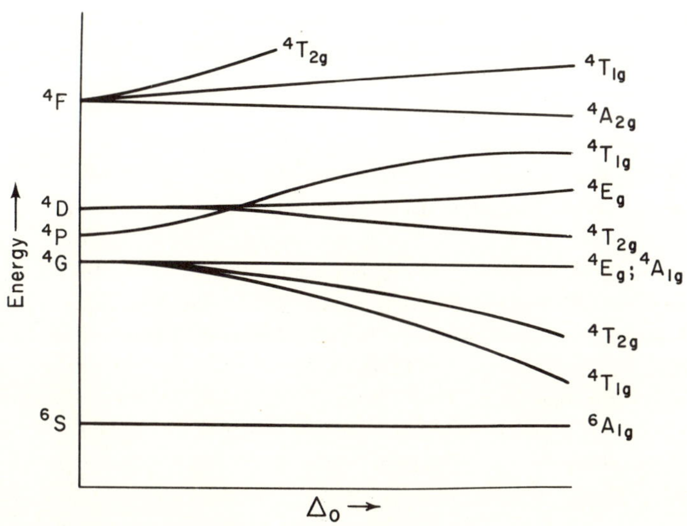

FIG. 4. Energy level diagram for Fe^{3+}. The diagram also applies to Mn^{2+} ions.

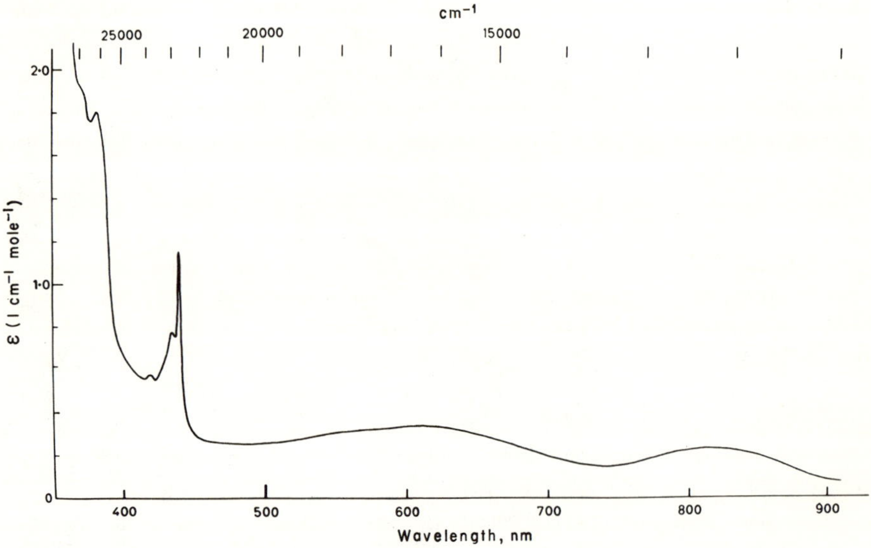

FIG. 5. Absorption spectrum of andradite, $Ca_3Fe_2(SiO_4)_3$. The spectrum shows the typical profile of a crystal field transition in Fe^{3+} ions. (From Burns, 1972.)

12,000 cm^{-1} to about 3,000 cm^{-1}. In pyroxenes, for example, there are three prominent spin-allowed bands at 11,000–10,000 cm^{-1}, 5,500–4,000 cm^{-1}, and 3,100 cm^{-1} (Runciman *et al.*, 1973a; Burns, 1970a, pp. 88-89).

2. The spectra usually contain contributions from Fe^{2+} ions in two or more sites in the ferromagnesian silicate crystal structure. Different site occupancies lead to changing spectral profiles. For example, the spectra of enstatite are dominated by intense bands at 10,800 cm^{-1}, 5,400 cm^{-1} and 3,100 cm^{-1} from Fe^{2+} ions in the noncentrosymmetric M2 sites, in which the Fe^{2+} ions are strongly enriched. Less prominent features at 8,500 cm^{-1} and around 10,000 cm^{-1} gain intensity in the spectra of iron-rich orthopyroxenes, in which there are greater occupancies of Fe^{2+} ions in the M1 site. The spectra of olivines are also composed of contributions from Fe^{2+} ions in M1 sites (11,700–11,400 cm^{-1} and 8,100–9,200 cm^{-1}) and M2 sites (9,600–9,300 cm^{-1}) (Burns, 1970b).

The relative intensities of these bands remain approximately constant across the forsterite–fayalite series because there is no strong enrichment of Fe^{2+} ions in either M1 or M2 sites (Birle *et al.*, 1968; Finger, 1970, 1971; Brown, 1971; Wenk and Raymond, 1973). The relative intensities change in the spectra of Fe_2SiO_4–Mn_2SiO_4 series as Mn^{2+} ions displace Fe^{2+} ions from the M2 sites (Burns, 1970b, Fig. 7). Although there has been debate recently about site assignment of the spin-allowed bands of olivine (Burns, 1970b; Runciman *et al.*, 1973b), it is now generally agreed (Burns, 1974; Runciman *et al.*, 1974; Mao and Bell, 1973) that Fe^{2+} ions in both M1 and M2 sites contribute to the spectra.

TABLE II

POSITIONS OF CRYSTAL FIELD BANDS IN THE SPECTRA OF Fe^{3+} OXIDE AND SILICATE MINERALS

	ABSORPTION BANDS (cm^{-1})					
MINERAL	$^4T_{1g}$	$^4T_{2g}$	^4E_g, $^4A_{1g}$	$^4T_{2g}$	^4E_g	REFERENCE
Sapphire (Al_2O_3/Fe^{3+})	9,700	18,700	22,200	25,600	26,700	Ferguson and Fielding (1971)
Kyanite (Al_2SiO_5/Fe^{3+})	(16,000–17,000)		22,400 23,400	26,500	27,000	Faye and Nickel (1969)
Pyroxene	14,000	16,700	22,200	—	—	Bell and Mao, (1972a,b) R. G. Burns (unpublished)
Epidote	9,470	16,480	{21,200 {22,000	24,800		Burns and Strens (1967)
Vesuvianite			21,600	26,000		Manning (1968b)
Andradite	12,453	16,650	{22,701 {22,999	24,000	27,000	Moore and White (1972)
Grossularite	13,111		{22,865 {23,121	23,592	27,040	Moore and White (1972)
Grossularite[a]	18,000	19,700	21,758	26,400		Moore and White (1972)
Orthoclase[a]		20,700	{24,000 {22,650	26,500		Faye (1969)
Phlogopite[a]	19,200	20,300	22,700	25,000		Faye and Hogarth (1969)

[a] Tetrahedral Fe^{3+}.

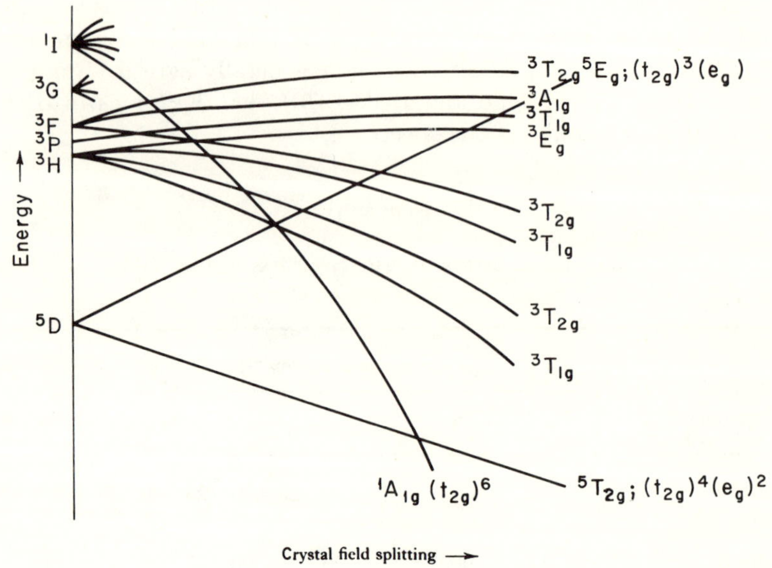

FIG. 6. Partial energy-level diagram for Fe^{2+} in octahedral coordination. (From Burns *et al.*, 1972a.)

3. Band maxima change with increasing Fe^{2+} content of a ferromagnesian silicate. The shifts have been well documented for olivines (Burns, 1970b), orthopyroxenes (Burns, 1965; Adams *et al.*, 1973; also see Chapter 4, Section III,C), and other silicate series (Burns, 1965). In the olivine spectra, the variations of spectral profiles around 10,000 cm^{-1} across the forsterite–fayalite series is a direct consequence of the greater compositional variation for the 11,700–11,400 cm^{-1} and 8,100–9,200 cm^{-1} bands relative to the band at 9,600–9,300 cm^{-1} (Burns, 1970b).

B. Intervalence Transitions

1. Homonuclear Intervalence Transfer

a. Iron–Iron Charge Transfer. The spectra, and consequently the color, of many iron-bearing minerals are strongly influenced by absorption bands in the 13,000–19,000 cm^{-1} region of the spectrum, which are attributed to electron transfer between neighboring Fe^{2+} and Fe^{3+} ions (Allen and Hush, 1967; Hush, 1967; Faye, 1968; Faye *et al.*, 1968; Faye and Nickel, 1969, 1970). The charge transfer process occurs in structures in which adjacent Fe^{2+} and Fe^{3+} ions share octahedral edges and there is possible overlap of t_{2g} orbitals. The directional nature of these interactions produces a marked polarization dependence of the absorption bands, and intense broad spectral features are observed when light is polarized along the direction of the Fe^{2+}–Fe^{3+} bond (Faye *et al.*, 1968; Burns, 1973b). Frequently, crystal field transitions in Fe^{2+} and Fe^{3+} may be seen superimposed upon the intervalence charge-transfer transitions in polarized absorption spectra. This is well illustrated by the spectrum of glaucophane shown in Fig. 7. For light polarized in the band of cations in the (001) plane of the amphibole structure (β and γ), the spectrum shows a broad, intense charge-transfer band centered around 18,520 and 16,130 cm^{-1}. However, the diagnostic spin-forbidden peak for Fe^{3+} at 22,850 cm^{-1} and spin-allowed bands for Fe^{2+} at 10,100 and 8,400 cm^{-1} are clearly visible. The measured energies of Fe^{2+}–Fe^{3+} intervalence

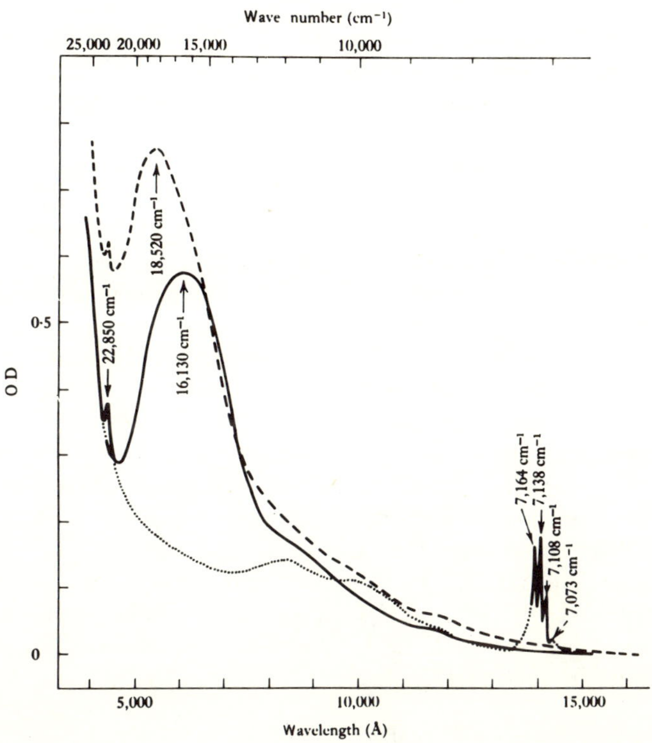

FIG. 7. Polarized absorption spectra of glaucophane, $Na_2(Mg, Fe^{2+})_3(Al, Fe^{3+})_2$-$Si_8O_{22}(OH)_2$. The band centered around 18,500–16,000 cm^{-1} represents a $Fe^{2+} \rightarrow Fe^{3+}$ intervalence charge transfer transition. .... α spectrum; - - - - β spectrum; ——— γ spectrum. (From Bancroft and Burns, 1969.)

transitions, which have been suggested for a variety of iron-bearing minerals, are summarized in Table III. Note that the proposed energies cluster around a relatively narrow range (13,000–19,000 cm^{-1}). In spite of the copious spectral and crystallographic data on the host phases, no quantitative correlations between metal–metal distances and the energy and intensity of Fe^{2+}–Fe^{3+} interactions have been proposed, although it was suggested that the probability (and hence spectral intensity) of an intervalence transition decreases with increasing internuclear separation (Allen and Hush, 1967, p. 359). Molecular orbital calculations using the SCF-X_α scattered wave cluster method have been performed on regular octahedral $(FeO_6)^{-9}$ and $(FeO_6)^{-10}$ clusters (Tossell *et al.*, 1974). Using the results of these calculations, preliminary values have been obtained for the energies of $Fe^{2+} \rightarrow Fe^{3+}$ charge transfer in silicates (Table III).

b. Titanium–Titanium Charge Transfer. The best-documented example of a homonuclear intervalence transition in titanium involves $TiCl_3$ in concentrated HCl solutions (Hush, 1967, p. 398). Jorgensen (1957) observed that if a sky-blue solution of Ti(III) in strong HCl is slighty oxidized, it becomes purplish black in color. The broad band centered at 16,750 cm^{-1} in $TiCl_3$ solutions, attributed to the $^2T_{2g} \rightarrow {}^2E_g$ crystal field transition in Ti^{3+} ions, is intensified as a prominent inflexion of a broad band centered at 20,400 cm^{-1} for Ti(III, IV) chloride solutions. The 20,400 cm^{-1} band represents a $Ti^{3+} \rightarrow Ti^{4+}$ intervalence transition in titanium ions coordinated to Cl^- ions.

In silicates, however, $Ti^{3+} \rightarrow Ti^{4+}$ charge transfer transitions have been used to assign the spectra of a variety of minerals over a wide range of energies. These are summarized in Table IV. Note that in some cases, alternative interpretations have been proposed for these absorption bands. For example, the 16,500 cm^{-1} band for the titanium pyroxene from the Allende meteorite was assigned to a crystal field transition in Ti^{3+} ions (Burns and Huggins, 1973; see Section III, A, 1b), as was the band at 15,600 cm^{-1} for synthetic $NaTiSi_2O_6$. Furthermore, in iron-bearing minerals, the bands attributed to $Ti^{3+} \rightarrow Ti^{4+}$ charge transfer may represent heteronuclear intervalence transitions between Fe and Ti ions. These are discussed further in Section III, B, 2.

2. Heteronuclear Electron Transfer

Iron–Titanium Charge Transfer. Many silicate minerals contain appreciable amounts of both iron and titanium. Since each element can

TABLE III

POSITIONS OF SUGGESTED Fe^{2+}–Fe^{3+} CHARGE TRANSFER BANDS IN IRON-BEARING MINERALS

MINERAL	M–M DISTANCE (Å)	INTERVALENCE TRANSITION (cm^{-1})	REFERENCE
Vivianite	2.85	15,200	Hush (1967); Faye (1968); Faye *et al.* (1968); Townsend and Faye (1970)
Sapphire	2.65, 2.79	16,400	Faye (1971a,b)
Kyanite	2.75	16,700	Faye and Nickel (1969)
Orthopyroxene	3.20, 3.25	15,500–16,500	Burns (1970a, p. 72)
Augite	3.15, 3.25	13,000	Burns (1970a, p. 93)
Crocidolite	3.09, 3.10, 3.22, 3.21	15,000 and 18,000	Littler and Williams (1965); Faye and Nickel (1970); Bancroft and Burns (1969); Hush (1967)
Cordierite	2.85	17,500	Faye *et al.* (1968)
Tourmaline	3.00	14,000 17,000	Faye *et al.* (1968) Burns and Simon (1973)
Calculated[a]	3.20	13,600	J. A. Tossell, R. G. Burns, D. J. Vaughan, and K. H. Johnson (unpublished)

[a] MO Calculation on Fe^{2+} and Fe^{3+} in regular octahedra of oxygens ($Fe^{3+} - O = 2.06$ Å; $Fe^{2+} - O = 2.17$ Å; $Fe^{3+} - Fe^{2+} = 3.20$ Å).

occur in two or more oxidation states, a variety of heteronuclear charge transfer processes are possible. These include

$$(1)\quad Fe^{2+} + Ti^{4+} \rightleftarrows Fe^{3+} + Ti^{3+}$$

$$(2)\quad Fe^{3+} + Ti^{3+} \rightleftarrows Fe^{2+} + Ti^{4+}$$

$$(3)\quad Fe^{2+} + Ti^{3+} \rightleftarrows Fe^{3+} + Ti^{2+}$$

TABLE IV

ENERGIES SUGGESTED FOR Ti^{3+}–Ti^{4+} CHARGE TRANSFER TRANSITIONS

COMPOUND OR MINERAL	ENERGY OF CHARGE TRANSFER	REFERENCE
Ti(III, IV) chloride	20,400	Jorgensen (1957)
Melanite (Ti-garnet)	5,280	Moore and White (1971)
Reduced rutile (TiO_{2-x})	6,700	Johnson *et al.* (1968)
Magneli phases (Ti_nO_{2n-1})	12,000–15,000	See Moore and White (1971)
$SrTiO_{3-x}$	4,500	McCarthy *et al.* (1969)
Synthetic $NaTiSi_2O_6$	15,600	Prewitt *et al.* (1972)
Allende pyroxene $Ca(Mg, Ti^{3+}, Ti^{4+})(Si, Al)_2O_6$	16,500	Dowty and Clark (1973a,b)
Andalusite	20,800	Faye and Harris (1969)
Clintonite	22,000	Manning (1968c)
Astrophyllite	23,000	Manning (1968c)
Tourmaline	22,000	Manning (1969b)

In each of these examples, the species on the left-hand side of the equilibria represent coexisting ions in the ground state, which are capable of being identified by physical (e.g., Mössbauer spectroscopy, ESR, electronic spectra) and chemical techniques. The species on the right represent transient ions in the excited state formed by photochemical oxidation-reduction. By analogy with electron-hopping phenomena in magnetite, an oscillation of a valence may be more rapid than 10^8 sec^{-1} (Robin and Day, 1967) when Fe^{2+} and Fe^{3+} ions, for example, are independently characterized by Mössbauer spectroscopy.

Heteronuclear intervalence transitions between Fe^{2+} and Ti^{4+} ions have been suggested for terrestrial titanaugites, a typical set of polarized spectra for which are shown in Fig. 8. There are clearly a large number of spectral features portrayed in these spectra. Interest centers on the absorption maximum around 22,000 cm^{-1}. Originally, this feature was attributed to a crystal field transition in Ti^{3+} ions (Chesnokov, 1959; Burns *et al.*, 1964; Burns, 1970a). However, Manning and Nickel (1969) curve-fitted the spectrum of a titanaugite to four broad bands at 22,000, 19,000, 14,000, and 10,000 cm^{-1}, and proposed the following assignments: 22,000 cm^{-1} to the $^6A_{1g} \rightarrow {}^4A_{1g}, {}^4E_g$ spin-forbidden transition in octahedral Fe^{3+} (compare Table II); 19,000 cm^{-1} to the $^2T_{2g} \rightarrow {}^2E_g$ crystal field transition in Ti^{3+} (see Burns and Huggins, 1973); 14,000 cm^{-1} to $Fe^{2+} \rightarrow Fe^{3+}$ charge transfer (compare Table III); and 10,000 cm^{-1} to $^5T_{2g} \rightarrow {}^5E_g$ spin-allowed transition in Fe^{2+}. Mössbauer spectroscopy of titanaugites (Burns *et al.*, 1972b; Virgo, 1972) has revealed considerable amounts of tetrahedrally coordinated Fe^{3+} ions (in addition to octahedral Fe^{3+} ions) replacing silicon in the linked tetrahedra in the pyroxene structure, leading to the suggestion that the $^6A_{1(g)} \rightarrow {}^4A_{1(g)}, {}^4E_{(g)}$ transition in octahedrally and tetrahedrally coordinated Fe^{3+} ions causes the observed intensification and broadening of the band at 22,000 cm^{-1} (Burns *et al.*, 1972b). Additional inflexions in the spectra shown in Fig. 8 may be attributed to spin-forbidden transitions in Fe^{2+} and Fe^{3+} ions. Recently, however, Dowty and Clark (1973b) have proposed that the band at 22,000 cm^{-1} represents charge transfer between Fe^{2+} and Ti^{4+} ions. Clearly, further investigations of the spectra of Fe–Ti pyroxenes are needed to resolve these multiple interpretations of the electronic spectra. Some predictions of $Fe^{2+} \rightarrow Ti^{4+}$ charge transfer energies are being made on the basis of molecular orbital calculations (Table V) and refinement of these data may clarify the assignments.

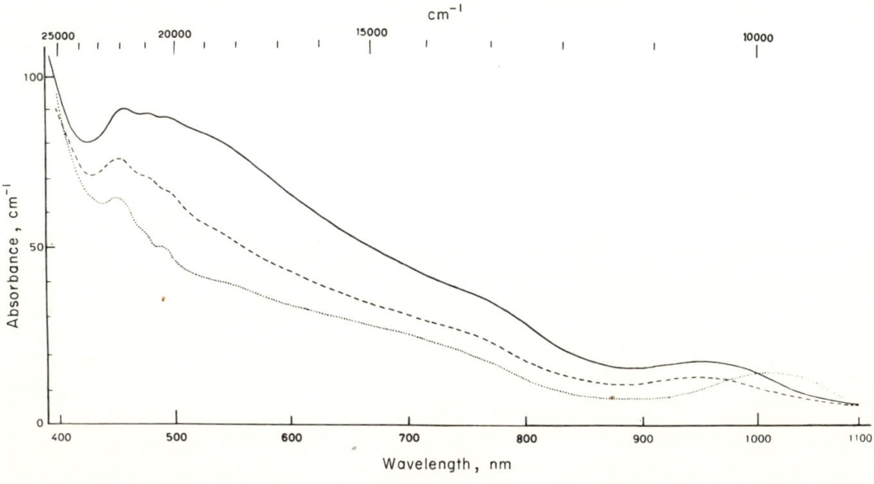

FIG. 8. Polarized absorption spectra of a terrestrial titanaugite. The specimen is from a techenite Caroline Is., and has the composition $Ca_{0.26}Na_{0.03}Mg_{0.64}Fe_{0.21}Ti_{0.15}Al_{0.47}$-$Si_{1.56}O_6$. .... α spectrum; ---- β spectrum; ——— γ spectrum.

TABLE V

ENERGIES SUGGESTED FOR IRON–TITANIUM CHARGE TRANSFER TRANSITIONS

COMPOUND OR MINERAL	ENERGY OF CHARGE TRANSFER	REFERENCE
Sapphire	14,200–17,800	Townsend (1968); Fergusen and Fielding (1971); Eigenmann *et al.* (1972)
Andalusite	20,800	Faye and Harris (1969); Dowty and Clark (1973a)
Titanaugite	22,000	Dowty and Clark (1973a)
Micas	25,000	Faye (1968)
Calculated[a]	18,800	J. A. Tossell, R. G. Burns, D. J. Vaughan, and K. H. Johnson (unpublished)

[a] MO Calculation on Fe^{2+} and Ti^{4+} in regular octahedral of oxygens ($Fe^{2+} - O = 2.17$ Å; $Ti^{4+} - O = 1.965$ Å; $Fe^{2+} - Ti^{4+} = 3.20$ Å).

IV. Spectra of Lunar Minerals

A. Pyroxenes

Several polarized absorption spectral measurements of lunar pyroxenes have been reported from rocks from the Apollo 11 (Burns *et al.*, 1972a,c), Apollo 12 (Mao and Bell, 1971, 1973; Burns *et al.* 1972c; Cohen, 1972; Bell and Mao, 1972a,b), Apollo 14 (Bell and Mao, 1972b), Apollo 15 (Burns *et al.*, 1972c, 1973), and Luna 20 (Adams *et al.*, 1973) missions. A typical spectrum is shown in Fig. 9. The spectra contain prominent, intense absorption bands in the near-infrared region around 900–1000, 1200, and 1800–2000 nm, together with several weaker, sharp peaks in the visible region, which are superimposed on an absorption edge descending from the ultraviolet. The broad bands in the near-infrared represent crystal field transitions in Fe^{2+} ions located predominantly in the non-centrosymmetric M2 sites of the pyroxene structure (Hafner and Virgo,

1970; Hafner *et al.*, 1971a; Ghose *et al.*, 1972; Malysheva, 1972), although Fe^{2+} ions in M1 sites also contribute to the spectra at 1200 nm and around 1000 nm. The absorption maxima of these spin-allowed transitions shift with pyroxene-type and composition (Burns *et al.*, 1972c; Adams *et al.*, 1973, Fig. 9; also see Chapter 4, Section III, C).

There has been considerable debate over assignments of the peaks in the visible region (Burns *et al.*, 1973), particularly the doublet or asymmetric peak around 19,800 cm^{-1}, which has been variously assigned to charge transfer processes involving Fe or Ti, spin-forbidden peaks in Fe^{3+}, or spin-forbidden peaks in Fe^{2+}. Since intervalence transitions lead to absorption bands which are generally polarization dependent (e.g., see Fig. 7) and broader than the sharp peaks in the visible region, such charge-transfer transitions appear to be unlikely origins of the prominent peak at 19,800 cm^{-1} in pyroxene spectra. Ferric ions are of such low concentrations in lunar pyroxenes so as to be undetectable by Mössbauer spectroscopy, so that there are insufficient amounts of Fe^{3+} ions to account for the intensity of this spectral feature. Furthermore, the sharp spin-forbidden Fe^{3+} peak in silicates lies at higher energies (Table II). Thus, a spin-forbidden transition in Fe^{2+} ions appears to be the most plausible explanation for the peak at 19,800 cm^{-1} in pyroxenes (Runciman *et al.*, 1973a).

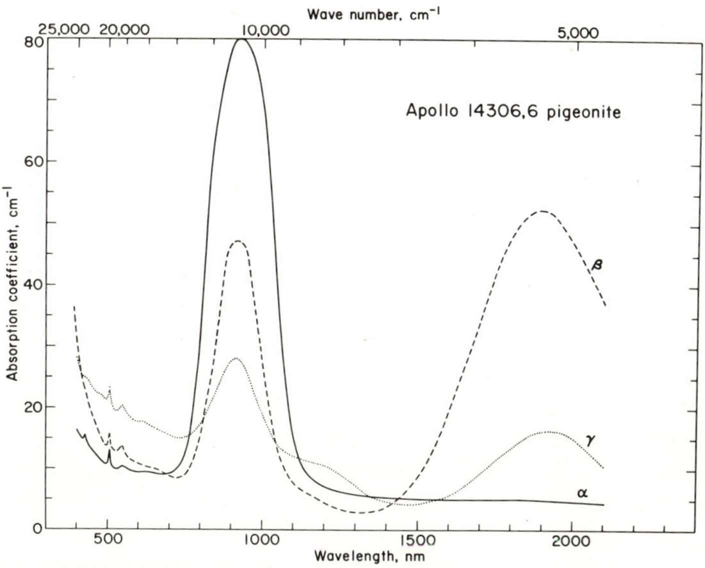

FIG. 9. Polarized absorption spectra of a lunar pigeonite from Apollo 14 rock 14306,6. (From Bell and Mao, 1972b.)

There has also been discussion over the weak, broad bands centered around 16,500 cm^{-1} and 21,000 cm^{-1} in the titaniferous lunar pyroxenes. These have been assigned to crystal field transitions in Ti^{3+} ions located in distorted M1 sites of the pyroxene structure (Burns *et al.*, 1972a,c), by analogy with the spectra of the Allende pyroxene (Burns and Huggins, 1973; see Section III, A, 1b). However, Dowty and Clark (1973a,b) have proposed alternative assignments for these two bands in lunar pyroxenes in terms of $Fe^{2+} \rightarrow Ti^{4+}$ (21,000 cm^{-1}) and $Ti^{3+} \rightarrow Ti^{4+}$ (16,500 cm^{-1}) intervalence transitions.

Further work is required to resolve these alternative band assignments. Note that although the 16,500 cm^{-1} band lies in the regions of Cr^{3+} ion crystal field transitions (Table I) and of Fe^{2+}–Fe^{3+} homonuclear intervalence transitions (Table III), the low concentrations of Cr relative to Ti and the undetectable Fe^{3+} ion contents in the Apollo 11 pyroxenes mitigate against contributions to the spectra from Cr^{3+} and Fe^{3+} ions.

Spectral evidence has been cited for Cr^{3+} ions in certain lunar pyroxenes from Apollo 15 rocks (Burns *et al.*, 1973). Thus, the relatively broad and weak bands at 16,670 and 22,200 cm^{-1} measured from the pigeonite cores of large phenocrysts in rock 15058 may be correlated with crystal field transitions in Cr^{3+} ions (Table I). The low Fe^{2+} ion concentration and high Cr/Fe ratios in these core pigeonites enabled the Cr^{3+} peaks to be resolved.

Spectral data for pyroxenes from Apollo 17 rocks have revealed high Ti^{3+}/Ti^{4+} ratios (Sung *et al.*, 1974).

B. *Pyroxferroite*

The new pyroxenoid mineral, pyroxferroite $CaFe_6(SiO_3)_7$, was reported in the groundmass or mantling pyroxene phenocrysts in several lunar rocks (Chao *et al.*, 1970). Large phenocrysts of pyroxferroite, containing negligible Cr, Mn, and Ti contents, occur in rock 15058 and are suitable for polarized spectral measurements (Burns *et al.*, 1973). The spectra are illustrated in Fig. 10. Broad, intense bands occur at 9,800, 7,300, and 4,750 cm^{-1}, while sharper, weak but prominent peaks occur at 18,200 and 19,800 cm^{-1}. There are obvious correlations with the spectra of isostructural pyroxmangite (Manning, 1968a), but peaks attributable to Mn^{2+} ions are absent in the lunar pyroxferroite. The crystal structure of pyroxferroite (Burnham, 1971) indicates that Fe^{2+} ions occupy at least five of the seven cation sites. Thus, the breadth of the bands in the near-infrared region may be attributed to closely overlapping spin-allowed transitions in Fe^{2+} ions located in several coordination sites of the pyroxferroite structure. In the absence of large amounts of other transition metal ions,

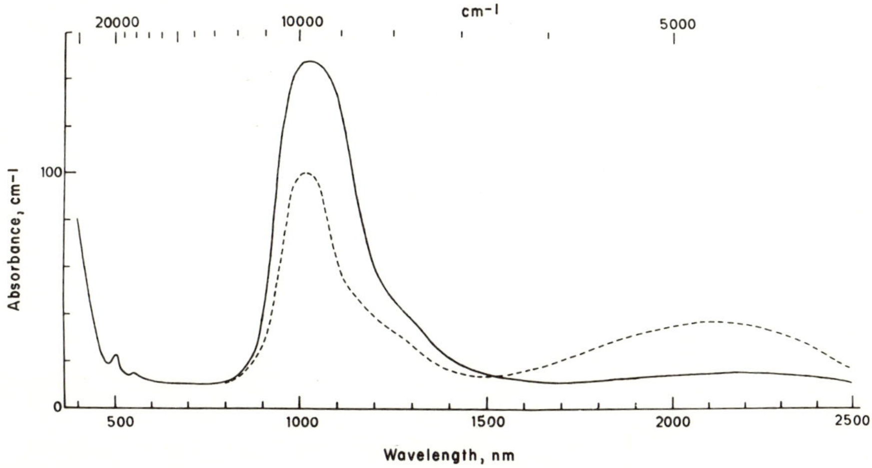

FIG. 10. Polarized absorption spectra of pyroxferroite from Apollo 15 rock 15058. The composition of the lunar pyroxenoid is $Ca_{0.13}Mg_{0.03}Fe_{0.83}Al_{0.02}Si_{0.98}O_6$. ——— β spectrum; - - - - α spectrum.

the spectrum of pyroxferroite serves to delineate clearly the positions of the weaker spin-forbidden bands due to Fe^{2+} ions in the visible region. The visible-region spectrum of pyroxferroite shown in Fig. 10 has thus helped in the assignment of Cr^{3+} and Ti^{3+} crystal field bands in other lunar pyroxenes (Burns *et al.*, 1973).

C. Olivine

Several polarized absorption spectra of olivines in lunar rocks have been described for Apollo 11 (Haggerty *et al.*, 1970; Burns *et al.*, 1972a), Apollo 12 (Mao and Bell, 1971; Bell and Mao, 1972a,b), Apollo 15 (Bell and Mao, 1972; Mao and Bell, 1973; Burns *et al.*, 1973), and Luna 20 (Adams *et al.*, 1973) samples. Typical spectra are reproduced in Fig. 11. There are remarkable similarities with the spectra of terrestrial olivines (Burns, 1970a,b; Runciman *et al.*, 1973b). The broad bands at 8680 and 11,210 cm^{-1} are attributed to Fe^{2+} ions in olivine M1 positions (Burns, 1970b, 1974; Mao and Bell, 1973; Runciman *et al.*, 1974), while the more intense 9500 cm^{-1} band originates from Fe^{2+} ions in noncentrosymmetric M2 positions. The band maxima vary with composition and conform well with the determinative curves established for terrestrial Mg–Fe olivines derived from polarized spectral measurements (Burns, 1970b; Burns *et al.*, 1972c); (also see Chapter 4, Section III, B.)

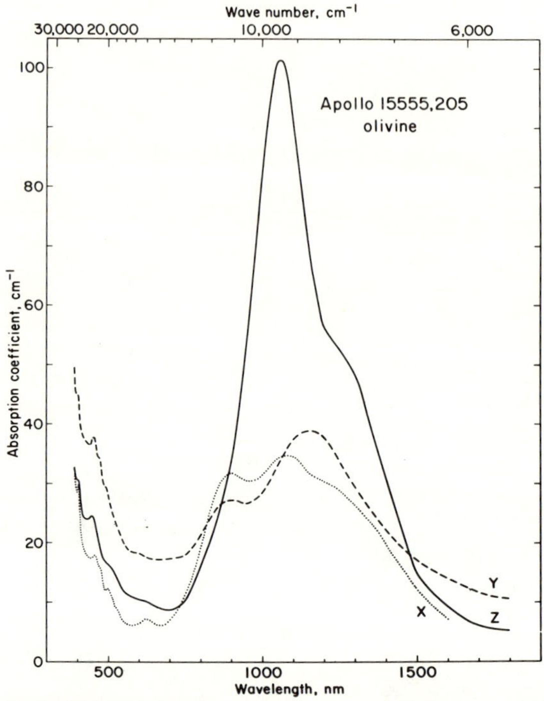

FIG. 11. Polarized absorption spectra of a lunar olivine from Apollo 15 rock 15555,205. (From Bell and Mao, 1972b; Mao and Bell, 1973.)

There is a debate over assignment of the abundant features in the visible region spectra of olivines. Most of the discussion centers around the peaks at 16,800 and 22,200 cm^{-1}, which have been variously assigned to $Fe^{2+} \rightarrow Fe^{3+}$ charge transfer (Mao and Bell, 1971; Bell and Mao, 1972a,b; compare Table III), Cr^{3+} spin-allowed transitions (Burns *et al.*, 1973; see Table II), Fe^{3+} spin-forbidden transitions (Burns *et al.*, 1973; compare Table II), and Fe^{2+} spin-forbidden transitions (Runciman *et al.*, 1973b; Burns, 1970b). Clearly further work is required to resolve these assignments.

Another area of discussion centers around spectral evidence for Cr^{2+} ions in lunar olivines, particularly from Apollo 11 rocks (Burns *et al.*, 1973). Spectral features reported at 720 nm (13,900 cm^{-1}) and 1,040 nm (9,600 cm^{-1}) have been attributed to crystal field transitions in Cr^{2+} ions (Haggerty *et al.*, 1970). Although Cr^{2+} ions have been predicted to be stabilized in the olivine structure (Burns, 1969, 1970a, p. 124) and indeed Cr^{2+}-bearing forsterites were synthesized (Scheetz and White, 1972; see Section III, A, 2, b), it has been suggested that Fe/Cr ratios in lunar olivines are unfavorable for resolution of Cr^{2+} spectral features among the numerous spin-allowed and spin-forbidden bands arising from the more abundant Fe^{2+} ions (Burns *et al.*, 1973). Again, additional work is required to resolve this problem.

D. Plagioclase Feldspars

The spectra of plagioclase feldspars in rocks from Apollo 14 (Bell and Mao, 1972a,b; Mao and Bell, 1973) and Luna (Adams *et al.*, 1973) missions show a broad, strongly polarized band centered around 8000 cm^{-1} with pronounced asymmetry on the long wavelength side. The absorption band may be attributed to Fe^{2+} ions in a variety of sites of the plagioclase structure. Although small amounts of Fe^{3+} ions have been reported in ESR measurements of lunar plagioclase feldspars (Kolopus *et al.*, 1971; Weeks, 1972, 1973; Hafner *et al.*, 1973), it is extremely unlikely that the low concentrations of Fe^{3+} ions would be sufficient to produce the weak spin-forbidden bands in this region of the near-infrared spectrum (compare Table II). For the role of Fe^{3+} ions in the near-infrared luminescence of lunar and terrestrial feldspars see Chapter 3, Section III, C. Mössbauer spectroscopy of lunar plagioclase feldspars has demonstrated that Fe^{2+} ions occur in the irregular 8–12 coordinated calcium sites (Appleman *et al.*, 1971; Hafner *et al.*, 1971b; Schürmann and Hafner, 1972) and in the tetrahedral sites (Appleman *et al.*, 1971) by analogy with other framework silicates (Burns and Burns, 1974). The large width of the absorption band centered at 8000 cm^{-1} in the crystal field spectrum may thus be assigned to overlapping contributions from Fe^{2+} ions in two or more of the four calcium positions of the plagioclase structure (Megaw *et al.*, 1962). Tetrahedrally coordinated Fe^{2+} ions, which produce crystal field bands at lower energies (Slack, 1964; Bates *et al.*, 1966), are probably responsible for the asymmetry on the long wavelength shoulder of this absorption band.

V. Applications to Lunar Science

A. The Oxidation States of Transition Metal Ions in Lunar Minerals and the Redox Potentials at the Time of Their Crystallization

There has been considerable debate in the literature regarding the oxidation states of Fe, Ti, and Cr present in the Apollo lunar samples, as determined from electronic absorption spectra (Burns *et al.*, 1973). However, these debates, many of which have been discussed in Section

IV, have centered around subtleties of spectral interpretation, and tend to obscure the large extent of agreement between different workers. That almost all of the iron in lunar minerals occurs as Fe^{2+} has been established independently by Mössbauer spectroscopy (Hafner and Virgo, 1970; Hafner *et al.*, 1971a; Ghose *et al.*, 1972; Malysheva, 1972). Mössbauer studies have failed to detect Fe^{3+} ions in the ferromagnesian silicate minerals, and the generally accepted detectability limit for this technique lies in the range 0.3–1% of the total iron as Fe^{3+}. However, ESR measurements of lunar plagioclase feldspars have clearly shown the presence of several ppm Fe^{3+} (e.g., Weeks, 1972; Hafner *et al.*, 1973), although this may have been formed after crystallization by shock and radiation effects (Weeks, 1973).

The occurence of Ti^{3+} in addition to Ti^{4+} in lunar pyroxenes was postulated on analytical and crystal chemical grounds from the earliest studies (e.g., Agrell *et al.*, 1970; Lovering and Ware, 1970; Dence *et al.*, 1970; Bence and Papike, 1972; Boyd and Smith, 1971). The subsequent identification of peaks attributed to Ti^{3+} in spectra of lunar pyroxenes (Burns *et al.*, 1972a,c, 1973) has established the presence of Ti^{3+}. The presence of Cr^{2+} in lunar olivines, postulated from spectral evidence in Apollo 11 samples (Haggerty *et al.*, 1970) and subsequently used to explain the crystal chemistry of Apollo 12 (Boyd and Smith, 1971; Weill *et al.*, 1971) and Apollo 15 (Boyd, 1972) pyroxenes, remains unresolved. However, the presence of Cr^{3+} ions in lunar olivines was proposed by Burns *et al.* (1973) on the basis of reproducible spectral evidence.

Thus, to summarize, the presence of appreciable amounts of Fe^{2+}, Ti^{4+}, and Ti^{3+} in lunar oxide and silicate minerals is well established. It is also likely that Cr^{3+} ions are present in olivines, although the evidence for Cr^{2+} ions requires further support. Fe^{3+} ions, certainly present at ppm concentrations in some minerals, are unlikely to exceed these concentrations except under very unusual circumstances, such as in the Apollo 16 "rusty rock" (66095) which may have oxidized after removal from the lunar surface (Taylor *et al.*, 1973). By contrast, metallic iron is widespread throughout the Apollo and Luna samples.

The occurrence of low oxidation states in the transition elements of lunar rocks indicates the low oxygen fugacities prevalent during the formation of the lunar crust. On the basis of a significant Ti^{3+}/Ti^{4+} ratio and the possible presence of Cr^{2+}, estimates of f_{O_2} at 1000°C in the ranges 10^{-13}–10^{-14} (Brett *et al.*, 1971) and $10^{-15}10-^{-16}$(Haggerty, 1972) have been made. Ideally, synthesis experiments determining Ti^{3+}/Ti^{4+} distribution as a function of f_{O_2} should be performed to confirm these estimates.

B. Electronic Spectra of Lunar Samples and Telescopic Observations of Lunar Surface Reflectance Spectra

Information on the gross mineralogy and petrology of different regions of the moon's surface may be obtained from remote reflectivity measurements of the lunar surface (Adams and Jones, 1970; Adams and McCord, 1970, 1972; Pieters *et al.*, 1973). The interpretation of the reflectivity measurements depends on the laboratory spectral studies of single crystals and whole rock powder samples using both the techniques described in Section II, B and diffuse reflectance methods (Wendlandt and Hecht, 1966; White and Keester, 1966). Also see Chapter 4, Section I, for diffuse reflectance technique in the visible and near-infrared.

Not surprisingly, major features in the remote reflectance profiles arise from absorption by the pyroxenes (Adams and Jones, 1970), and Adams and McCord (1972) have related laboratory and remote sensing measurements in order to classify lunar rocks and soils in terms of average pyroxene composition. Since the remote measurements are of reflectance rather than absorption, the features which appear as peaks in an absorption spectrum are troughs in the reflectance spectrum. Comparison of the reflectance spectra with the polarized absorption spectra shown in Fig. 9 indicates that the two major features around 900–1000 nm (the 1 μm band) and 1800–2200 nm (the 2 μm band) are due to spin-allowed transitions in Fe^{2+} ions. Compositional variations of the 1 μm band are useful for determining the average pyroxene compositions for areas of the lunar surface. Furthermore, intense absorption toward the ultraviolet-blue end of the spectrum has been used to distinguish areas of the lunar surface rich in iron and titanium, such as at the Apollo 17 landing site (Pieters *et al.*, 1973). This absorption at the blue end of the spectrum has generally been attributed to "charge transfer processes" involving Fe^{2+}, Ti^{3+}, Ti^{4+} (Pieters *et al.*, 1973). Further laboratory and theoretical studies of charge transfer spectra could clarify these assignments. In this regard, the results of recent SCF-X_α calculations (Section II, D, Vaughan *et al.*, 1973 Loeffler *et al.*, 1974) suggest that oxygen → metal charge transfer processes are of greatest importance in absorption in the near ultraviolet.

ACKNOWLEDGMENTS

We thank Dr. P. M. Bell and Professor W. B. White for providing some of the spectra illustrated in this review. Bibliographic research was carried out by Mrs. Virginia Mee Burns, and Roxanne Regan assisted in the preparation of the manuscript. The lunar-research is supported by grants from the National Aeronautics and Space Administration (NGR 22-009-551 and NGL 22-009-187).

References

Abu-Eid, R. M., Mao, H. K., and Burns, R. G. (1973). *Annu. Rep. Geophys. Lab., Yearb.* **72**, 564–567.

Adams, J. A., and Jones, R. L. (1970). *Science* **167**, 737–739.

Adams, J. A., and McCord, T. B. (1970). *Proc. Lunar Sci. Conf.*, Geochim. Cosmochim. *Acta, 1st, 1970,* Vol. 3, Suppl. 3, pp. 1937–1945.

Adams, J. B., and McCord, T. B. (1972). *Proc. Lunar Sci. Conf., Geochim. Cosmochim. Acta, 3rd, 1972,* Vol. 3, Suppl. 3, pp. 3021–3034.

Adams, J. B., Bell, P. M., Conel, J. E., Mao, H. K., McCord, T. B., and Nash, D. B. (1973). *Geochim. Cosmochim. Acta* **37**, 731–744.

Agrell, S. O., Scoon, J. H., Muir, I. D., Long, J. V. P., McConnel, J. D. C., and Peckett, A. (1970). *Proc. Lunar Sci. Conf., Geochim. Cosmochim. Acta, 1st, 1970,* Vol. 1, Suppl. 1, 93–128.

Allen, G. C., and Hush, N. S. (1967). *Progr. Inorg. Chem.* **8**, 357–389.

Appleman, D. E., Steward, D. B., Clark, J. R., Dowty, E., and Huebner, J. S. (1971). *Proc. Lunar Sci. Conf., Geochim. Cosmochim. Acta, 2nd, 1971,* Vol. 1, Suppl. 2, pp. 117–133.

Ballhausen, C. J. (1962). "Introduction to Ligand Field Theory." McGraw-Hill, New York.

Bancroft, G. M. (1973). "Mössbauer Spectroscopy: An Introduction for Inorganic Chemists and Geochemists." McGraw-Hill, New York.

Bancroft, G. M., and Burns, R. G. (1969). *Mineral. Soc. Amer., Spec. Paper* **2**, 137–148.

Bates, C., White, W. B., and Roy, R. (1966). *J. Inorg. Nucl. Chem.* **28**, 397–405.

Bell, P. M., and Mao, H. K. (1969). *Annu. Rep. Geophys. Lab., Yearb.* **68**, 253–256.

Bell, P. M., and Mao, H. K. (1972a). *Proc. Lunar Sci. Conf., Geochim. Cosmochim. Acta, 3rd, 1972,* Vol. 1, Suppl. 3, pp. 545–553.

Bell, P. M., and Mao, H. K. (1972b). *Annu. Rep. Geophys. Lab., Yearb.* **71**, 480–489.

Bell, P. M., and Mao, H. K. (1972c). *Annu. Rep. Geophys. Lab., Yearb.* **71**, 531–534.

Bence, A. E., and Papike, J. J. (1972). *Proc. Lunar Sci. Conf., Geochim. Cosmochim. Acta, 3rd, 1972,* Vol. 1, Suppl. 3, pp. 431–469.

Bethe, H. (1929). *Ann. Phys.* (*Leipzig*) [5] **3**, 133–206.

Birle, J. D., Gibbs, G. V., Moore, P. B., and Smith, J. V. (1968). *Amer. Mineral.* **53**, 807–824.

Boyd, F. R. (1972). *Annu. Rep. Geophys. Lab., Yearb.* **71**, 459–463.

Boyd, F. R., and Smith, D. (1971). *J. Petrol.* **12**, 439–464.

Brett, P. R., Butler, P., Jr., Meyer, C., Jr., Reid, A. M., Takeda, H., and Williams, R. (1971). *Proc. Lunar Sci. Conf., Geochim. Cosmochim. Acta, 2nd, 1971,* Vol. 1, Suppl. 1, pp. 301–317.

Brown, G. E. (1971). *Trans. Amer. Geophys. Union* **53**, 188.

Burnham, C. W. (1971). *Proc. Lunar Sci. Conf., Geochim. Cosmochim. Acta, 2nd, 1971,* Vol. 1, Suppl. 2, pp. 47–57.

Burns, R. G. (1965). Ph.D. Dissertation, University of California, Berkeley.

Burns, R. G. (1966). *J. Sci. Instrum.* **43**, 58–60.

Burns, R. G. (1969). *Chem. Geol.* **5**, 275–283.

Burns, R. G. (1970a). "Mineralogical Applications of Crystal Field Theory." Cambridge Univ. Press, London and New York.

Burns, R. G. (1970b). *Amer. Mineral.* **55**, 1608–1632.

Burns, R. G. (1972). *Can. Spectroscopy,* **17**, 51–59.

Burns, R. G. (1973a). *Abstr., Geol. Soc. Amer., Annu. Meet.*, Vol. 5, pp. 562–563.
Burns, R. G. (1973b). *Trans. Amer. Geophys. Union* **54,** 910–912.
Burns, R. G. (1974). *Amer. Mineral.* **59,** 625–629.
Burns, R. G., and Burns, V. M. (1974). *In* "Handbook of Geochemistry" (K. H. Wedephol, ed.), Chapter 12A.
Burns, R. G., and Fyfe, W. S. (1967). *Res. Geochem.* **2,** 259–285.
Burns, R. G., and Huggins, F. E. (1973). *Amer. Mineral.* **58,** 955–961.
Burns, R. G., and Simon, H. F. (1973). *Abstr., Geol. Soc. Amer., Annu. Meet.*, Vol. 5, pp. 563–564.
Burns, R. G., and Strens, R. G. J. (1967). *Mineral. Magazine,* **36,** 204–226.
Burns, R. G., Clark, R. H., and Fyfe, W. S. (1964). *In* "Chemistry of the Earth's Crust" (A. P. Vinogradov, ed), *Proc. Vernadsky Centenn. Symp.*, Vol. 2, pp. 88–106 (Transl.: Vol. 2, pp. 93–112. Isr. Program Sci. Transl., Jerusalem, 1967).
Burns, R. G., Abu-Eid, R. M., and Huggins, F. E. (1972a). *Proc. Lunar Sci. Conf., Geochim. Cosmochim. Acta, 3rd, 1972,* Vol. 1, Suppl. 3, pp. 533–543.
Burns, R. G., Abu-Eid, R. M., and Leung, I. S. (1972b). *Abstr., Geol. Soc. Amer., Annu. Meet.* Vol. 4, pp. 463–464.
Burns, R. G., Huggins, F. E., and Abu-Eid, R. M. (1972c). *Moon* **4,** 93–102.
Burns, R. G., Vaughan, D. J., Abu-Eid, R. M., Witner, M., and Morawski, A. (1973). *Proc. Lunar Sci. Conf., Geochim. Cosmochim. Acta, 4th, 1973,* Vol. 1, Suppl. 4, pp. 983–994.
Chao, E. C. T., Minkin, J. A., Frondel, C., Klein, C., Jr., Drake, J. C., Fuchs, L., Tani, B., Smith, J. V., Anderson, A. T., Moore, P. B., Zechman, G. R., Jr., Traill, R. J., Plant, A. G., Douglas, J. A. V., and Dence, M. R. (1970). *Proc. Lunar Sci. Conf., Geochim. Cosmochim. Acta, 1st, 1970,* Vol. 1, Suppl. 1, pp. 65–79.
Chesnokov, B. V. (1959). *Dokl. Akad. Nauk SSSR* **129,** 647–649.
Clark, J. R., Appleman, D. E., and Papike, J. J. (1969). *Mineral Soc. Amer., Spec. Pap.* **2,** 31–50.
Clark, R. J. H. (1968). "The Chemistry of Titanium and Vanadium." Elsevier, Amsterdam.
Cohen, A. J. (1972). *Moon* **4,** 141–154.
Cotton, F. A. (1963). "Chemical Applications of Group Theory." Wiley (Interscience), New York.
Cotton, F. A., and Meyers, M. D. (1960). *J. Amer. Chem. Soc.* **82,** 5023–5026.
Dence, M. R., Douglas, J. A. V., Plant, A. G., and Traill, R. J. (1970). *Proc. Lunar Sci. Conf., Geochim. Cosmochim. Acta, 1st, 1970,* Vol. 1, Suppl. 1, pp. 315–340.
Dowty, E., and Clark, J. R. (1973a). *Amer. Mineral.* **58,** 230–242.
Dowty, E., and Clark, J. R. (1973b). *Amer. Mineral.* **58,** 962–964.
Dunn, T. M., McClure, D. S., and Pearson, R. G. (1965). "Some Aspects of Crystal Field Theory." Harper, New York.
Eigenmann, K., Kurtz, K., and Gunthard, H. H. (1972). *Helv. Phys. Acta* **45,** 452–480.
Faye, G. H. (1968). *Can. Mineral.* **9,** 403–425.
Faye, G. H. (1969). *Can. Mineral.* **10,** 112–117.
Faye, G. H. (1971a). *Can. Mineral.* **10,** 889–895.
Faye, G. H. (1971b). *Amer. Mineral.* **56,** 344–350.
Faye, G. H., and Harris, D. C. (1969). *Can. Mineral.* **10,** 47–56.
Faye, G. H., and Hogarth, D. D. (1969). *Can. Mineral.* **10,** 25–34.
Faye, G. H., and Nickel, E. H. (1969). *Can. Mineral.* **10,** 35–46.
Faye, G. H., and Nickel, E. H. (1970). *Can. Mineral.* **10,** 616–635.

Faye, G. H., Manning, P. G., and Nickel, E. H. (1968). *Amer. Mineral.* **53,** 1174–1201.
Ferguson, J., and Fielding, P. E. (1971). *Chem. Phys. Lett.* **10,** 262–265.
Figgis, B. N. (1966). "Introduction to Ligand Fields." Wiley (Interscience), New York.
Finger, L. W. (1970). *Annu. Rep. Geophys. Lab., Yearb.* **69,** 302–305.
Finger, L. W. (1971). *Geol. Soc. Amer., Abstr.* **3,** 310.
Ghose, S., Ng, G., and Walter, L. S. (1972). *Proc. Lunar Sci. Conf., Geochim. Cosmochim. Acta, 3rd, 1972,* Vol. 1, Suppl. 1, pp. 507–531.
Greskovich, C., and Stubican, V. S. (1966). *J. Phys. Chem. Solids* **27,** 1379–1384.
Hafner, S. S., and Virgo, D. (1970). *Proc. Lunar Sci. Conf., Geochim. Cosmochim. Acta, 1st, 1970,* Vol. 1, Suppl. 3, pp. 2183–2198.
Hafner, S. S., Virgo, D., and Warburton, D. (1971a). *Proc. Lunar Sci. Conf., Geochim. Cosmochim. Acta, 2nd, 1971,* Vol. 1, Suppl. 1, pp. 91–108.
Hafner, S. S. Virgo, D., and Warburton, D. (1971b). *Earth Planet. Sci. Lett.* **12,** 159–166.
Hafner, S. S., Niebuhr, H. H., and Zeira, S. (1973). *Proc. Lunar Sci. Conf., Geochim. Cosmochim Acta, 4th, 1973,* pp. 326–328.
Haggerty, S. E. (1972). *Proc. Lunar Sci. Conf., Geochim. Cosmochim. Acta, 3rd, 1972,* Vol. 1, Suppl. 3, pp. 305–332.
Haggerty, S. E., Boyd, F. R., Bell, P. M., Finger, L. W., and Bryan, W. B. (1970). *Proc. Lunar Sci. Conf., Geochim. Cosmochim. Acta, 1st, 1970,* Vol. 1, Suppl. 1, pp. 541–550.
Hartmann, H., and Schläfer, H. L. (1951). *Z. Phys. Chem.* **197,** 116–142.
Hartmann, H., Schläfer, H. L., and Hansen, K. H. (1957). *Z. Anorg. Allg. Chem.* **289,** 40–65.
Healy, G. W. and Schottmiller, J. C. (1964). *Trans. AIME* **230,** 420–425.
Hoff, G. R., and Brubaker, C. H., Jr. (1971). *Inorg. Chem.* **10,** 2063–2066.
Holmes, O. G., and McClure, D. S. (1957). *J. Chem. Phys.* **26,** 1686–1694.
Hush, N. S. (1967). *Progr. Inorg. Chem.* **8,** 357–389.
Johnson, K. H. (1973). *Advan. Quantum Chem.* **7,** 143–185.
Johnson, O. W., Ohlsen, W. D., and Kingsbury, P.I., Jr. (1968). *Phys. Rev.* **175,** 1102–1109.
Jorgensen, C. K. (1957). *Acta Chem. Scand.* **11,** 73.
Kolopus, J. L., Kline, D., Chatelain, A., and Weeks, R. A. (1971). *Proc. Lunar Sci. Conf., Geochim. Cosmochim. Acta, 2nd, 1971,* Vol. 3, Suppl. 2, pp. 2501–2514.
Lever, A. B. P. (1968). "Inorganic Electronic Spectroscopy." Elsevier, Amsterdam.
Littler, J. G. F., and Williams, R. J. P. (1965). *J. Chem. Soc., London,* pp. 6368–6371.
Loeffler, B. M., Burns, R. G., Tossell, J. A., Vaughan, D. J., and Johnson, K. H. (1974) *Proc. Lunar Sci. Conf., Geochim. Cosmochim., Acta, 5th, Suppl. 5,* Vol. 3, 3007–3016.
Lovering, J. F., and Ware, N. G. (1970). *Proc. Lunar Sci. Conf., Geochim. Cosmochim. Acta, 1st, 1970,* Vol. 1, Suppl. 1, pp. 633–654.
McCarthy, G. J., White, W. B., and Roy, R. (1969). *J. Amer. Ceram. Soc.* **52,** 463–467.
Malysheva, T. V. (1972). *Proc. Lunar Sci. Conf., Geochim. Cosmochim. Acta, 3rd, 1972,* Vol. 1, Suppl. 1, pp. 105–114.
Manning, P. G. (1968a). *Can. Mineral.* **9,** 348–357.
Manning, P. G. (1968b). *Can. J. Earth Sci.* **5,** 89–92.
Manning, P. G. (1968c) *Can. Mineral.* **9,** 663–667.
Manning, P. G. (1969a). *Can. Mineral.* **10,** 57–70.
Manning, P. G. (1969b). *Can. Mineral.* **10,** 678–690.
Manning, P. G. (1970). *Can. Mineral.* **10,** 677–688.

Manning, P. G., and Nickel, E. H. (1969). *Can. Mineral.* **10,** 71–83.

Mao, H. K., and Bell, P. M. (1971). *Annu. Rep. Geophys. Lab., Yearb.* **70,** 207–215.

Mao, H. K., and Bell, P. M. (1972). *Annu. Rep. Geophys. Lab., Yearb.* **71,** 520–524 and 524–527.

Mao, H. K., and Bell, P. M. (1973). *Amer. Soc. Test. Mater., Spec. Tech. Publ.* **539,** 100–119.

Mao, H. K., Bell, P. M., and Dickey, J. S., Jr. (1972). *Annu. Rep. Geophys. Lab., Yearb.* **71,** 538–541.

Megaw, H. D., Kempster, C. J. E., and Radoslovich, E. W. (1962). *Acta Crystallogr.* **15,** 1017–1035.

Melankholin, N. M. (1956). *Zap. Vses. Mineral. Obshchest.* **85,** 218–223.

Moore, R. K., and White, W. B. (1971). *Amer. Mineral.* **56,** 826–840.

Moore, R. K., and White, W. B. (1972). *Can. Mineral.* **11,** 791–811.

Neuhaus, A. (1960). *Z. Kristallog, Kristallgeometric, Kristallphys., Kristallchem.,* **113,** 195–233.

Orchin, M., and Jaffé, H. H. (1971). "Symmetry, Orbitals and Spectra." Wiley (Interscience), New York.

Orgel, L. E. (1952). *J. Chem. Soc., London,* pp. 4756–4761.

Orgel, L. E. (1966). "An Introduction to Transition Metal Chemistry: Ligand Field Theory," 2nd ed. Methuen, London.

Pieters, C., McCord, T. B., Zisk, S., and Adams, J. B. (1973). *J. Geophys. Res.* **78,** 5867–5875.

Poole, C. P., Jr. (1964). *J. Phys. Chem. Solids* **25,** 1169–1182.

Prewitt, C. T., Shannon, R. D., and White, W. B. (1972). *Contrib. Mineral. Petrol.* **35,** 77–82.

Robin, M. B., and Day, P. (1967). *Advan. Inorg. Chem. Radiochem.* **10,** 248–422.

Runciman, W. A., Sengupta, D., and Marshall, M. (1973a). *Amer. Mineral.* **58,** 444–450.

Runciman, W. A., Sengupta, D., and Gourley, J. T. (1973b). *Amer. Mineral.* **58,** 451–456.

Runciman, W. A., Sengupta, D., and Gourley, J. T. (1974). *Amer. Mineral.* **59,** 630–631

Scheetz, B. E., and White, W. B. (1972). *Contrib. Mineral. Petrol.* **37,** 221–227.

Schläfer, H. L. (1964). *In* "Theory and Structure of Complex Compounds (B. Jezowska-Trzebiatowska, ed.), *Proc. Symp. Theory Struct. Complex. Comp., Wroclaw, 1962,* pp. 181–193.

Schläfer, H. L., and Gliemann, G. (1968). "Basic Principles of Ligand Field Theory." Wiley (Interscience), New York.

Schürmann, K., and Hafner, S. S. (1972). *Proc. Lunar Sci. Conf., Geochim. Cosmochim. Acta, 3rd, 1972,* Vol. 1, Suppl. 3, pp. 615–621.

Slack, G. A. (1964). *Phys. Rev.* **134,** A1268–A1280.

Slater, J. C., and Johnson, K. H. (1972). *Phys. Rev. B* **5,** 844–853.

Sung, C-M., Abu-Eid, R. M., and Burns, R. G. (1974). *Proc. Lunar Sci. Conf., Geochim. Acta, 5th, 1974,* Suppl. 5, Vol. 1, 717–726.

Taylor, L. A., Mao, H. K., and Bell, P. M. (1973). *Proc. Lunar Sci. Conf., Geochim. Cosmochim. Acta, 4th, 1973,* Vol. 1 pp. 715–716.

Tossell, J. A., Vaughan, D. J., and Johnson, K. H. (1973). *Nature* (*London*) **244,** 42–45.

Tossell, J. A., Vaughan, D. J., and Johnson, K. H. (1974). *Amer. Mineral.* **59,** 319–334.

Townsend, M. G. (1968). *Solid State Commun.* **6,** 81–83.

Townsend, M. G., and Faye, G. H. (1970). *Phys. Status Solid,* **38,** K57–K60.

Tsuchida, R., and Kobayashi, M. (1938). *Bull. Chem. Soc. Jap.* **13,** 618–623.

Ulmer, G. C., and White, W. B. (1966). *J. Amer. Ceram. Soc.* **49,** 50–51.

Vaughan, D. J., Burns, R. G., Loeffler, B., Tossell, J. A., and Johnson, K. H. (1973). *EOS, Trans. Amer. Geophys. Union,* **54,** 1229.

Virgo, D. (1972). *Annu Rep. Geophys. Lab., Yearb.* **71,** 534–538.

Vishnevskii, V. N., and Klimovskaya, L. K. (1958). *Ukr. Fiz. Zh.* **3,** 239–244.

Weeks, R. A. (1972). *Proc. Lunar Sci. Conf., Geochim. Cosmochim. Acta, 3rd, 1972,* Vol. 3, Suppl. 3, pp. 2503–2517.

Weeks, R. A. (1973). *In* "Lunar Science-IV, Abstracts" (J. W. Chamberlain and C. Watkins, eds.), pp. 772–774 and 775–778. The Lunar Science Institute, Houston.

Weill, D. F., Greive, R. A., McCallum, I. S., and Bottinga, Y. (1971). *Proc. Lunar Sci. Conf., Geochim. Cosmochim. Acta, 2nd, 1971,* Vol. 1, Suppl. 1, pp. 413–430.

Wendlandt, W. W., and Hecht, H. G. (1966). "Reflectance Spectroscopy." Wiley (Interscience), New York.

Wenk, H. R., and Raymond, K. N. (1973). *Z. Kristallogr., Kristallgeometric, Kristallphys., Kristallchem.* **137,** 86–105.

White, W. B., and Keester, K. L. (1966). *Amer. Mineral.* **51,** 774–791.

White, W. B., McCarthy, G. J., and Scheetz, B. E. (1971). *Amer. Mineral.* **56,** 72–89.

3

LUMINESCENCE OF MINERALS IN THE NEAR-INFRARED

J. E. Geake and G. Walker

I. Introduction

Luminescence occurs when a substance absorbs energy in any form and reemits it as light after a delay that varies from nanoseconds to hours. If the delay is less than about 1ms the effect is described as fluorescence; if longer it is called phosphorescence, and this persistence of the emission after the source of excitation has been removed is termed afterglow. The emission may be in any part of the spectrum, except that by Stokes' law it cannot be at a shorter wavelength than that of the source of excitation. If the excitation is by fast particles, e.g., electrons, protons, or α-particles, then it is their effective quantum wavelength which counts, given by $\lambda = hc/E$, where E is the particle's energy, and h is Planck's constant. However, whereas a photon of radiation gives up all or none of its energy to a luminescence center, a particle may give up part of its energy, so the effective quantum wavelength is merely the shortest possible effective wavelength of the interaction.

The emission wavelength is also given by $\lambda = hc/E$, where E is now the energy difference between the energy levels involved; the emission will be considered to be in the IR if its wavelength* is greater than about 7000 Å, corresponding to an energy difference of less than 2.8×10^{-12} ergs or 1.8 eV. Whether there is luminescence emission at all depends on whether the radiative dissipation of the stored energy has a higher probability than the many competing nonradiative processes, e.g., lattice vibrations.

The types of crystal field situation that favor radiative transfers, and which give energy levels sufficiently close for the emission to be in the IR, will be discussed in Section II.

Several common minerals and other crystalline solids exhibit IR luminescence, usually with transition-metal ions as activators; for example, it is caused by Fe^{3+} ions in feldspars and spinels, and in alumina and sodalite. Plagioclase feldspar has proved especially interesting as it is the main luminescent mineral found on the moon, and the strength of the IR luminescence emission is found to be one of the major differences between lunar and corresponding terrestrial materials. The properties of all these IR luminescent materials will be discussed in Section III.

II. Crystal Field Considerations

The effect of the crystal environment on the energy levels of transition-metal ions in solids is outlined in Chapters 1 and 2. An excellent account of the application of crystal field theory to transition-metal mineralogy and geochemistry has been given by Burns (1970). Since about 1960, crystal or ligand field theory has also been used extensively to explain the spectral emission characteristics of well-known luminescence centers such as Mn^{2+} and Cr^{3+} in many different solids.

Any luminescence emission is usually (but not always) the result of a radiative transition from the lowest excited crystal field state of the ion to the ground state. Obviously, if the energy of this electronic transition is low enough, the emission will be in the IR. Excitation into higher states also usually produces the same luminescence (but again not always) because the center undergoes nonradiative transitions to the lowest excited state from which emission occurs. Often the lowest excited crystal field state arises from the splitting of the free ion ground state and therefore has the same multiplicity (spin). However, for the most common transition-metal luminescence centers such as Cr^{3+} and Mn^{2+}, the lowest excited

*A wavelength of 10,000 Å = 1,000 nm = 1μm. To convert wave length in μm to wave number in cm^{-1} divide 10,000 by the wavelength.

crystal field state is derived from the lowest excited state of the free ion which in both cases has different multiplicity to the ground state. The consequence is that in the former situation the transition energy becomes greater, and therefore the emission wavelength shorter, with increasing crystal field, whereas in the latter case this is not so; indeed, for d^5 species such as Mn^{2+}, the transition energy decreases with increasing crystal field, as can be seen from Fig. 1. The crystal field splitting parameter Δ is, according to the simple point charge model, inversely proportional to the fifth power of the average cation–anion distance, and therefore very sensitive to changes in this variable. It also obviously depends on the number, nature, and position of the surrounding anions or ligands (see Burns, 1970). Because the magnitude of the crystal field in a tetrahedral site is about half that in an octahedral site, if all other parameters such as cation–anion distances, etc., remain the same, the emission of a d^5 ion in tetrahedral coordination is usually at a shorter wavelength than that of the same ion in an octrahedral site. Nevertheless, it is theoretically possible, if unlikely, for the reverse situation to occur. For a d^5 ion, such as Mn^{2+} or Fe^{3+}, the lowest excited state in any environment of cubic symmetry is the $^4T_1$ state, or some component thereof, this state being triply degenerate in any environment with simple cubic symmetry. However, cation sites in minerals seldom have such simple symmetry, and distortions from simple octahedral or tetrahedral symmetry cause the $^4T_1$ level to split into either two or three components. The energy gap between these levels and the ground state is quite sensitive to variations in the crystal field (see Fig. 1) and therefore the emission band is usually broad. In addition, since the transition is between states $^4T_1$ and $^6A_1$, which have different multiplicities, the transition is spin-forbidden, and therefore the emission usually has a decay time of several milliseconds; afterglow may also be observed in systems where metastable trapping states are numerous and the concentration of quenchers such as Fe^{2+} is small, but such effects are

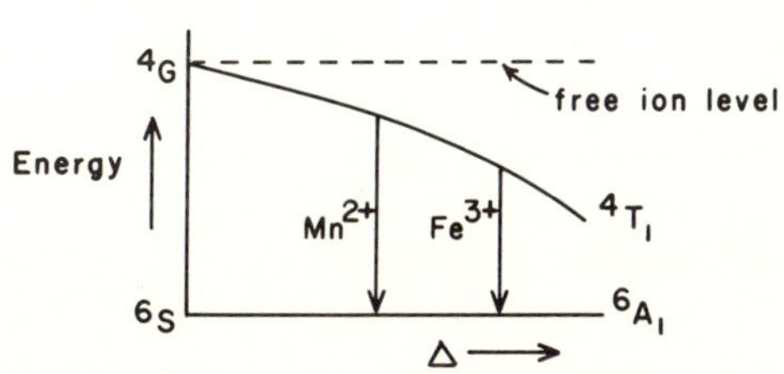

FIG. 1. Schematic energy level diagram for transition metals of d^5 electron configuration in a cubic crystal field. The $^4T_1$ level is actually triple; only the emitting state is shown.

only relevant to thermoluminescence and will not be discussed here. Mn^{2+}, which is the most ubiquitous luminescence center in solids, usually gives emission between red and green in color, depending on the size and symmetry of the crystal field in which it finds itself; furthermore, it is known to substitute easily for ions such as Mg^{2+}, Ca^{2+}, and Zn^{2+} in various structures. Fe^{3+}, which is isoelectronic with Mn^{2+}, has been found to produce luminescence emission which is usually deep red or near-IR. In a similar environment Fe^{3+} should produce emission at a much longer wavelength (i.e., lower energy) than Mn^{2+}, because of the larger crystal field splitting parameter Δ caused by its higher charge (see Fig. 1). However, because of its different ionic size and charge, Fe^{3+} will probably tend to substitute in different sites to those preferred by Mn^{2+}. There are now many reports, which will be discussed in the next section, of Fe^{3+} emission on the verge of the IR, and most of these concern solids containing aluminum. Since d^5 ions have zero crystal field stabilization energy, the crystal field itself should not influence site preference.

Luminescence is, of course, also caused by ions other than d^5 ions; for example, Cr^{3+} (a d^3 ion) has already been mentioned. However, the emission of this ion in solids is usually red (e.g., the well-known ruby laser emission) and its wavelength is almost independent of the magnitude of the crystal field. Other transition metals, such as Cu^{2+}, Co^{2+}, V^{2+}, V^{3+}, and also Ni^{2+} at low temperatures, have been shown to be responsible for IR emission in oxides, sulfides and selenides; however, there do not appear to be any reports of such luminescence emission centers in important minerals, such as silicates, although this is theoretically possible.

III. Crystalline Solids Showing IR Luminescence

A. Fe^{3+} Centers

Perhaps one of the earliest reports of Fe^{3+} as a luminescence center was that of Hummel and Sarver (1964). These authors measured the emission spectrum of Mn^{2+}- and Fe^{3+}-activated spinel ($MgAl_2O_4$) and found that Fe^{3+} activation gave rise to an emission band at about 7000 Å (14,300 cm^{-1}), which they attributed to octahedrally coordinated Fe^{3+}; another emission band very similar to that of the Mn^{2+}-activated spinel was attributed to tetrahedrally coordinated ferric iron. However, it is probable that this latter band was either due to manganese impurity or to a defect center, and that the 7000 Å emission band is that of tetrahedrally coordinated Fe^{3+} centers. More recently, Fe^{3+} emission has been reported in another spinel structure, that of $LiAl_5O_8$ (Melamed *et al.*, 1970; Pott and McNicol, 1971a, 1972). Pott and McNicol found that the emission band, which was again at about 7000 Å, was broad and structureless at room temperature but showed fine structure at 77°K; this fine structure they interpreted, not unreasonably, as vibrational fine structure (Fig. 2). Figure 3 shows the luminescence excitation spectrum (and the emission spectrum) of $LiAl_5O_8$ at 77°K. Excitation spectra, like absorption spectra, give information about higher excited states provided that the center in question undergoes nonradiative "internal conversion" to the emitting state and that direct excitation into higher excited states is possible. Figure 4 shows the excitation spectrum of Fe^{3+}-activated β-$NaAlO_2$ which also exhibits luminescence in the region of 7000 Å. From these spectra, Pott and McNicol calculate Δ (which they refer to as 10 Dq) values and the Racah parameters B and C, and from these deduce that the ferric iron is again tetrahedrally coordinated.

γ-Alumina can also be activated by Fe^{3+} to produce an emission band at about 8000 Å (12,500 cm^{-1}), as shown in Fig. 5 (Pott and McNicol,

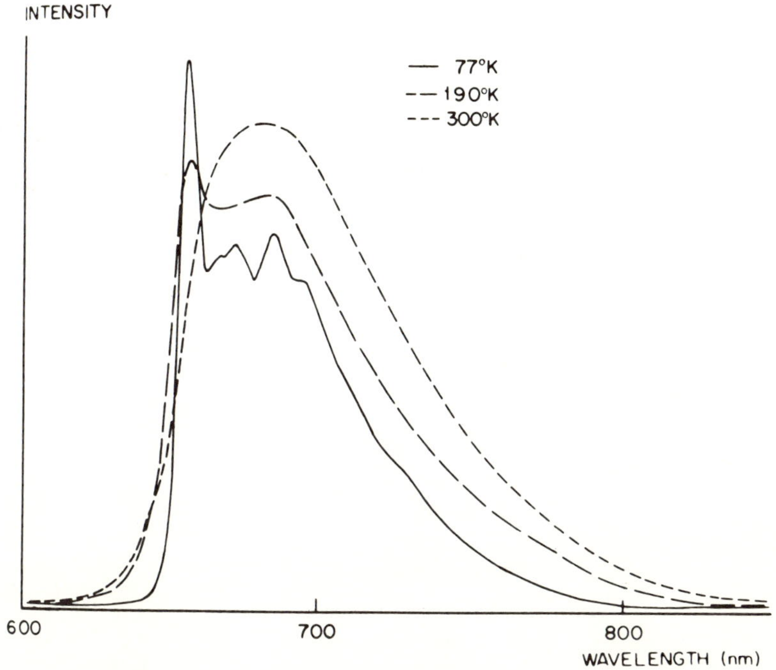

FIG. 2. Luminescence spectrum of Fe^{3+}-activated $LiAl_5O_8$, measured at various temperatures. (From Pott and McNicol, 1972.)

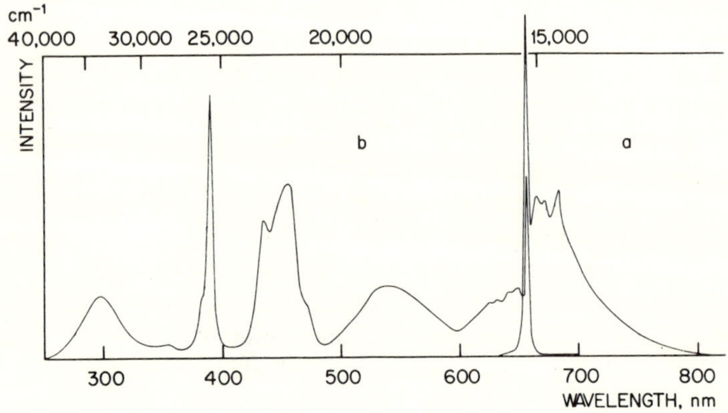

FIG. 3. Luminescence excitation and emission spectra of $LiAl_5O_8$: Fe^{3+}, measured at 77°K. (From Pott and McNicol, 1972.)

1971b); γ-alumina has a defect spinel structure, i.e., vacant cation sites occur. This luminescence emission is ascribed to Fe^{3+} in tetrahedral coordination.

Both Pott and McNicol (1971a) and Van Doorn and Schipper (1971) have reported luminescence emission from Fe^{3+}-activated sodalite ($Na_4Al_3Si_3O_{12}Cl$ or $Na_4Al_3Si_3O_{12}Br$); the emission is again in the region of 7000 Å and is shown in Fig. 6. A full discussion of the luminescence properties of sodalite has been given by Van Doorn, Schipper, and Bolwijn (1972). Bergstein and White (1969) have listed no less than nine synthesized compounds which show luminescence emission in the region of 7000 Å when activated by Fe^{3+}, which they state is in tetrahedral sites.

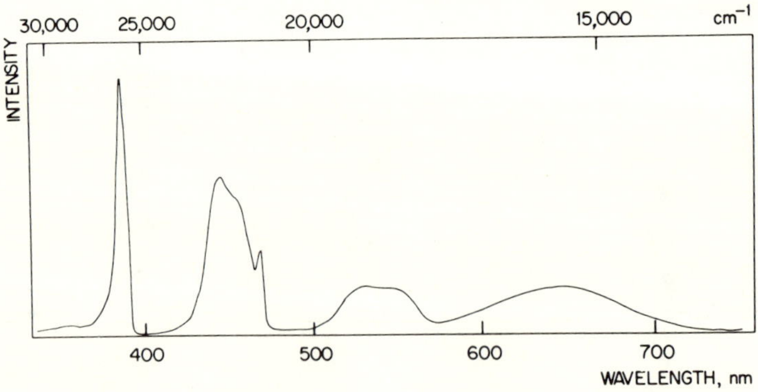

FIG. 4. Luminescence excitation spectrum of β-$NaAlO_2$: Fe^{3+}, measured at 77°K. (From Pott and McNicol, 1971a.)

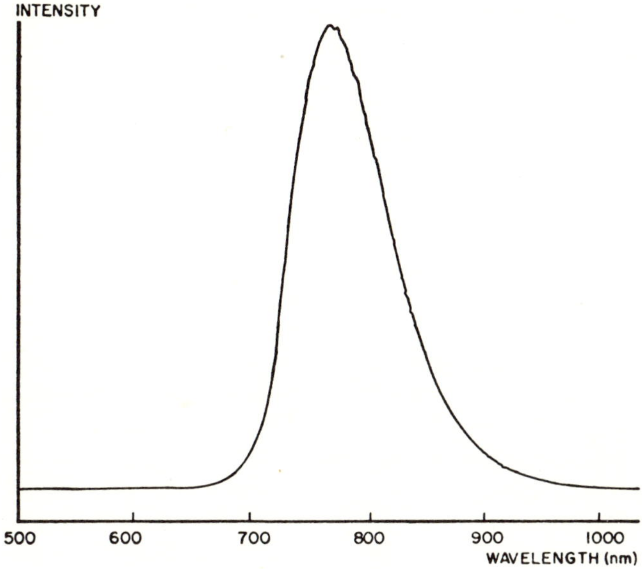

FIG. 5. Luminescence emission spectra of γ-Al_2O_3: Fe^{3+}, with about 100 ppm Fe^{3+}. (From Pott and McNicol, 1971b.)

There is also now reasonably conclusive evidence that Fe^{3+} is the cause of the strong emission band in the very near-infrared exhibited by most plagioclases (and probably orthoclases also) of terrestrial origin (Geake *et al.*, 1972, 1973). It is interesting to note that plagioclases of extraterrestrial origin do not usually show significant Fe^{3+} emission. The luminescence of Fe^{3+} in both natural and synthetic samples of this very important mineral will be discussed in Section III, C.

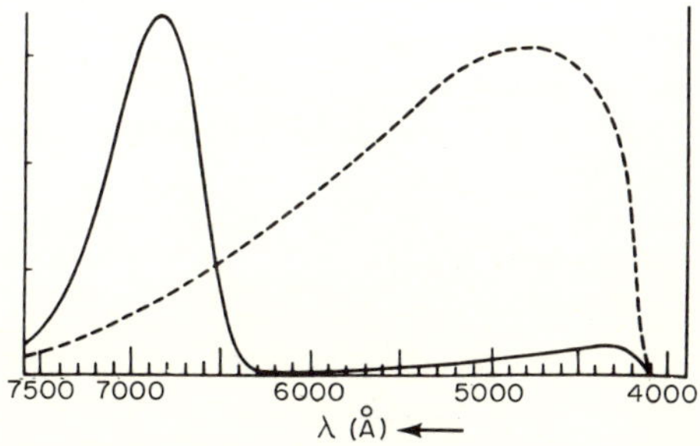

FIG. 6. Luminescence emission spectrum of Fe^{3+}-doped bromide sodalite at 77°K, uncorrected for the spectral response of the instrument which is indicated by the dashed curve. (From Van Doorn and Schipper, 1971.)

B. Other IR Luminescence Centers

It has already been mentioned in Section II that transition-metal ions other than Fe^{3+} can cause IR luminescence. However, reports of such centers have been limited mainly to their occurrence in group II–VI compounds, particularly ZnS and CdS, which are probably the most well-known and most extensively investigated of all phosphors.

It has been known for nearly twenty years that copper-activated ZnS exhibited IR luminescence in the region 1.3–2.1 μm (7,700 to 4,800 cm^{-1}). Broser *et al.* (1965) have measured the luminescence spectra of copper-activated ZnS and CdS at temperatures in the vicinity of 4°K, which revealed vibrational fine structure in addition to the structure which is evident at room temperature. They attribute this luminescence to transitions between crystal-field states of the Cu^{2+} ion (d^9 configuration). However, this interpretation has been questioned by Cox, Hagston and Radford (1968), who argue that this emission could equally well be caused by cation vacancies.

Divalent cobalt also causes IR emission in ZnS in the region of 3 μm (3300 cm^{-1}). Figure 7 shows the luminescence spectrum of Co^{2+} in ZnS, and the probable transitions between crystal-field states of Co^{2+} which are responsible for the emission (Garlick, 1964). More recently, Reynolds and Garlick (1967) have found that Ni^{2+} produces IR luminescence in various phosphors including ZnS and ZnO, but only at low temperatures; no luminescence is observed at room temperature.

Avinor and Meijer (1960) found that ZnS, and some other II–VI compounds, activated with vanadium exhibited IR luminescence. The spectrum of ZnS:V consists of a broad emission band at about 2 μm (5000 cm^{-1}) with little or no structure. Allen (1963) has discussed whether V^{3+} or V^{2+} is responsible for the emission and concluded that V^{3+} (d^2 configuration) is more likely.

Since such IR luminescence emission bands in the wavelength region 1–3 μm are observed in well-known phosphor systems, it is possible that some minerals may be found to show similar emission bands in this region.

C. Lunar and Terrestrial Feldspars

Feldspar is a general term for framework aluminum silicate minerals in which divalent or monovalent metal cations provide charge compensation; one of these is plagioclase [(Na, Ca) (Al, Si) $AlSi_2O_8$], which is an isomorphous mixture of albite ($NaAlSi_3O_8$) and anorthite ($CaAl_2Si_2O_8$). It occurs with a full range of Na:Ca ratios; in practice the names used are albite, 0–10% Ca; oligoclase, 10–30% Ca; andesite, 30–50% Ca; labra-

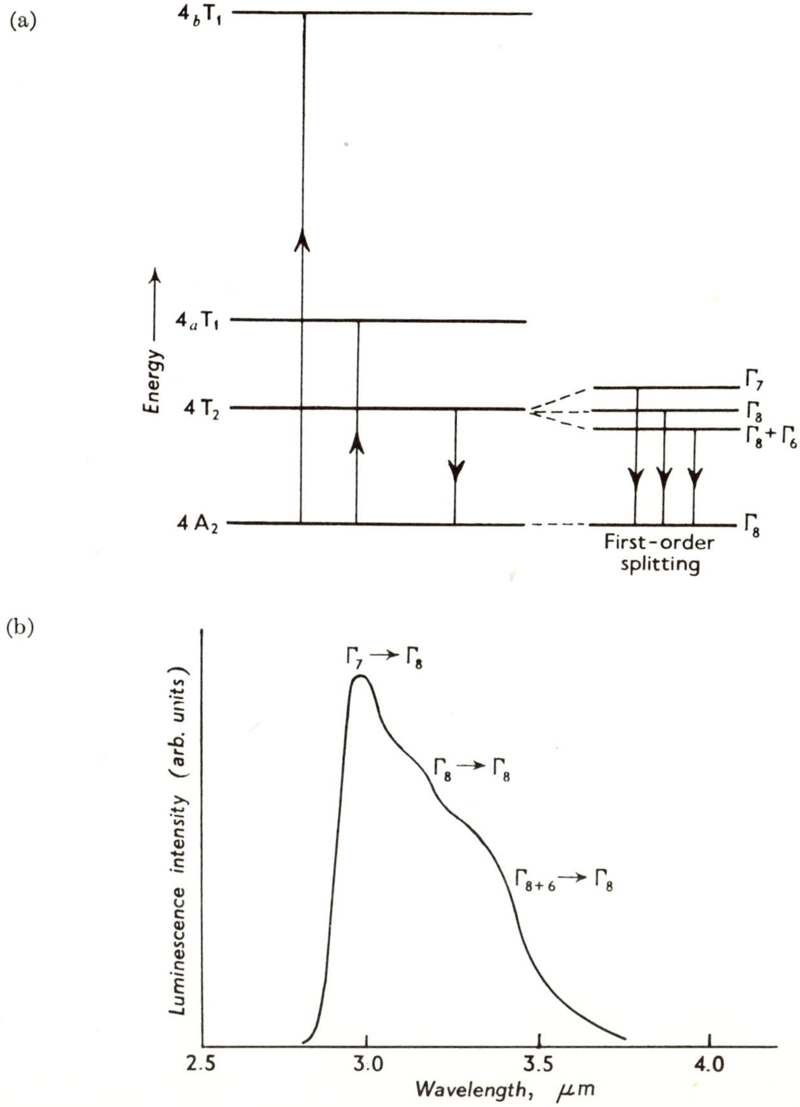

FIG. 7. (a) Energy level scheme for absorption and emission of the Co^{2+} ion center in zinc sulfide. (b) Emission spectrum of ZnS:Co. (From Garlick, 1964.)

dorite, 50–70% Ca; bytownite, 70–90% Ca and anorthite, 90–100% Ca. The structure of anorthite is known with high precision (Megaw *et al.*, 1962).

Plagioclase, in the form of almost pure anorthite, is a common constituent of lunar samples; it is the main mineral of lunar highlands, and being white it causes the relatively high albedo of these regions. Plagioclase is also fairly common on the earth, where it occurs in basic igneous rocks such as basalts and gabbros. It is found with the full range of Na:Ca ratios, but the almost pure anorthite so common on the moon is rather rare on the earth.

Plagioclase is usually weakly luminescent under electron or proton bombardment or x-ray irradiation, but UV irradiation (down to 2537 Å) is ineffective; indeed, it is the most luminescent constituent of the lunar samples, with an energy conversion efficiency as high as 0.1%. However, there are interesting differences between the emission spectra of lunar and terrestrial plagioclases, and these differences are mainly in the near IR part of the spectrum. Some typical luminescence emission spectra for lunar samples under electron excitation are shown in Fig. 8; (a) represents four fines (dust) samples, from four different landing sites; (b) represents a crystalline rock (10058) and a breccia—a rock made from fragments of other rocks welded together by partial melting, usually by impact; (c) shows two more rocks, from different missions. All these spectra contain the same features, at different relative intensities, and several workers identified plagioclase as the constituent mainly responsible (Geake *et al.*, 1970; Nash and Greer, 1970; Sippel and Spencer, 1970). This has been confirmed by electron microprobe analysis (Geake *et al.*, 1973). The samples vary in overall efficiency, probably because of differing plagioclase contents; also, the efficiency of the fines material is probably reduced by radiation damage effects while on the lunar surface. The fines efficiencies cover the range 10^{-5} to 10^{-4} (in the same sequence as their albedos, which also depend on plagioclase content); some high-plagioclase rocks have efficiencies nearly as high as that for pure plagioclase—about 10^{-3}. The main features of the lunar plagioclase spectrum are a blue band, with its peak at about 4500 Å (22,200 cm^{-1}), and a yellow-green band with its peak at about 5600 Å (17,900 cm^{-1}); there is some weak emission in the near IR, but none of the spectra in Fig. 8 shows a peak there. However, the spectra for terrestrial plagioclases, shown in Fig. 9 (b) and (c), usually show a strong, and often a dominant, peak in the near IR, between about 7300 and 7800 Å (13,700 and 12,800 cm^{-1}). The only lunar sample to show an IR peak is the separated lunar plagioclase [Fig. 9 (a)], which was separated by density from the sawdust of a lunar rock. Another extra-terrestrial plagioclase sample, from the meteorite Juvinas [Fig. 9 (a)] does not show an IR peak (Geake and Walker, 1966).

The spectral features common to all the plagioclases investigated are thus a blue and a green peak; the blue one is probably due to lattice defects rather than to any particular activator, and may be enhanced by shock

(a)

(b)

(c)

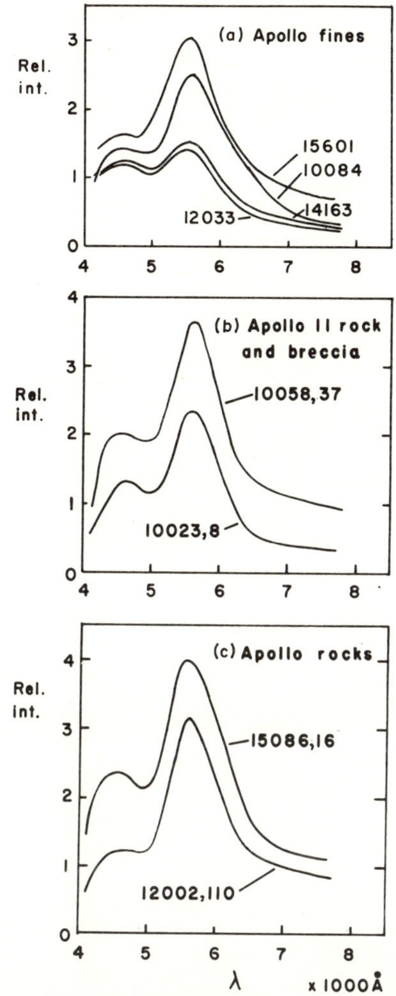

FIG. 8. Luminescence emission spectra for lunar materials, corrected for the spectral response of the instrument. The relative intensity scales for different samples are unrelated. (From Geake *et al.*, 1973.)

effects, such as impacts (Sippel and Spencer, 1970). The green peak is found to be caused by Mn^{2+} as an activator, in Ca^{2+} sites (Geake *et al.*, 1971, 1973). However, it is in the strength of the IR peak that the main difference is found between the emission spectra for terrestrial plagioclases, where it is usually dominant, and lunar plagioclase, where it is weak or absent. An explanation of this difference, and of the cause of this peak,

(a)

(b)

(c)

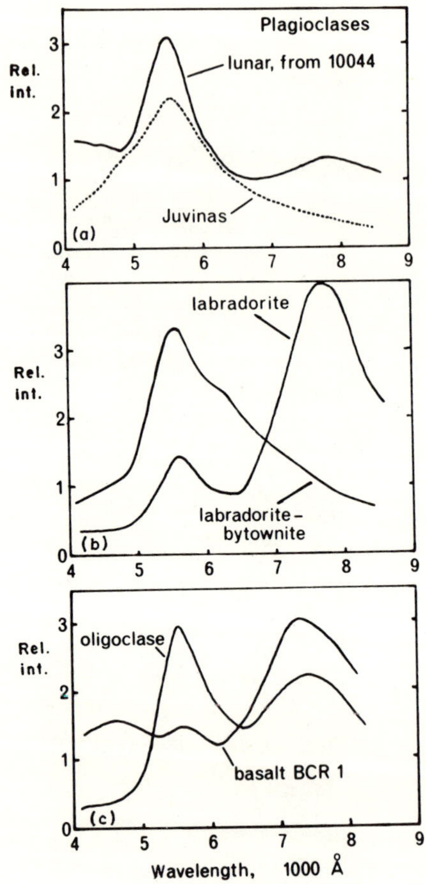

FIG. 9. Corrected luminescence emission spectra for (a) a plagioclase-rich meteorite (the eucrite Juvinas), and for separated plagioclase from a lunar rock; (b) and (c) terrestrial plagioclases. (From Geake *et al.*, 1973.)

provides a clue as to the different conditions prevailing during the formation of the lunar and terrestrial surface materials.

Circumstantial evidence suggested that this IR peak band might be caused by Fe^{3+} present as an impurity in the plagioclase (Geake *et al.*, 1972); this suggestion was supported by luminescence decay-time measurements and by crystal field theory considerations. A marginal enhancement of the IR peak was indeed produced by doping terrestrial labradorite with Fe^{3+}, but this was difficult to interpret as the material already showed a large IR peak before doping. In order to elucidate the cause, a series of synthetic anorthites, labradorites, and albites were produced from gels, using the hydrothermal techniques described by Hamilton and Henderson (1968). Some of these synthetic samples were selectively doped during gelling with Mn^{2+} or Fe^{3+}, or both. One of the samples produced was a synthetic anorthite of sufficient purity to show a negligible IR peak before doping, but when this material was doped with 1000 ppm of Fe^{3+} a large peak at about 7000 Å (14,300 cm^{-1}) was produced, as shown in Fig. 10.

In the synthesis of these materials a major difficulty was contamination by Cr^{3+}; this occurred by the diffusion of chromium into the sample from the metal of the hydrothermal bomb during the firing process, and it gave rise to an emission band in the near IR rather near to the Fe^{3+} band. No actual confusion arose as the chromium band was easily recognized by its fine structure; also, the presence of Fe^{3+} tended to quench it. However, it was better not to have it there, and this contamination was largely prevented by using double-layer encapsulation with a silver foil barrier between two layers of gold, and by firing at a slightly lower temperature ($\sim$800°C for anorthites). Labradorites and albites were also synthesized; up to 1% Fe^{3+} was added, and in each case strong IR emission was induced.

In both natural and synthetic plagioclase samples there is a considerable variation in the IR emission peak wavelength for different Na:Ca ratios (expressed as anorthite percentages). Figure 11 shows this variation over the full range of composition from 0 to 100% anorthite (Geake *et al.*, 1973). The peak wavelengths were obtained from spectral scans uncorrected for the spectral response of the instrument; this adequately shows the variation, although correction would shift all the peaks to slightly longer wavelengths. There appears to be a trend toward shorter wavelengths with increasing anorthite percentage, and a total range of variation of

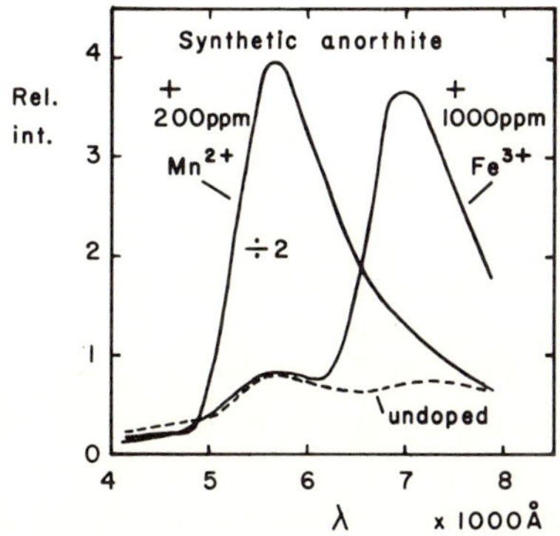

FIG. 10. Luminescence emission spectra for synthetic anorthite, before doping, and after adding either Mn^{2+} or Fe^{3+}. (From Geake *et al.*, 1973.)

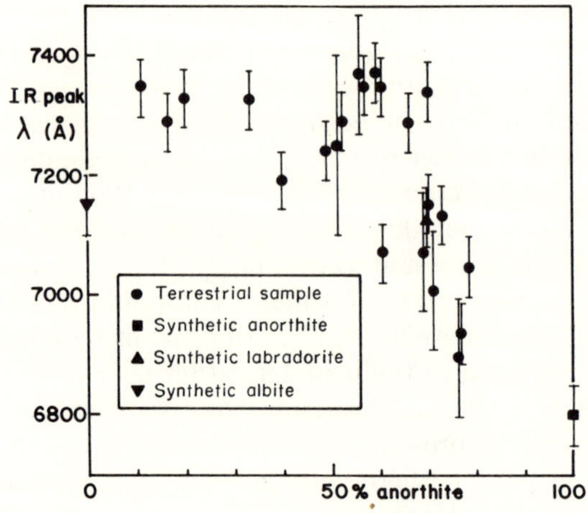

FIG. 11. The wavelength of the near-IR emission peak in uncorrected spectra plotted against percentage anorthite, for a suite of terrestrial plagioclases of known composition. (From Geake *et al.*, 1973.)

about 600 Å. However, the picture is evidently not a simple one, as there is one group of natural samples in the 55–70% anorthite range that show anomalously long wavelengths, and the synthetic albite shows a rather short wavelength. However, the wide overall range of variation shown is in contrast with the much smaller range, of about 100 Å, found in the position of the green Mn^{2+} peak for the same samples.

The position of the Fe^{3+} emission peak should depend on its environment in the crystal; if Fe^{3+} were in a similar lattice site to Mn^{2+} (i.e., the Ca^{2+} site) then a positive correlation of spectral shifts of the emission bands of the isoelectronic species Mn^{2+} and Fe^{3+} would be expected, although the converse is not necessarily true. However, no clear correlation is found; one group of samples does show a correlated shift of both peaks, suggesting that the Mn^{2+} and Fe^{3+} might be in similar sites (i.e., Ca^{2+}) but for another group there is no correlation, suggesting that the Fe^{3+} may not always be in the same site; indeed there is much evidence in the literature to suggest that Fe^{3+} would tend to be in Al^{3+} sites. In agreement with Pott and McNicol (1970) we have found that Fe^{3+} produces an IR emission band in γ-Al_2O_3 at about 8000 Å (12,500 cm^{-1}).

Although the *average* size of a tetrahedral site in plagioclases is smaller for albite than for anorthite, the aluminum-rich tetrahedral sites are not appreciably different in size. Thus there does not appear to be a simple correlation between the size of the largest available tetrahedral site (Fe^{3+} is larger than Al^{3+}) and the IR peak position. Further investigation of this problem is required and more evidence as to the environment of the Fe^{3+} ion may be found by the measurement of excitation spectra; e.g., those for the IR emission of both natural and synthetic Fe^{3+}-doped plagioclases have been determined recently, and found to show similar features to the excitation spectrum of $LiAl_5O_8$:Fe^{3+} (Telfer, D. J., and Walker, G., unpublished).

It is still not entirely clear why terrestrial plagioclases show such strong IR emission, whereas similar lunar plagioclases usually show hardly any. It may simply be evidence of the extreme scarcity of oxygen when the lunar surface materials were formed, which could result in most of the iron in lunar materials going in as Fe^{2+} which quenches luminescence, rather than as Fe^{3+} which produces it, and which is the dominant form in terrestrial materials, formed under oxygen-rich conditions. Nevertheless, Fe^{3+} has been detected in lunar plagioclases by using electron paramagnetic resonance (EPR) techniques (Weeks, 1972), although the concentration does not appear to be known (see Chapter 2, Section IV, D). As regards the concentration of Fe^{3+} required to produce emission, Geake *et al.* (1973) have found that it is still detectable at only 100 ppm; also that the intensity of this band increases with the amount of Fe^{3+} added, up to about 1%, beyond which there is no further intensity increase, probably because self-quenching becomes important. Furthermore, the proportion of the added Fe^{3+} that actually enters the appropriate lattice sites to produce luminescence centers will depend on the conditions during preparation, in a way that is not yet understood. Probably the most important factor is the Fe^{2+}:Fe^{3+} ratio, since Fe^{2+} is likely to quench Fe^{3+} emission; there is no means of knowing what this ratio is for lunar plagioclase, although as mentioned above one might expect less Fe^{3+} and more Fe^{2+} than in terrestrial samples, in view of the scarcity of free oxygen when the lunar surface materials were formed, for which there is independent mineralogical evidence. It may therefore be possible to use the strength of the IR emission peak as an indicator of the amount of oxygen available during formation.

IV. Applications

Luminescence spectroscopy, whether in the visible or the infrared, is a very useful complement to absorption spectroscopy and in particular instances has great advantages over other analytical techniques. It does not usually require either good mineral separation or special sample preparation. For example, it is possible to study centers in lunar plagioclase, using rock or soil samples, without the need for thin polished sections or very flat surfaces as would be required for electron microprobe work, and without the sometimes difficult and laborious mineral separation necessary for EPR or x-ray fluorescence work. Furthermore, thin-section preparation and mineral separation both require much larger samples than are necessary for luminescence analysis. However, luminescence spectroscopy is usually confined to minerals which are relatively iron-free, for, although Fe^{3+} often behaves as a luminescence center, Fe^{2+} nearly always quenches luminescence. If luminescence emission of appreciable intensity is shown by an iron-rich mineral, it is usually indicative of poor equilibration, with the luminescence coming from areas that are relatively iron-free.

Luminescence emission spectra and decay times are valuable parameters for discerning the nature of the center responsible; once this is known, the distribution and concentration of the centers in the mineral can often give information about the history of the sample.

The measurement of the excitation spectrum of a particular emission can give the same information as an absorption spectrum, even in conditions where the absorption spectrum would be either difficult to obtain because of the state of the sample, or impossible because absorption by other species masks the relevant one. There is often ambiguity in the assignment of bands in an absorption spectrum, because the suspected absorbing species may not be responsible for all the bands observed. However, if any absorbing species is luminescent, and can be directly excited, then an excitation spectrum can give information which is unambiguous in this respect. It is therefore clear that there are situations in which luminescence spectroscopy, including that in the IR region, can provide data of importance to transition-metal geochemistry that would be difficult or impossible to obtain by other techniques.

References

Allen, J. W. (1963). *Physica* (*Utrecht*) **29,** 764–768.

Avinor, M., and Meijer, G. (1960). *J. Phys. Chem. Solids* **12,** 211–215.

Bergstein, A., and White, W. B. (1969). In *Extended Abstr. Spring Meet. Electrochem. Soc.* p. 142.

Broser, I., Maier, H., and Schulz, H. J. (1965). *Phys. Rev.* **140,** A2135–A2138.

Burns, R. G. (1970). "Mineralogical Applications of Crystal Field Theory" Cambridge Univ. Press, London and New York.

Cox, A. F. J., Hagston, W. E., and Radford, C. J. (1968). *J. Phys. C* **1,** 1746–1758.

Garlick, G. F. J. (1964). *Sci. Progr.* (*London*) **52,** 3–25.

Geake, J. E., and Walker, G. (1966). *Geochim. Cosmochim. Acta* **30,** 929–937.

Geake, J. E., Dollfus, A., Garlick, G. F. J., Lamb, W., Walker, G., Steigmann, G. A., and Titulaer, C. (1970). *In Proc. 1st Lunar Sci. Conf.*, *Geochim. Cosmochim. Acta*, Suppl. 1, Vol. 3, pp. 2127–2147.

Geake, J. E., Walker, G., Mills, A. A., and Garlick, G. F. J. (1971). *Proc. 2nd Lunar Sci. Conf., Geochim. Cosmochim. Acta,* Suppl. 2, Vol. 3, pp. 2265–2275.

Geake, J. E., Walker, G., Mills, A. A., and Garlick, G. F. J. (1972). *Proc. 3rd Lunar Sci. Conf., Geochim. Cosmochim. Acta,* Suppl. 3, Vol. 3, pp. 2972–2979.

Geake, J. E., Walker, G., Telfer, D. J., Mills, A. A., and Garlick, G. F. J. (1973). *Proc. 4th Lunar Sci. Conf., Geochim. Cosmochim. Acta,* Suppl. 4, Vol. 3, pp. 3181–3189.

Hamilton, D., and Henderson, C. M. B. (1968). *Mineral. Mag.* **36,** 832–838.

Hummel, F. A., and Sarver, J. F. (1964). *J. Electrochem. Soc.* **111,** 252–253.

Megaw, H. D., Kempster, C. J. E., and Radoslovich, E. (1962). *Acta Crystallogr.* **15,** 1017–1035.

Melamed, N. T., Vicaro, P. J., Artman, J. O., and de S. Barros, F. (1970). *J. Luminescence* **1, 2,** 348.

Nash, D. B., and Greer, R. T. (1970). *Proc. 1st Lunar Sci. Conf., Geochim. Cosmochim. Acta,* Suppl. 1, Vol. 3, pp. 2341–2350.

Pott, G. T., and McNicol, B. D. (1971a). *Chem. Phys. Leit.* **12,** 63–64.

Pott, G. T., and McNicol, B. D. (1971b). *Discuss. Faraday Soc.* **52,** 121–131.

Pott, G. T., and McNicol, B. D. (1972). *J. Chem. Phys.* **56,** 5246–5254.

Reynolds, M. R., and Garlick, G. F. J. (1967). *Infrared Phys.* **7,** 151–159.

Sippel, R. J. and Spencer, A. B. (1970). *Proc. 1st Lunar Sci. Conf., Geochim. Cosmochim. Acta,* Suppl. 1, Vol. 3, pp. 2413–2426.

Van Doorn, C. Z., and Schipper, D. J. (1971). *Phys. Lett. A* **34,** 139–140.

Van Doorn, C. Z., Schipper, D. J., and Bolwijn, P. T. (1972). *J. Electrochem. Soc.* **119,** 85–92.

Weeks, R. A. (1972). *Proc. 3rd Lunar Sci. Conf., Geochim. Cosmochim. Acta,* Suppl. 3, Vol. 3, pp. 2503–2517.

4

INTERPRETATION OF VISIBLE AND NEAR-INFRARED DIFFUSE REFLECTANCE SPECTRA OF PYROXENES AND OTHER ROCK-FORMING MINERALS

John B. Adams

I. Introduction

The purpose of this chapter is to discuss the uniqueness of visible and near-infrared diffuse reflectance spectra of pyroxenes and other common rock-forming minerals, and the extent to which telescopic spectra of unexplored surfaces can be used to identify the constituent mineralogy. Visible and near-infrared spectra are now available of the surfaces of a variety of solar system objects, including the moon, Mercury, Mars, asteroids, and Galilean satellites. To interpret these spectra it is necessary to understand the optical absorption properties of minerals and rocks.

There is an enormous number of terrestrial, meteoritic, and lunar minerals available for study, but fortunately the number of important rock-forming minerals is small. Certain silicate minerals such as plagioclase feldspar, pyroxene, and olivine in various combinations comprise the bulk of many terrestrial and extraterrestrial rocks. This chapter especially addresses the reflectance properties of minerals and rocks that are known to occur on extraterrestrial surfaces, or that are possibly important constituents of these objects.

The laboratory diffuse reflectance data were obtained in the range 0.35–2.5 μm with a Beckman DK-2A ratio-recording spectroreflectometer. Samples were powdered, mounted in aluminum dishes, and covered with polished silica cover glasses. A freshly prepared MgO reference was covered with a matched cover glass. Adsorbed water on the MgO standard was monitored with the spectrometer by placing the MgO standard in the reference position on the integrating sphere and mounting a sample of sand-blasted gold foil in the sample position. Spectra were recorded simultaneously on chart paper and on digital paper-tape.

II. Background

The theoretical basis for interpretation of visible and near-infrared spectra of minerals was established in a series of papers by Burns (1965, 1970a), White and Keester (1966, 1967), and Bancroft and Burns (1967). These authors pioneered in the application of crystal-field theory to the interpretation of optical spectra of natural materials. White and Keester (1966) and Adams and Filice (1967) showed that diffuse reflectance spectra of powdered minerals retain the main spectral details that are observed in the transmission curves. Adams (1968) presented evidence that absorption bands in the reflectance spectra of planetary surfaces could be used to obtain information on remote mineralogy and petrology. Several authors have studied the reflectance properties of minerals and rocks from the point of view of interpreting remotely sensed spectra (see, for example, Salisbury, 1972, and Chapter 5), however, a review of this extensive literature is not possible here.

Burns (1970a) provided a detailed discussion of the application of crystal-field theory to the study of minerals. Chapters 1 and 2 report on recent results in this rapidly expanding area of research where the spectra of single mineral grains are measured in polarized light. Readers are refered to these chapters for further background on theory and absorption-band assignments.

To apply the detailed information now available on the optical properties of single mineral crystals to the diffuse reflectance spectra of powders it is necessary to know how the crushing of minerals affects their spectral reflectance. The effects of particle size and shape and packing densities on the spectral reflectance properties of minerals and rocks were investigated by Adams and Filice (1967), and it was shown that variations in these physical parameters change the albedo of the powders but do not change the wavelengths of the electronic or vibrational absorption bands. For further information on the effects of variations in these parameters on spectral reflectance, see Chapter 6.

Because albedo is a function of the physical state of a material it has been possible to use the photometric properties of powders to interpret telescopic measurements of the photometric functions of the moon and other objects (Hapke and VanHorn, 1963; Hapke, 1968). These and other measurements such as of the polarization properties of powders (e.g., Dollfus and Titulaer 1971; Pieters, 1973) are outside the scope of this chapter. However, it should be noted that these measurements can yield very useful information for remote sensing that compliments the spectral reflectance data.

III. Reflectance Spectra of Rock-Forming Minerals

A. Introduction

In this section, diffuse reflectance spectra are presented for some principal groups of rock-forming minerals. Spectra also are given for a variety of naturally occurring glasses. The attempt here is to illustrate the important spectral characteristics of materials that are known to be common constituents of the earth, of the moon, and of meteorites. The emphasis, therefore, is on common silicate minerals. The main point being addressed is the uniqueness of the spectral features of each mineral species. It will become clear that some spectra are highly unique, whereas others are ambiguous. From a knowledge of the spectral properties of the various mineral types it is possible to develop a sound basis for interpretation of reflectance curves of unexplored surfaces.

In the following discussion and in Figs. 1–9 the minerals are arranged according to structural type, following the format of mineralogy texts such as Deer, Howie, and Zussman (1966). Glasses (Fig. 10) are included because they exhibit regular spectral features. No attempt has been made to include all minerals. For a survey of the reflectance spectra of a wide variety of minerals the reader is referred to the series of papers by Hunt and Salisbury (1970, 1971) and Hunt *et al.* (1971a,b, 1972).

B. Olivine Group

The olivines are an important group of minerals in the scheme of interpreting telescopic spectra of solid bodies. Olivines (except pure forsterite) show strong absorption near 1 μm; furthermore, they are a common constituent of meteorites, they occur in some lunar samples, and they are abundant in many terrestrial mafic and ultramafic rocks.

The strong absorption slightly beyond 1 μm (Fig. 1) arises from Fe^{2+} ions in sixfold coordination. The critical role of the Fe^{2+} ions in the absorption process is emphasized by the monotonus spectrum of pure (synthetic) forsterite (Mg_2SiO_4). Very pure forsterite occurs in some meteorites, and as such does not contribute any absorption bands.

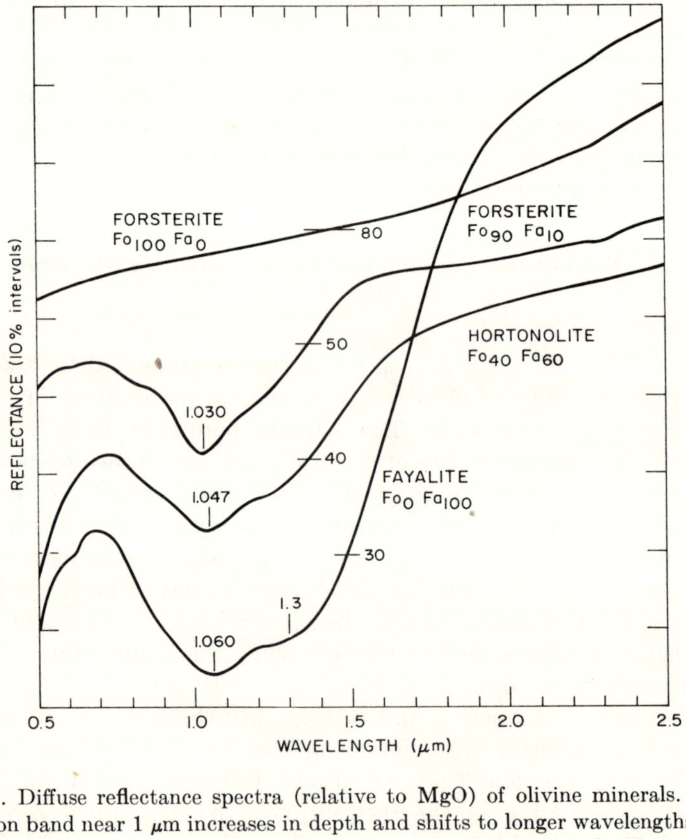

FIG. 1. Diffuse reflectance spectra (relative to MgO) of olivine minerals. The Fe^{2+} absorption band near 1 μm increases in depth and shifts to longer wavelengths, and the 1.3 μm inflection enlarges with increasing iron content of the minerals. Horizontal lines crossing the curves that are labeled with a two-digit number indicate percent reflectance.

A few percent of the fayalite molecule induces a characteristic single band centered at about 1.03 μm (9710 cm^{-1}). This absorption arises mainly from Fe^{2+} on the noncentrosymmetric M2 site (Burns, 1970a,b). With increasing iron content the band depth increases and the band shifts to longer wavelengths and develops a pronounced inflection near 1.3 μm (7700 cm^{-1}). This behavior is explained by referring to polarized absorption spectra (see Chapter 2, Section IV, C). Burns (1970a) assigns the 1.3 μm feature to Fe^{2+} on M1 sites. Runciman *et al.* (1973b), however, do not agree, and thus this assignment is the subject of debate in the literature. Burns (1970b) shows that the shift of the absorption band to longer wavelengths is linear (in cm^{-1} units) with respect to Mg^{2+}–Fe^{2+} compositions, based on polarized transmission spectra. The change in band depth and band position in reflectance spectra with composition was shown by Hunt and Salisbury (1970).

The 1.03–1.05 μm (9710–9520 cm^{-1}) band of the forsteritic olivines, by itself, is not a unique spectral feature. Several other minerals, including calcic pyroxenes, display this band (as will be discussed). However, the other minerals (and some glasses) also display other absorption bands at longer or shorter wavelengths besides the one at 1.03–1.05 μm (Fig. 11). Thus, the spectrum of olivine can be considered unique if the presence of the other bands can be ruled out. The spectra of the fayalite-rich olivines have a unique shape, owing to the exceptionally strong absorption, and to the prominent inflection near 1.3 μm (7700 cm^{-1}).

C. Pyroxene Group

The pyroxenes appear to be the most important group of minerals for the purpose of interpreting the spectra of rock and rock-derived surfaces in the solar system. Their importance arises from their relative abundance in terrestrial rocks, meteorites, lunar rocks, and asteroids, and their probable presence in the surface materials of Mars and Mercury. Of the common rock-forming silicates, the iron-bearing pyroxenes typically have the deepest and best-defined absorption bands. Furthermore, information on the composition and structure of a pyroxene commonly can be derived from the spectral curve.

Figure 2 illustrates the characteristic double-band structure of the low-calcium pyroxenes. Both absorption bands are caused by Fe^{2+} on highly distorted octahedral (M2) sites. A discussion of the assignments of the major bands is given in White and Keester (1966, 1967), Burns (1970a), and Runciman *et al.* (1973a) (also see Chapter 2, Section IV, A). In the examples given in Fig. 2, note that the short wavelength band ranges

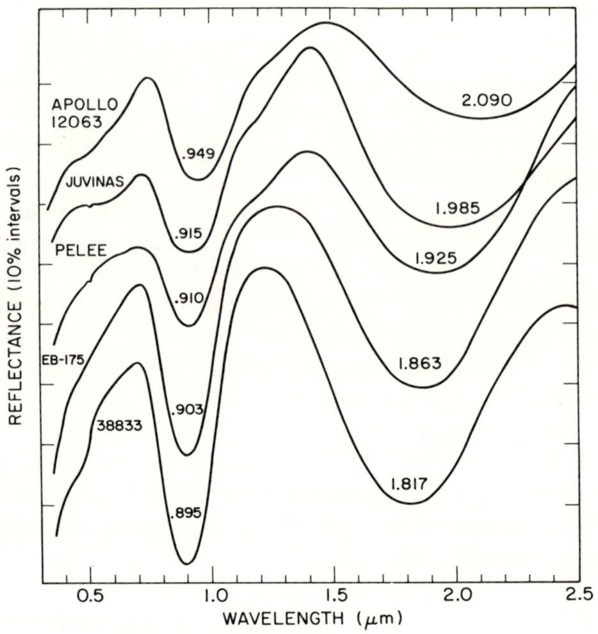

FIG. 2. Diffuse reflectance spectra (relative to MgO) of pyroxenes. Samples 38833, EB-175, and Pelee are orthopyroxenes and increase in Fe in that order. Juvinas is a pigeonite, and Apollo 12063 is a subcalcic augite. Chemical data for these samples are given in Table I. Note the two prominent Fe^{2+} absorption bands for all samples, and the shift of both bands to longer wave-lengths from bottom to top of the figure.

from 0.895 (11,200 cm^{-1}) to 0.949 μm (10,500 cm^{-1}), and the long wavelength band ranges from 1.817 (5,500 cm^{-1}) to 2.090 μm (4,780 cm^{-1}). The shifts in both bands to longer wavelengths are related to increasing iron and to increasing calcium.

Several details of the curves in Fig. 2 are worth noting. First, the 0.9 μm (11,000 cm^{-1}) band is consistently narrower and deeper than the 1.8 (5,600 cm^{-1}) to 2.0 μm (5,000 cm^{-1}) band. Second, several of the curves show weak structure at the shorter wavelengths. And third, there is strong absorption in the ultraviolet portion of the spectrum. Burns *et al.* (1972a) consider broad, weak bands near 0.44 (23,000 cm^{-1}) and 0.65 μm (15,000 cm^{-1}) to be the result of charge-transfers involving Fe^{2+}–Fe^{3+}, and Fe^{2+}–Ti^{3+}, whereas the very narrow band at 0.505 μm (19,800 cm^{-1}) (see samples Juvinas and Pelee of Fig. 2) is considered to be an Fe^{2+} spin-forbidden band. Runciman *et al.* (1973a) assign both the 0.65 and the 0.505 μm bands as spin-forbidden. The sharp drop in the UV reflectance is largely the result of charge-transfer absorptions. Chemical analyses of the pyroxenes in Fig. 2 are listed in Table I.

The weak inflection near 1.25 μm (8000 cm^{-1}) in the top three curves of Fig. 2 is contributed by minor amounts of plagioclase that remained in the separates of the respective pyroxenes. The plagioclase contributes a band near 1.25 μm which is due to Fe^{2+} (see Fig. 6).

In terms of uniqueness, the band at 0.90 μm (11,100 cm^{-1}) is almost exclusively an indicator of low Ca pyroxene. Possible confusion could arise with the 0.89 μm (11,200 cm^{-1}) Fe^{3+} band of goethite (Fig. 8), and with weak Fe^{2+} and Fe^{3+} bands in the micas (Fig. 5); however, other parts of these curves would serve to distinguish them. From about 0.93 (10,800 cm^{-1}) to 1 μm (10,000 cm^{-1}) the Fe^{2+} absorption also occurs in several of the amphiboles (Fig. 4), and in magnetite (Fig. 7), although the band depths in these minerals are very shallow relative to those of the pyroxenes.

The long wavelength band (1.78–2.2 μm) (5620–4500 cm^{-1}) is perhaps the most diagnostic single feature of the pyroxenes (Lewis and White, 1975). No other common silicates exhibit this strong band. The only

TABLE I

ANALYSES OF PYROXENES SHOWN IN FIG. 2

	ENSTATITE 38833[a]	ENSTATITE EB-175[a]	HYPERSTHENE PELEE[a]	PIGEONITE JUVINAS[b]	SUBCALCIC AUGITE APOLLO 12063[c]
SiO_2	57.38	55.17	52.26	48.49	51.61
TiO_2	0.02	0.22	0.17	1.22	1.15
Al_2O_3	1.12	1.00	1.00	2.03	1.82
Fe_2O_3	(as FeO)	(as FeO)	(as FeO)	—	(as FeO)
FeO	8.08	15.59	24.45	32.34	16.43
MnO	—	—	—	0.80	0.37
CaO	0.27	1.61	1.32	4.14	8.45
MgO	33.98	27.59	20.56	11.43	18.86
Na_2O	—	—	—	0.09	—
K_2O	—	—	—	0.06	—
Cr_2O_3	—	—	—	—	0.72
	100.85	101.18	99.76	100.60	99.41
	$Wo_0En_{88}Fs_{12}$	$Wo_3En_{74}Fs_{23}$	$Wo_3En_{58}Fs_{39}$	$Wo_8En_{36}Fs_{56}$	$Wo_{18}En_{55}Fs_{27}$

[a] Microprobe analysis by A. Chodos.
[b] Duke and Silver (1967).
[c] Microprobe analysis by P. Bell.

possible sources of confusion would be spinel (Fig. 7) and the silicate glasses (Fig. 10), which typically show a weak Fe^{2+} absorption structure in the same spectral region.

When taken together, the two major absorption bands of the types shown in Fig. 2 provide a unique identification of pyroxenes. Furthermore, there is a regular shift in the wavelength positions of the two bands with composition. The wavelength of the short band when plotted against the wavelength of the long band defines a curving line (Fig. 11) along which pyroxenes change in composition. This pyroxene trend line is discussed in detail elsewhere (Adams and McCord, 1972; Adams, 1974).

The absorption features of the calcic pyroxenes shown in Figs. 3 and 4 differ somewhat from those of Fig. 2. The (low-Fe) diopsides shown in Fig. 4 still possess two very prominent bands; however, these occur longward of 1 μm (10,000 cm^{-1}), and near 2.3 μm (4,300 cm^{-1}). The bands are related to the Fe^{2+} bands of the low-calcium pyroxenes discussed above, but they reflect the different crystal structure (and therefore crystal field) in which the Fe^{2+} occurs in diopside. In the orthopyroxenes and low-Ca clinopyroxenes Fe^{2+} occurs largely on the M2 sites. In the diopside structure, although it is similar to the orthopyroxene structure, Ca^{2+} ions occupy the M2 sites and Fe^{2+} ions preferentially fill the M1 sites (Burns, 1970a). Burns *et al.* (1972a) assign the two prominent bands of the diopside–hedenbergite series to Fe^{2+} ions on the M1 sites. The splitting of the two bands, as in the orthopyroxenes, probably is determined in part by the large distortion of the coordination site on which the Fe^{2+} occurs. Burns *et al.* (1972a) illustrate, on the basis of polarized absorption spectra of single crystals, that the absorption-band shift from 0.9 (11,000 cm^{-1}) to 1.04 μm (9,620 cm^{-1}) from orthopyroxenes to diopside-hedenbergite is a measure of the decreasing Fe^{2+} site occupancy of the M2 positions. The same qualitative measure of M2 site occupancy in the Ca-rich pyroxenes appears in the diffuse reflectance data (Figs. 3 and 4), and is expressed in Fig. 11.

The pyroxene curves shown in Fig. 3 lack the simple two-band structure, and instead, typically display three or more shallow bands. The bands at 1.01–1.02 μm (9900-9800 cm^{-1}) and at 2.2–2.3 μm (4500–4300 cm^{-1}) are still present in the augites and diopside-salites (labeled type B), but two of the curves (labeled A) instead have bands near 0.95 μm (10500 cm^{-1}) and at 1.155 μm (8660 cm^{-1}) in addition to the one near 2.3 μm. Apparently in the type A calcic pyroxenes of Fig. 3 the 1.03 μm (9710 cm^{-1}) band has been replaced by two bands. From the data of Burns (1970a) it appears that a separation between the 0.96 and the 1.16 μm bands occurs in the Eg orbital group in Fe^{2+} on the M1 site. Type B pyroxenes in Fig. 3

do not show clearly developed bands near 0.95 and 1.155 μm; thus, these bands are exclusive of the absorption near 1.03 μm. Analyses of the pyroxenes from Fig. 3 appear in Table II. There is no obvious chemical difference that accounts for the differences between type A and type B pyroxenes.

Most of the curves shown in Fig. 3 also have a clearly developed absorption feature between 0.75 (13,300 cm^{-1}) and 0.80 μm (12,500 cm^{-1}). This band appears to be a charge-transfer absorption involving Fe^{2+} and

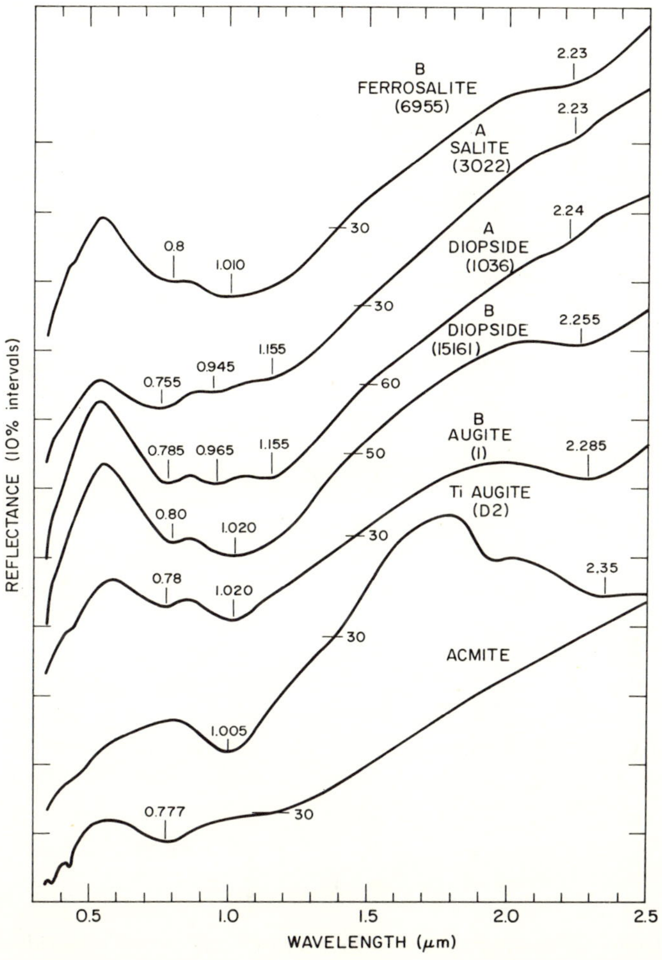

FIG. 3. Diffuse reflectance spectra of clinopyroxenes. Iron-rich calcic pyroxenes typically have more complex Fe^{2+} and Fe^{2+}-Fe^{3+} band structure than do the orthopyroxenes and pigeonites. See text for discussion. Chemical analyses are given in Table II.

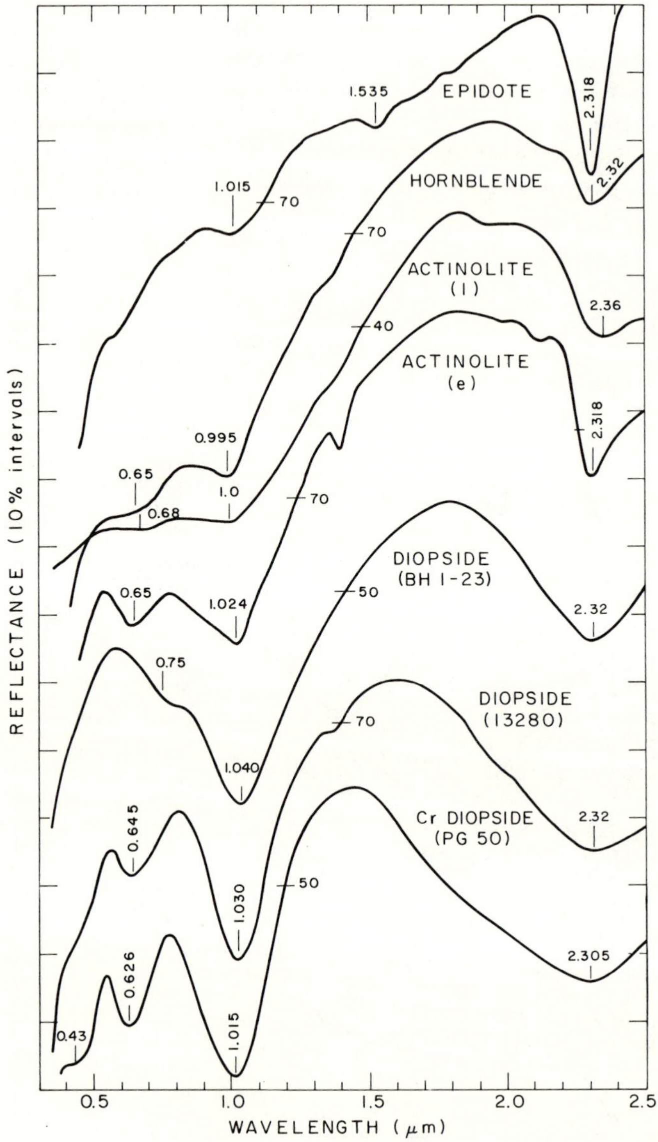

FIG. 4. Diffuse reflectance spectra of Fe-poor calcic pyroxenes (diopsides) compared with spectra of amphiboles and of epidote. Diopsides have well-developed Fe^{2+} bands near 1 and 2.3 μm. The amphiboles have a narrower and shallower band near 2.3 μm, and a characteristic charge-transfer absorption near 0.65 μm. See text for discussion. Chemical analyses are given in Table III.

TABLE II

ANALYSES OF PYROXENES SHOWN IN FIG. 3

	TITANIUM-AUGITE D2[a]	AUGITE 1[b]	DIOPSIDE 15161[c]	DIOPSIDE 1036[c]	SALITE 3022[c]	FERROSALITE 6955[c]
SiO_2	50.85	52.76	54.54	54.84	49.70	51.54
TiO_2	1.05	0.16	0.01	0.14	0.12	0.02
Al_2O_3	2.70	1.12	0.64	0.38	4.74	0.23
Fe_2O_3	1.11	1.73	(as FeO)	(as FeO)	(as FeO)	(as FeO)
FeO	10.36	8.92	4.00	3.71	10.13	16.46
MnO	0.25	0.47	—	—	—	—
CaO	19.20	20.48	25.52	24.96	23.69	23.78
MgO	14.09	13.43	15.58	16.11	10.64	7.55
Na_2O	0.30	0.35	—	—	—	—
K_2O	0.00	0.05	—	—	—	—
Cr_2O_3	0.13	—	—	—	—	—
H_2O+	0.01	0.45	—	—	—	—
H_2O-	0.09	—	—	—	—	—
NiO	0.02	—	—	—	—	—
	100.16	99.92	100.27	100.14	99.02	99.58
	$Wo_{40}En_{41}Fs_{19}$	$Wo_{44}En_{40}Fs_{16}$	$Wo_{51}En_{43}Fs_{6}$	$Wo_{50}En_{44}Fs_{6}$	$Wo_{51}En_{32}Fs_{17}$	$Wo_{51}En_{22}Fs_{27}$

[a] Hess (1949).
[b] Allen and Fahey (1951).
[c] Microprobe analyses by A. Chodos.

Fe^{3+} ions (Burns *et al.*, 1972b). In diffuse reflectance spectra the presence of the 0.75–0.80 μm band is characteristic of those pyroxenes that contain at least a few tenths of a percent of Fe_2O_3. Fe^{3+} ions are more easily accommodated in the augite-type structures than, for example, in the orthopyroxenes, and the charge-transfer absorptions in which Fe^{3+} participate account in large part for the overall differences between the pyroxene curves in Figs. 2, 3, and 4: Not only do most of the pyroxenes in Fig. 3 show the 0.75–0.80 μm band, but they also exhibit a steep overall slope and masking of the major Fe^{2+} spin-allowed bands. Lunar augites (see Figs. 2 and 11), having little or no Fe^{3+} (Burns *et al.*, 1973), do not show the masking of the main Fe^{2+} bands near 1 (10,000 cm^{-1}) and 2 μm (5,000 cm^{-1}). With increasing Fe^{3+}/Fe^{2+} ratio, the charge-transfer absorptions at 0.75–0.80 μm and in the ultraviolet become dominant, and the Fe^{2+} absorption bands are entirely masked, as is shown in the spectrum for acmite in Fig. 3. Note also that the acmite spectrum has sharp narrow bands at 0.38 (26,300 cm^{-1}) and 0.43 μm (23,300 cm^{-1}) which probably are Fe^{2+} spin-forbidden absorptions. A similar weak band near 0.45 μm (22,200 cm^{-1}) occurs in the augite spectrum (Fig. 3).

The titan-augite spectrum (Fig. 3) is of interest because of the notable absence of a band at 0.75–0.80 μm (13,300–12,500 cm^{-1}). Lewis and White (1975) remarked that the presence of titanium in augite appears to suppress the 0.77 μm (13,000 cm^{-1}) charge transfer band without masking the 1.0 μm band. The spectra of several Ti-augites investigated in this study suggest that this is a general rule, which, in turn, implies that Ti-augites have a unique reflectance curve. Note also that the Ti-augite spectrum in Fig. 3 has weak hydroxyl and water bands at 1.4 (7100 cm^{-1}) and 1.9 μm (5300 cm^{-1}), probably due to traces of an alteration phase (amphibole?) in the augite.

In Fig. 4 the diopside spectra have well-developed absorption features near 0.43 (23,200 cm^{-1}), 0.62 (16,100 cm^{-1}), 0.64 (15,600 cm^{-1}), and 0.75 μm (13,300 cm^{-1}). The bands near 0.45 (22,200 cm^{-1}) and 0.62 μm in the spectrum of the Cr-diopside probably arise from Cr^{3+} on the M1 site (Burns *et al.*, 1973). This pyroxene contains 1.21% Cr_2O_3 (Table III). The Cr^{3+} bands in diopside are shifted slightly to longer wavelengths relative to those in Cr_2O_3 (Fig. 7). The absorption near 0.645 μm (15,500 cm^{-1}) in diopside 13280 may also be due to Cr^{3+}; however, the companion band near 0.45 μm is indistinct, perhaps owing to the very steep slope of the curve in that spectral region. The 0.645 μm absorption is too strong to be a Fe^{2+} spin-forbidden band (Runciman *et al.*, 1973a). It is also unlikely that this band is due to Ti^{3+}, owing to the very low (0.04%) content of TiO_2. (see Table III). The weak feature near 0.75 μm in diopside BH 1-23 is probably the result of Fe^{2+}–Fe^{3+} charge transfers.

TABLE III

ANALYSES OF PYROXENES SHOWN IN FIG. 4

	Cr-DIOPSIDE PG 50[a]	DIOPSIDE 13280[b]	DIOPSIDE BH 1.23[a]
SiO_2	51.98	54.27	54.51
TiO_2	0.15	0.04	0.04
Al_2O_3	3.67	1.37	0.40
Fe_2O_3	0.62	(as FeO)	0.40
FeO	3.62	2.69	2.35
MnO	0.14	—	0.16
CaO	20.13	24.47	24.90
MgO	18.28	17.65	17.19
Na_2O	0.17	—	0.06
K_2O	0.08	—	0.05
Cr_2O_3	1.21	—	—
H_2O+	0.11	—	0.03
H_2O-	0.03	—	0.03
NiO	0.03	—	—
	100.22	100.50	100.12
	$Wo_{41}En_{52}Fs_7$	$Wo_{48}En_{48}Fs_4$	$Wo_{49}En_{47}Fs_4$

[a] Hess (1949).
[b] Microprobe analyses by A. Chodos.

The unique features of the Ca-pyroxene spectra can be summarized as follows: The two main Fe^{2+} bands near 1.03 (9710 cm^{-1}) and 2.3 μm (4300 cm^{-1}) are typical of the calcic pyroxenes; however, these bands also appear in the spectra of actinolite (see Fig. 4). Bands near 0.95 (10,500 cm^{-1}) and 1.15 μm (8,700 cm^{-1}) may replace the single band near 1.03 μm, and in conjunction with the 2.3 μm band, uniquely characterize the type A calcic pyroxenes (see Fig. 11). The band at 0.75–0.80 μm (13,300–12,500 cm^{-1}) also is characteristic of the augites and other pyroxenes that contain both Fe^{2+} and Fe^{3+}; however, this band by itself must be treated with caution for it does not always allow a unique identification of a pyroxene (for example, Hunt and Salisbury, 1970). In combination with the other, spin-allowed Fe^{2+} bands the 0.75–0.80 μm band appears to be a unique indicator of calcic pyroxene containing some Fe^{3+}. The bands at 0.43–0.45 μm (23,300–22,200 cm^{-1}) and 0.62–0.65 μm (16,100–15,400 cm^{-1}) are good indicators of Cr^{3+} when seen in conjunction with the other bands discussed above. The absence of any other weak bands below 1.0 μm in a calcic pyroxene curve implies the presence of titanium. The relationships between the several bands discussed above are shown in Fig. 11.

D. Amphiboles

The diffuse reflectance spectra of three common amphiboles are shown in Fig. 4. The curves are displayed along with spectra of diopsides and epidote to allow comparison. There is a striking overall similarity among the curves of Fig. 4. All spectra have Fe^{2+} spin-allowed bands near 1.0 μm. The amphiboles, however, display a band near 0.65 μm (15,400 cm^{-1}) that is probably due to Fe^{2+}–Fe^{3+} charge transfer (Burns, 1970b). This assignment is supported by the observation that in amphiboles such as crossite (spectrum not shown) having a high Fe^{3+}/Fe^{2+} ratio, the bands near 1.0 (10,000 cm^{-1}) and 2.3 μm (4,300 cm^{-1}) are very faint and the 0.65 μm band is the principal remaining feature. In contrast, tremolite and gedrite (not shown), having a trace of FeO but essentially no Fe_2O_3, have two exceedingly strong bands at 0.93 (10,800 cm^{-1}) and 2.3 μm, but no band at 0.65 μm. The assignment of the band at 2.3 μm is uncertain; however, it appears to be a combination of two overlapping bands, one a broad band contributed by Fe^{2+} (analogous to the 2.3 μm band in the calcic pyroxenes), and the other due to combination tones of the OH stretch with lattice modes (White and Keester, 1967; Hunt and Salisbury, 1970). The broad band at 2.3 μm in the curves of hornblende and of actinolite (1) in Fig. 4 probably is due largely to Fe^{2+}, as evidenced by the very weak to absent OH and water bands at 1.4 (7,100 cm^{-1}) and 1.9 μm (5300 cm^{-1}) in these curves. The curve of actinolite (e), in contrast, has a clear 1.4 μm OH band, a weak 1.9 μm H_2O band, and the 2.3 μm band is distinctly deeper and narrower than in the other curves. Amphibole spectra sometimes display the hydroxyl and water bands at 1.4 and 1.9 μm; however, these bands tend to be suppressed by strong Fe^{2+} bands.

In terms of uniqueness, the 0.65 μm band along with the wavelengths of the bands near 1 and 2.3 μm serve to separate the amphiboles from all but the Cr-diopsides. Comparison of band depths and band shapes (Fig. 4) usually allows the spectra of these minerals to be discriminated. For example, the 2.3 μm band in Cr-diopsides is much broader than the 2.3 μm band in the amphiboles. Figure 11 compares the wavelengths of the amphibole bands with bands in the spectra of other minerals.

E. Layer Silicates

The diffuse reflectance spectra of several layer silicates are shown in Fig. 5. It is useful to separate the vibrational features arising from OH

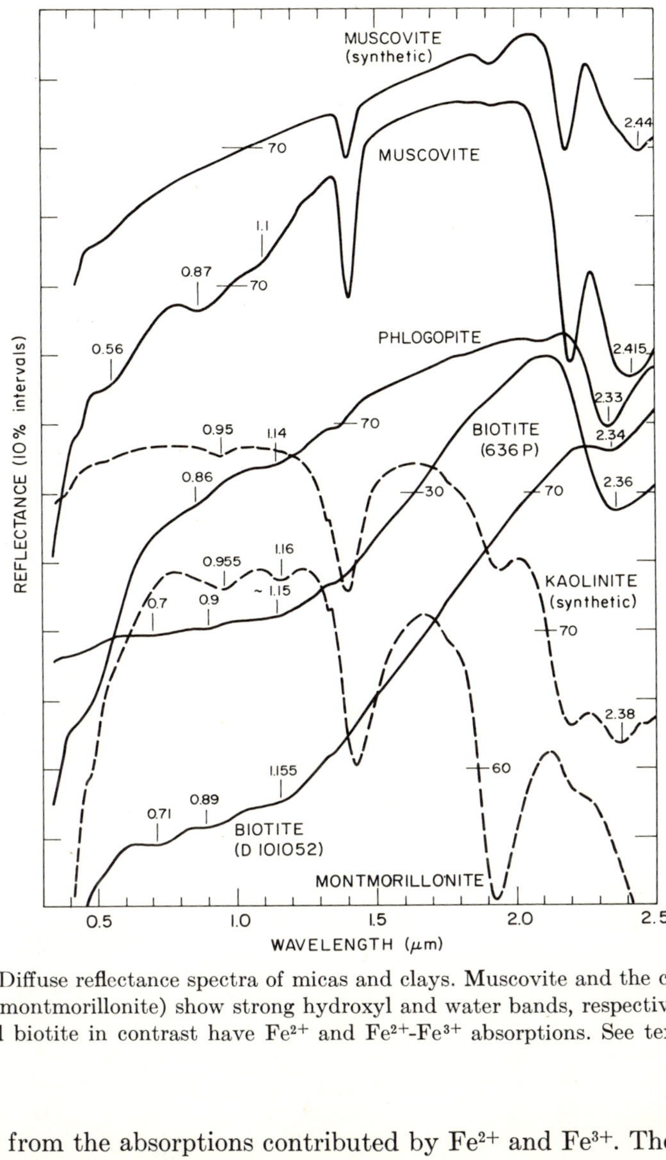

FIG. 5. Diffuse reflectance spectra of micas and clays. Muscovite and the clays (kaolinite and montmorillonite) show strong hydroxyl and water bands, respectively. Phlogopite and biotite in contrast have Fe^{2+} and Fe^{2+}-Fe^{3+} absorptions. See text for discussion.

and H_2O from the absorptions contributed by Fe^{2+} and Fe^{3+}. The spectra of muscovite, and of kaolinite and montmorillonite illustrate a variety of bands, which are indicative of the different sites on which OH and water molecules are located. Muscovites have a strong narrow absorption feature at 1.4 μm (7100 cm^{-1}) which is the OH ($2\nu_3$) overtone, and similar narrow bands at 2.3 (4300 cm^{-1}) and 2.4 μm (4200 cm^{-1}), which are combination tones involving the OH-stretching mode with lattice or librational modes. The narrowness of these bands indicates that the OH is on well-defined, ordered sites (Hunt and Salisbury, 1970).

The 1.9 μm (5300 cm^{-1}) H_2O band ($\nu_2 + \nu_3$) is weak or absent in the muscovites, whereas the clays—especially those with adsorbed water—typically have a well-developed band near 1.9 μm. These clays also have water bands near 1.15 μm (8,700 cm^{-1}) ($\nu_1 + \nu_2 + \nu_3$) and near 0.95 μm (10,500 cm^{-1}) ($2\nu_1 + \nu_3$). It is important to note that the water bands near 0.95 and 1.15 μm are very weak features, and that they overlap the positions of Fe^{2+} bands such as occur in the biotites and in the A-type diopsides. Ambiguity in the assignments of these bands is possible; however, in practice, the water bands are much too weak to show up in the reflectance spectra of iron-bearing minerals (even when the mineral powders are water-soaked); if water is present, the bands at 1.4 and 1.9 μm should be about a factor of 10 deeper than those at 0.95 and 1.15 μm.

The spectra of phlogopite and biotite in Fig. 5 illustrate the strong effect of electronic absorptions (Fe^{2+} and Fe^{3+}) in surpressing the hydroxyl and water bands. In the phlogopite the band at 0.86 μm (11,600 cm^{-1}) and the inflection at 0.5 μm (20,000 cm^{-1}) probably are due to Fe^{3+}, by analogy with the spectrum of Fe_2O_3 (Fig. 7). The bands at 1.14 (8770 cm^{-1}) and 2.34 μm (4270 cm^{-1}) are most likely Fe^{2+} spin-allowed bands, although there is probably some contribution of OH to the band near 2.3 μm. Similar bands occur in the reflectance spectra of muscovite and biotite, and the biotites have an additional absorption near 0.7 μm (14,000 cm^{-1}) which is probably a Fe^{2+} - Fe^{3+} charge transfer feature.

F. Feldspars

Reflectance spectra of common plagioclase and K-feldspars are shown in Fig. 6. Aside from the usual presence of water bands [1.4 (7100 cm^{-1}) and 1.9 μm (5,300 cm^{-1})] owing to trapped and/or adsorbed water, there are significant absorptions arising from trace amounts of Fe^{2+} and Fe^{3+}. The plagioclase feldspars (excepting the synthetic sample) show an absorption band ranging from 1.1 (9100 cm^{-1}) to 1.3 μm (7700 cm^{-1}) that is due to Fe^{2+} (Mao and Bell, 1972). In two of the spectra shown (bytownite and anorthite) there is a very weak band at 0.86 μm (11,600 cm^{-1}) that probably is due to Fe^{3+}. The Fe^{2+} band appears to vary in wavelength as a predictable function of the crystal structure, and of the $NaAlSi_3O_8$ to $CaAl_2Si_2O_8$ ratio. The wavelength of the Fe^{2+} absorption band plotted (vertically) against the percent anorthite (horizontally) defines a curve that is the inverse of the 2V vs. percent anorthite curve (Deer *et al.*, 1966; Smith, 1958). There is a 2V maximum for low temper-

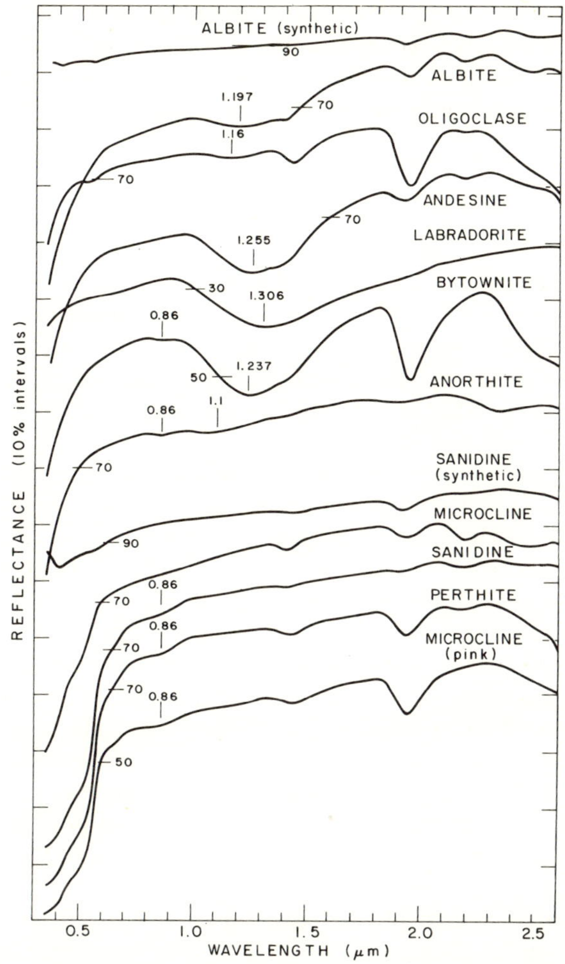

FIG. 6. Diffuse reflectance spectra of plagioclase and K-feldspars. Plagioclase minerals show a characteristic Fe^{2+} band near 1.2 μm, whereas K-feldspars typically have Fe^{3+} absorptions near 0.86 μm, and below 0.5 μm.

ature plagioclase at about 20% An, corresponding to the shortest wavelength position [1.15 μm (8700 cm^{-1})] for the Fe^{2+} band, and a 2V minimum at about 55% An, corresponding to a maximum wavelength of about 1.35 μm (7410 cm^{-1}) for the Fe^{2+} band. These results will be discussed in more detail in a separate paper.

The K-feldspars differ in spectral properties from plagioclase in that there is no detectable Fe^{2+} absorption feature, but there typically is absorption by Fe^{3+}. The Fe^{3+} features appear at 0.86 μm (11,600 cm^{-1}), and in the sharp absorption edge near 0.5 μm (20,000 cm^{-1}). The dominance of Fe^{3+} absorptions in K-feldspar is expectable in view of the fact that Fe^{3+} is most easily accommodated on the Al sites, whereas the Fe^{2+} readily substitutes on Ca sites.

G. Oxides

Diffuse reflectance spectra of several pure (synthetic) oxides and natural minerals are given in Fig. 7. Pure TiO_2 lacks absorption bands (since Ti^{4+} is without d-orbital electrons); therefore, a natural anatase with some iron impurity was used. The sample of Fe_3O_4 is a nearly stoichiometric magnetite, and the $MgAl_2O_4$ (spinel) sample apparently contains some iron.

The very strong and narrow absorption bands in the spectra of Cr_2O_3 and Fe_2O_3 arise from the respective Cr^{3+} and Fe^{3+} ions, and already have been discussed in the context of silicate minerals. The impure-anatase spectrum appears to be dominated by Fe^{2+} [0.90 (11,100 cm^{-1}), 1.12 (8,920 cm^{-1}), and 2.34 μm (4,270 cm^{-1})] and Fe^{2+}-Fe^{3+} [0.71 μm (14,100 cm^{-1})] and is very similar to the biotite spectra (Fig. 5).

There is little spectral detail in the strongly absorbing Fe_3O_4 (magnetite) except for a broad depression centered around 1.0 μm (10,000 cm^{-1}). The 1.0 μm feature is probably due to Fe^{2+}, but the overall spectrum appears to be dominated by intense Fe^{2+}–Fe^{3+} charge-transfer absorptions. The curve for $FeTiO_3$ has a broad band centered near 0.5 μm (20,000 cm^{-1}) that is at least in part due to Ti^{3+}. The exact origins of the other very broad structures in the curves of $FeTiO_3$, MnO_2, and $MgAl_2O_4$ are not known.

The reflectance spectra of two naturally occurring ferric oxide phases are given in Fig. 8. Goethite (FeO·OH) shows the characteristic Fe^{3+} band at 0.89 μm (11,200 cm^{-1}) and the sharp absorption edge in the blue ultraviolet region, with a slight inflection near 0.64 μm (15,600 cm^{-1}) Note that there there are only very weak OH bands near 1.4 (7100 cm^{-1}), 2.2 (4500 cm^{-1}), and 2.4 μm (4200 cm^{-1}).

Hematite (Fe_2O_3) is the anhydrous equivalent of goethite. The spectrum is characterized by the Fe^{3+} band at 0.85 μm (11,800 cm^{-1})—distinctly shorter in wavelength than in goethite—and by the broad saucer-shaped curve from 1.0 (10,000 cm^{-1}) to 2.5 μm (4,000 cm^{-1}). The differences in shapes between the two curves are important in considering possible interpretations of telescopic curves of the surface of Mars.

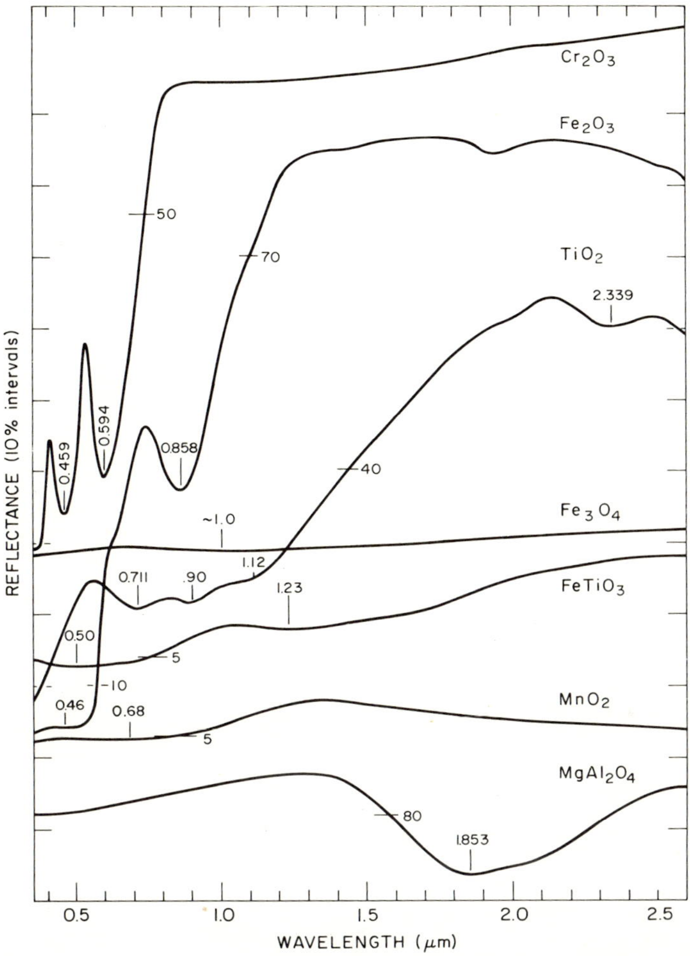

FIG. 7. Diffuse reflectance spectra of oxides. A variety of electronic and charge-transfer absorption features occur in the oxides. See text for discussion.

H. Carbonates

Four carbonate mineral spectra are shown in Fig. 9. The numerous bands between 1.4 (7,100 cm^{-1}) and 2.5 μm (4,000 cm^{-1}) are the result either of water or of the carbonate ion. The assignments of the CO_3 bands are discussed in Hunt and Salisbury (1971), and the same paper covers the reflectance properties of carbonate minerals in detail. In Fig. 9 the samples

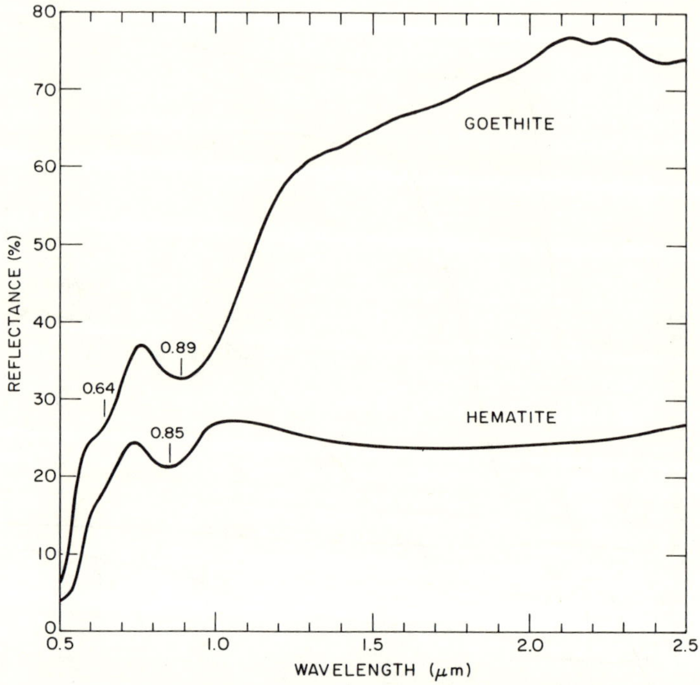

FIG. 8. Diffuse reflectance spectra of the common ferric oxides, goethite and hematite. Although the main elements of the spectra of pure Fe_2O_3 (Fig. 8) are retained, the natural ferric oxide minerals have distinctive curves.

of calcite and aragonite are quite pure and therefore do not show Fe^{2+} absorption bands. However, the 1.0 μm (10,000 cm^{-1}) band in dolomite, and the broad absorption between 0.8 (12,500 cm^{-1}) and 1.5 μm (6,700 cm^{-1}) in siderite, are due to Fe^{2+}. The broad absorption feature of siderite appears to be unique, whereas the 1.0 μm band in dolomite, by itself, is not diagnostic.

I. Glasses

The spectra of eight rock glasses are shown in Fig. 10. These glasses, along with artificially prepared iron-bearing silicate glasses, typically have two broad absorption bands near 1 (10,000 cm^{-1}) and 2 μm (5,000 cm^{-1}). Both bands are the result of Fe^{2+}. The band near 1 μm is due to Fe^{2+} on octahedral sites, whereas the band near 2 μm appears to be due to Fe^{2+} on tetrahedral sites (Boon and Fyfe, 1972). Typically the 1 μm band is somewhat variable in wavelength, ranging between 1.0 and 1.15

μm (8700 cm^{-1}). The 2 μm band is usually broad and weak. Boon and Fyfe (1972) show that less than 0.5% of the Fe^{2+} occupies the tetrahedral sites. Even though the extinction coefficient is 10 to 100 times stronger for Fe^{2+} on the tetrahedral (in contrast to the octahedral) sites (Burns, 1970a), the low concentration of tetrahedral iron in the iron-poor glasses (such as obsidian) may account for a very weak, or absent, 2 μm band.

Although the two absorption bands in glass appear to be similar to pyroxene bands, it is important to note that the glass bands are closer

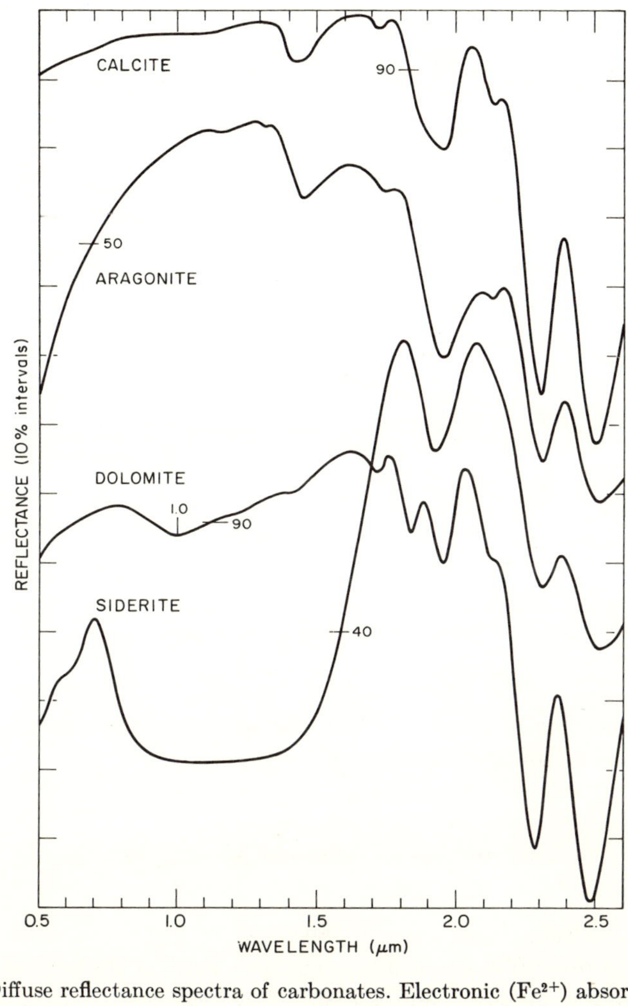

FIG. 9. Diffuse reflectance spectra of carbonates. Electronic (Fe^{2+}) absorptions occur near 1 μm and strong water and carbonate-ion bands appear between 1.4 and 2.5 μm.

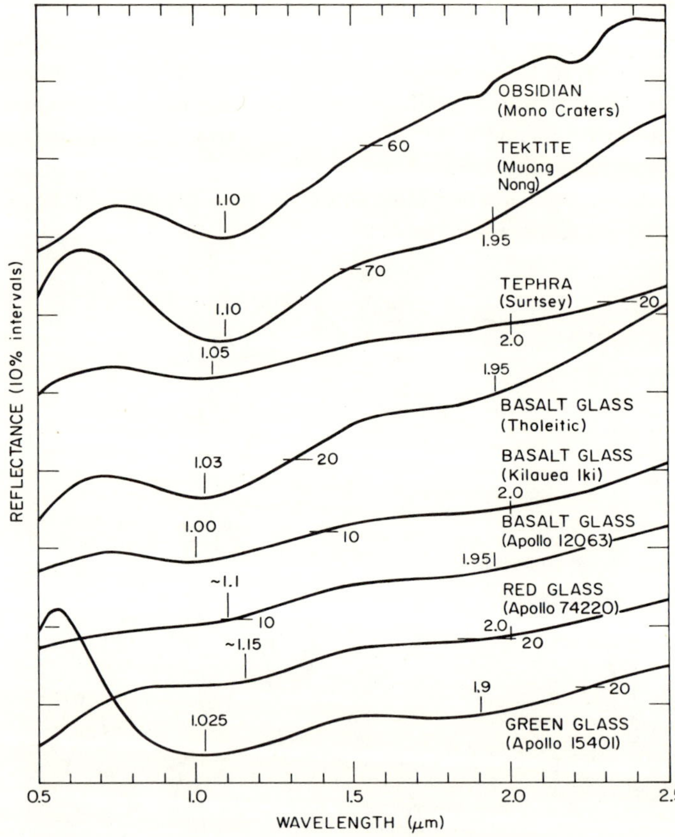

FIG. 10. Diffuse reflectance spectra of iron-bearing glasses. The broad Fe^{2+} bands above 1 μm and below 2 μm are characteristic of glasses.

together. Therefore the two bands in the spectra of glasses allow a unique identification as can be seen in Fig. 11.

IV. Uniqueness of Mineral Absorption Bands

The several reflectance curves presented in Fig. 1–10 were chosen to illustrate the main features in the spectra of some of the important classes of rock-forming minerals and glasses. The wavelength positions of the bands and the band assignments reveal details of the structure and the chemical make-up of most of the minerals. The same abundance of information in the spectral curves can lead to confusion in attempting to interpret the spectrum of an unknown substance. This task necessitates comparing the unknown spectrum with the spectra of all possible minerals or combinations of minerals. One way to illustrate the relationships between the principal absorption bands of minerals and glass is shown in Fig. 11. The diagram incorporates the absorption band data from Figs. 1–10 and from over 100 additional spectra of analysed minerals and glasses.

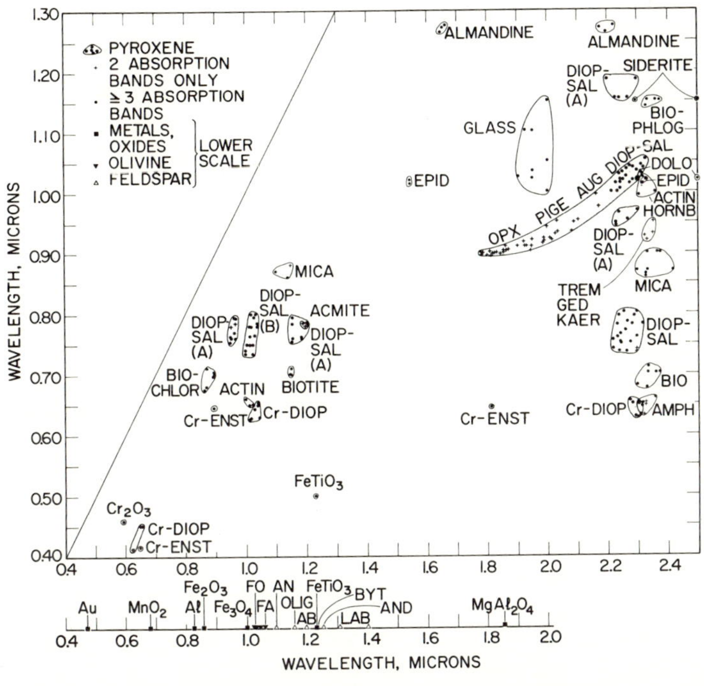

FIG. 11. Plot of centers of absorption bands in the diffuse reflectance spectra of several common minerals. Minerals having only one main band are plotted on the lower scale. Minerals with two or more bands in their spectra are designated by points in the main upper field. Hydroxyl and water bands at 1.4 and 1.9 μm are not used. Bands for minerals species and for glasses plot in separate fields. See text for discussion. Diop-sal = disoside-salite; bio-phlog = biotite-phlogopite; aug = augite; pige = pigeonite; opx = orthopyroxene; dolo = dolomite; epid = epidote; actin-hornb = actinolite-hornblende; trem-ged-kaer = tremolite-gedrite-kaersutite; amph = amphibole; enst = enstatite; chlor = chlorite; Fo = forsterite; Fa = fayalite; Ab = albite; Olig = oligoclase; And = andesine; Lab = labradorite; Byt = bytownite; An = anorthite.

In Fig. 11 the centers of absorption bands in the range 0.40 (25,000 cm^{-1}) to 1.30 μm (7,690 cm^{-1}) are plotted on the vertical axis, and the centers of absorption bands in the range 0.40–2.5 μm (4000 cm^{-1}) are plotted on the horizontal axis. Minerals having only one band (exclusive of identifiable water bands or carbonate bands) are plotted on a separate scale from 0.40 to 2.0 μm (5000 cm^{-1}) at the bottom of the diagram. Minerals having two or more bands (again, exclusive of identifiable water bands or carbonate bands) are plotted in the main field of the diagram.

A mineral having only two main absorption bands, for example, an orthopyroxene (Fig. 2), plots as a single point designated by a cross. Minerals with three or more bands are plotted as dots which appear at more than one place on the diagram. For example, an almandine garnet has bands near 1.27 (8870 cm^{-1}), 1.65 (6060 cm^{-1}), and 2.2 μm (4500 cm^{-1}); therefore, this sample plots as two points near the top of the diagram. A mineral having four bands plots as three points. An example is diopside A (Fig. 3) which has bands near 0.78 (12,800 cm^{-1}), 0.96 (10,400 cm^{-1}), 1.15 (8,700 cm^{-1}), and 2.24 μm (4,460 cm^{-1}).

Although the diagram excludes water and carbonate bands (which can usually be told by their wavelength position and narrow shape) there is some ambiguity with the band near 2.3 μm (4300 cm^{-1}) which can be due to Fe^{2+}, CO_3^{2-} or OH^-, or any combination of these. In this case the band is always plotted for the reason that on a degraded spectrum it is not always possible to identify OH^- or CO_3^- bands near 2.3 μm on the basis of shape alone. Thus, minerals such as carbonates, amphiboles, and micas plot along the 2.3 μm zone. The prominent CO_3^- band at 2.5 μm (4000 cm^{-1}) also is plotted for siderite and dolomite, to illustrate that even if the 2.3 μm band is mistaken for a Fe^{2+} band the presence of the 2.5 μm band should identify the carbonate mineral.

Perhaps the most important conclusion to be drawn from Fig. 11 is that the mineral species shown have absorption bands that are quite distinct from one another in wavelength. Even allowing for uncertainties in measuring the centers of broad weak absorption bands, there are well-defined clusters of points. Most of the common rock-forming minerals can be identified by their diffuse reflectance spectra, provided that the main absorption features are present. If any ambiguities remain they almost always can be resolved by comparing the overall curve shape, and the intensity of the reflected light (albedo) with known curves. For example, a single absorption band near 1.2 μm (8300 cm^{-1}) would suggest either ilmenite ($FeTiO_3$) or plagioclase feldspar. Comparison of the unknown curve with the curves in Figs. 6 and 7 should quickly resolve the question, based on the large difference in albedo.

ACKNOWLEDGMENTS

Mineral samples were kindly provided by several colleagues including Drs. J. Conel, P. Misch, D. Nash, and L. Silver. Dr. R. Speed made available many analyzed mineral separates, and Dr. W. Melson facilitated access to the collection at the U.S. National Museum. M. Charette and J. Kunin ably assisted with the laboratory measurements and with the data reduction. I have benefited greatly from discussions with Drs. P. Bell, R. Burns, M. Gaffey, R. Huguenin, T. McCord, and Ms. C. Pieters. Dr. W. White kindly provided a copy of an important study of the reflectance properties of pyroxenes in advance of publication. Ms J. Doane and Ms D. Rowe assisted with the preparation of the manuscript.

References

Adams, J. B. (1968). *Science* **159,** 1453–1455.

Adams, J. B. (1974). *J. Geophys. Res.* **79,** 4829–4836.

Adams, J. B., and Filice A. L. (1967). *J. Geophys. Res.* **72,** 5705–5715.

Adams, J. B., and McCord, T. B. (1972). *Proc. 3rd Lunar Sci. Conf., Geochim. Cosmochim. Acta,* Vol. 3, pp. 3021–3034.

Allen, V. T., and Fahey, J. J. (1951). *Geol. Soc. Amer., Bull.* **62,** 1420.

Bancroft, G. M., and Burns, R. G. (1967). *Amer. Mineral.* **52,** 1278–1287.

Boon, J. A., and Fyfe, W. S. (1972). *Chem. Geol.* **10,** 287–278.

Burns, R. G. (1965). Ph.D. Thesis, University of California, Berkeley (unpublished).

Burns, R. G. (1970a). "Mineralogical Applications of Crystal-Field Theory." Cambridge Univ. Press, London and New York.

Burns, R. G. (1970b). *Amer. Mineral.* **55,** 1608–1632.

Burns, R. G., Abu-Eid, R. M., and Huggins, F. E. (1972a). *Proc. 3rd Lunar Sci. Conf., Geochim. Cosmochim. Acta,* Vol. 1, pp. 533–543.

Burns, R. G., Huggins, F. E., and Abu-Eid, R. M. (1972b). *Moon* **4,** 93–102.

Burns, R. G., Vaughan, D. J., Abu-Eid, R. M., Witner, M., and Morawski, A. (1973). *Proc. 4th Lunar Sci. Conf., Geochim. Cosmochim. Acta,* Vol. 1, pp. 983–994.

Deer, W. A., Howie, R. A., and Zussman, J. (1966). "An Introduction to the Rock-Forming minerals." Wiley, New York.

Dollfus, A., and Titulaer, C. (1971). *Astron. & Astrophys.* **12,** 199–209.

Duke, M. B., and Silver, L. T. (1967). *Geochim. Cosmochim. Acta* **31,** 1637–1665.

Hapke, B. (1968). *Science* **159,** 76–79.

Hapke, B., and VanHorn, J. (1963). *J. Geophys. Res.* **63,** 4545–4570.

Hess, H. (1949). *Amer. Mineral.* **34,** 621–666.

Hunt, G. R., and Salisbury, J. W. (1970). *Mod. Geol.* **1,** 283–300.

Hunt, G. R., and Salisbury, J. W. (1971). *Mod. Geol.* **2,** 23–30.

Hunt, G. R., Salisbury, J. W., and Lenhoff, C. J. (1971a). *Mod. Geol.* **2,** 195–205.

Hunt, G. R., Salisbury, J. W., and Lenhoff, C. J. (1971b). *Mod. Geol.* **3,** 1–14.

Hunt, G. R., Salisbury, J. W., and Lenhoff, C. J. (1972). *Mod. Geol.* **3,** 121–132.

Lewis, J. F., and White, W. B. (1975). *J. Geophys. Res.* (in press).

Mao, H. K., and Bell, P. W. (1973). *Amer. Soc. Test. Mater., Spec. Tech. Publ.* 539, pp. 100–119.

Pieters, C. (1974). *In* "Planets, Stars and Nebulae Studied with Photo-polarimetry" (T. Gehrels, ed.). Univ. of Arizona Press, Tucson pp. 405–418.

Runciman, W. A., Sengupta, D., and Marshall, M. (1973a). *Amer. Mineral.* **58,** 444–450.

Runciman, W. A., Sengupta, D. and Marshall, M. (1973b). *Amer. Mineral.* **58,** 451–456.

Salisbury, J. W. (1972). *Moon* **5,** 332–347.

Smith, J. R. (1958). *Amer. Mineral.* **43,** 1179–1194.

White, W. B., and Keester, K. L. (1966). *Amer. Mineral.* **51,** 774–791.

White, W. B., and Keester, K. L. (1967). *Amer. Mineral.* **52,** 1508–1514.

5

THE USE OF MID-INFRARED SPECTROSCOPY IN REMOTE SENSING OF SPACE TARGETS

Lloyd M. Logan, Graham R. Hunt, and John W. Salisbury

I. Introduction

In this chapter we discuss the spectral signature information that is available in the mid-infrared from silicate materials, and describe the effects of changing environmental parameters on this information. We also evaluate the usefulness of infrared spectroscopy for determining the composition of such space targets as the moon, Mars, Mercury, and the asteroids, and point out the applicability of these data for study of circumstellar dusts. Additional details on remote sensing techniques in the mid-infrared region can be found in Chapter 7.

II. Vibrational Spectra of Inorganic Solids

The fundamental properties which determine the spectroscopic behavior of a material in absorption, transmission, reflection or emission are the refractive index, n, and the extinction coefficient k. Other parameters, such as particle size and temperature, have their effect on the appearance of the spectrum, but all of the compositionally characteristic information is contained in n and k.

The usefulness of spectroscopy for compositional measurements lies in the fact that n and k are wavelength dependent because of optically active transitions between various energy levels in the material. The energies of the vibrational levels, and hence the wavelength dependence of n and k are governed by three characteristic properties of the material. These are (*a*) the chemical composition, (*b*) the geometry of the equilibrium positions of the atoms relative to each other, and (*c*) the potential field of the interatomic forces representative of interatomic bonds and angles.

For silicates, the crystal structure is much too complex for a precise calculation of the vibrational band structure to be made. However, attempts have been made to identify vibrational features observed in the crystal spectra with the vibrational modes of free molecular groups. Despite the simplicity of this approach, it has been quite successful in leading to a qualitative understanding of spectra where bands indicative of "functional groups" can be identified. Further analysis of the finer details of the band envelope is generally not possible, but the theoretically expected sensitivity to composition is an experimentally substantiated fact (see Chapter 8, Tables VII–XII).

Figure 1 contains curves showing n and k as a function of wavelength for quartz. Two major groups of features are apparent. Those occurring near 10 μm (1000 cm^{-1}) are associated with Si–O bond-stretching modes of the silicate group, while those near 20 μm are predominantly due to O–Si–O bending modes. While other bands are evident in quartz, it is the two major groups of features mentioned above which have counterparts in all silicate spectra, and it is the variations in them from material to material which permits compositional distinctions to be made.

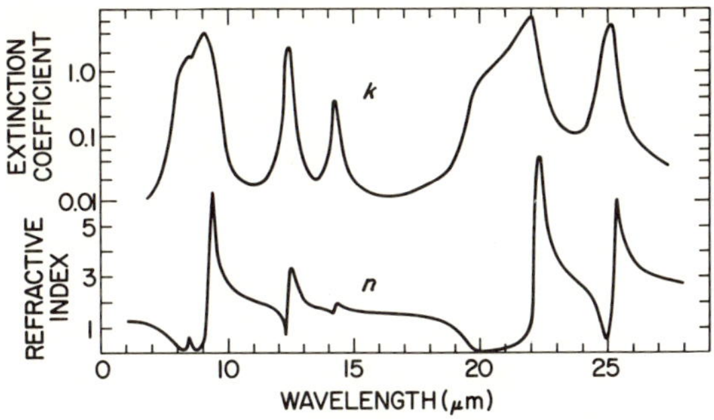

FIG. 1. Optical constants of quartz derived from the data of Spitzer and Kleinman (1961).

Large variations in n occur together with variations in k. Reflection spectra for small incidence angles to flat surfaces are directly related to n and k, by the Fresnel formula, namely

$$R = \frac{(n-1)^2 + k^2}{(n+1)^2 + k^2}$$

where R is the reflectivity.

Via some relatively complex treatment, such as the Kramers–Kronig analysis of reflection spectral data (see Chapter 11, Section III,A), n and k can be derived from the spectra. However, there is no simple way of obtaining n and k from a spectral measurement and the difficulty of attempting to do so is compounded by the anisotropy of silicate materials.

Thus the fundamental optical parameters, and thereby the chemical composition, may be derived from infrared spectra, but specific idealized experimental conditions must be employed. It is apparent that such evaluations cannot be made from the spectra of remote targets where the only information obtainable is a general reflection or emission spectrum in which bands are not singularly related to either fundamental parameters of the material. This leads to the important question of how compositionally characteristic the location of a band is in the spectrum of a remote target. The position of a band in the spectrum is often governed by environmental effects, so that an understanding of the compositional significance of a band location is best obtained from the study of a catalog of the spectra of materials recorded under conditions which simulate the environment of the target. We concentrate here on the transmission and emission spectra of silicate materials in the size ranges appropriate to space targets.

III. Optical Behavior of Fine Particles

When the diameter of the particles in the material approaches the wavelength of the radiation, the spectral properties become dominated by the scattering characteristics of the material. Several attempts have been made to describe the behavior theoretically. These include the multiple reflection model of Passell (1964); the qualitative theory of Emslie (1966) based on diffuse reflectance from thick clouds of absorbing scattering particles; a model based on Lorentz-Lorenz theory of dielectrics (Aronson *et al.*, 1967); the application of a diffuse reflectance theory based on the Kubelka-Munk method (Vincent and Hunt, 1968); and the application of a cloudy atmosphere model based on Mie diffraction theory (Conel, 1969) (also see Chapter 6).

An adequate description of scattering requires a rigorous solution of Maxwell's equations subject to the appropriate boundary conditions. For very small particles where r, the radius of the particle, is much smaller than the wavelength of the energy, λ, the particle is treated as though it were completely inside a homogeneous electric field, and Rayleigh scattering is adequate to describe the scattering process. For large particles, where $r \gg \lambda$, geometric optics and Fraunhofer diffraction theory provide an adequate description of the interaction of electromagnetic energy with the material. For the intermediate range, where $r \simeq \lambda$, the particle is too large to be considered as completely inside the field and too small to allow the radius of curvature of surface elements to be neglected. It was to provide specifically a solution to this latter case that Mie theory was developed; detailed treatments of the theory are available in a number of textbooks (e.g., Stratton, 1941; Van de Hulst, 1957; Samuelson, 1967).

Conel (1969) has made very effective use of Mie diffraction theory in conjunction with a cloudy atmosphere model to describe the spectral emission from condensed particulate mediums. In his paper, he presents detailed computations of the scattering parameters as functions of frequency for quartz and then applies these to the radiative transfer equations. The model accounts for many features observed in quartz. He predicted the occurrence of a maximum in the transmission spectrum of a powder film at the frequency where the refractive index of the individual particles equals that of the surrounding medium—the so-called Christiansen frequency (Henry, 1948). This occurs in the short wavelength side of the Si–O stretching absorption features near 10 μm (1000 cm^{-1}). The transmission maximum thus corresponds to a minimum in scattering. No quantitative calculations have been attempted for other silicates, but they behave

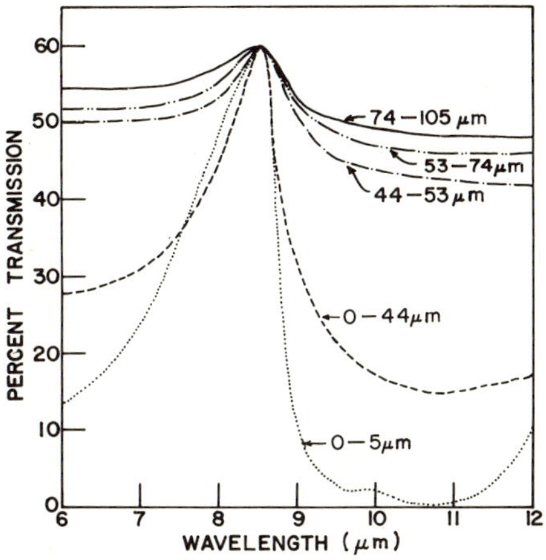

FIG. 2. Transmission spectra of various sized samples of hornblende.

similarly, displaying a prominent band which can be qualitatively related to the principal Christiansen frequency.

Figure 2 shows some transmission spectra of hornblende. The spectra were recorded using a Perkin Elmer Model 180 spectrophotometer fitted with a microreflection attachment which had been modified so that the incident beam impinged from above onto the sample supported on a mirror located in a horizontal plane. The light then traverses the sample twice, being reflected from the mirror surface. An air elutriation column was used in the preparation of underdense samples of 0–5 μm material; all other samples were prepared by depositing the material through a sieve of appropriate size. The thickness of the sample was controlled to provide 60% transmission at the peak.

The major features in the spectra are the maximum related to the Christiansen frequency and the minimum at longer wavelengths in the absorption band region. The contrast depends on particle size and there is some multiple band structure apparent in the transmission minimum of the finest sample. This general behavior is typical of all silicate materials, but the wavelength of the transmission maximum and the contrast for a given sample thickness and particle size depend on composition.

IV. Emission Spectra of Particulate Materials

Each of the theoretical studies of infrared optical behavior described in the previous section assumed an idealized isothermal environment. Quite clearly, such an idealized model does not apply to remote sensing conditions on space targets, and did not explain observational results. Logan *et al.* (1973) developed the first nonisothermal model to explain environmental effects on infrared emission behavior, a brief description of which follows.

The spectral dependence of radiation emitted from a particulate sample is complicated because the energy source is radiation from the particles themselves. All emitted energy is generated by thermal radiation from individual particles and the spectral character of this radiation is modified by interaction with other particles before it is viewed at the surface.

If the sample is optically thin, radiation can escape without significant interaction with other particles, and the spectrum records the emission characteristics of individual particles. In a thick sample the photons emitted by one particle interact with other particles before escaping through the surface; the resulting scattering and reabsorption affect the steady-state photon density within the sample, so that a spectrum of the radiation escaping through the surface is no longer a record of the emission properties of individual particles. This concept is the foundation of a description of the emission properties of thick samples in terms of the steady-state density of photons occupying the space between the particles, and the probability of photon escape through the surface.

The number of photons per unit time per micron, E_λ, emitted from the sample at wavelength λ can be expressed as

$$E_\lambda = \int_0^{z_0} f(\lambda, z) P(\lambda, z)\, dz \tag{1}$$

where $f(\lambda, z)$ is the steady-state photon density at depth z; $P(\lambda, z)$ is the probability of escape, and z_0 is the sample thickness.

For any volume element $f(\lambda, z)$ can be expressed as the rates of several photon mechanisms which are: generation by emitting particles; loss by reabsorption; escape through the sample surface; and net transfer with other volume elements. Because of photon escape through the surface, the steady-state photon density in the space between the particles is not directly related to the temperature of the particles. Such a relationship only exists deep within the sample, where surface effects can be ignored and there is zero net flow of photons out of or into the volume element. This results in a steady-state photon density at any wavelength which is that specified by the Planck function corresponding to the particle temperature. Associated with photon escape through the surface there is a net flow of thermal energy to the colder background. This radiatively cools the particles, and a steady-state temperature profile is established in which the radiative heat flow out is balanced by the inward flow of visible energy, which is absorbed and converted to heat inside the particles. We first discuss the steady-state photon density variation with depth for an assumed isothermal sample, and then the effects of a temperature gradient are introduced.

A. Isothermal Samples

Consider an isothermal sample surrounded by a colder shield. The assumption is that the temperature of the particles does not vary with depth, because visible energy is absorbed and internally converted to heat on just the right profile to correct for the depth dependence of the energy loss due to radiative transfer to the background. Then the generation rate will be independent of depth.

The energy balance that describes the dynamic equilibrium steady state photon density can be expressed as

$$Q(\lambda)S = k(\lambda)Q(\lambda)Sf(\lambda, z) + f(\lambda, z)P(\lambda, z) + D(\lambda, z) \tag{2}$$

$$\text{Emission} = \text{reabsorption} + \underset{\text{surface}}{\text{escape through}} + \underset{\text{transfer}}{\text{net photon}}$$

where $Q(\lambda)$ is the number of photons emitted per second per micron per unit particle surface area, S is the total surface area of the particles per unit volume, $P(\lambda, z)$ is the probability that a photon will escape from depth z through the sample surface per second, $D(\lambda, z)$ is the number of photons per second per micron per unit volume contributed to the photon density at depth z from the net photon transfer with all other parts of the sample, $k(\lambda)$ relates the probability of absorption to the probability of emission and can be evaluated by considering a region deep within any sample where the escape probability is zero, i.e., $P(\lambda, z) = 0$, and the steady-state photon density will be that of a blackbody $f_{\mathrm{BB}}(\lambda)$. From Eq. (2)

$$k(\lambda) = 1/f_{\mathrm{BB}}(\lambda) \tag{3}$$

and after substituting for $k(\lambda)$, Eq. (2) becomes

$$f(\lambda, z) = f_{\mathrm{BB}}(\lambda)\left\{\frac{1 - D(\lambda, z)/SQ(\lambda)}{1 + [f_{\mathrm{BB}}(\lambda)P(\lambda, z)]/SQ(\lambda)}\right\} \tag{4}$$

The term in braces in Eq. (4) describes how the steady-state photon density $f(\lambda, z)$ departs from $f_{\mathrm{BB}}(\lambda)$, the blackbody photon density prescribed by the temperature. Assuming that $D(\lambda, z)$ is small, for any given value of $P(\lambda, z)$, the influence of $Q(\lambda)$ on $f(\lambda, z)$ is emphasized when S is small, which occurs for very loosely packed samples of a given size range.

When $P(\lambda, z)$ is sufficiently small (deep within the sample), $f(\lambda, z)$ has essentially the wavelength dependence of a blackbody, $f_{BB}(\lambda)$. On the other hand, when $P(\lambda, z)$ is large (near the surface) $f(\lambda, z)$ is less than $f_{BB}(\lambda)$ and its actual value is influenced by $Q(\lambda)$. Thus as $P(\lambda, z)$ becomes progressively larger toward the surface, the wavelength dependence of $Q(\lambda)$ is gradually injected into $f(\lambda, z)$. The sharpness of the gradient in the depth dependence of $f(\lambda, z)$ will be reduced when the contribution from $D(\lambda, z)$ is considered in more detail, but that is the extent of its influence.

In the emission spectrum, which involves the integral of $f(\lambda, z) \cdot P(\lambda, z)$ over depth, the wavelength dependence character of $Q(\lambda)$ is most strongly injected into emission from regions of the sample where $P(\lambda, z)$ is large, i.e., the surface regions. Conversely, for emission from deep within the sample where $P(\lambda, z)$ is small, $f(\lambda, z)$ is essentially $f_{BB}(\lambda)$, and the departure from blackbody emission behavior is determined by the wavelength dependence of $P(\lambda, z)$.

B. Samples with Temperature Gradients

The above conclusions are equally valid when a temperature gradient is present in the sample. Temperature gradients are usually present in natural targets as a consequence of

1. The average depth scale on which energy is absorbed from the sun. Solar heating is achieved by absorption of the shorter wavelength (ultraviolet and visible) radiation where the sun radiates most of its energy, followed by internal conversion to heat. The heating efficiency drops with depth on a scale which depends on the average efficiency for solar energy absorption.
2. The overall conductivity of the sample. At atmospheric pressure the major contribution is from heat transfer through the interstitial gas. This mechanism is removed at a pressure of 10^{-6} torr, where the only mechanisms are radiative heat transfer and conduction through the small surface contacts between particles (Salisbury and Glazer, 1964).
3. The thermal environment outside the sample surface.

A thermal gradient with temperature increasing with depth is established by the overpowering effect of radiative cooling of the surface layers. The sharpness of the temperature fall off near the surface is governed by the heating profile, the depth from which the infrared radiation can escape, the background temperature to which it radiates, and the conductivity of the sample. The depth below the surface cooled via radiation is governed by the average transmissivity for thermal energy. For silicates, the transmission maximum and minimum occur in the 6–12 μm (833–1666 cm^{-1}) wavelength range, so radiative cooling between 2 and 40 μm (5000 and 250 cm^{-1}) will involve wavelengths having escape efficiencies intermediate between these extremes. Thus, significant variations in cooling efficiency and, therefore, in temperature occur on a depth scale of the same order as the optical depth of the photons in the 6–12 μm (833–1666 cm^{-1}) wavelength range of interest. This is one reason why the emissivity spectra are sensitive to the thermal environment of the sample.

A temperature gradient in the sample introduces a depth dependence to $Q(\lambda)$, the rate of emission from individual particles. Increasing the sharpness of the thermal gradient has the effect of increasing the average depth of emission. Thus, more energy comes from zones where the probability of escape is smaller; this causes a reduction in the $Q(\lambda)$ character in the emitted energy.

Ignoring the term in braces in Eq. (4), it is evident that at wavelengths where $P(\lambda, z)$ is small, little energy is emitted because the only energy that can escape originates in the cold low-radiance regions near the surface. Thus, a general correlation is expected between the transmission and emission spectra, the spectral contrast in the emission spectrum being related to the sharpness of the thermal gradient in the sample. Such a general correlation is substantiated by experiment, but it is not possible to explain the finer details of the spectra and the effect of the experimental parameters with this approximation, and so the term in braces in equation (4) must be considered. This term may be considered as a perturbation which injects some source function character into the overall emission from the sample. Changes in the emissivity spectrum can be correlated with the importance of this term for the experimental conditions involved.

C. Experimental Measurement of P(λ, z) *and* Q(λ)

Application of the above discussion to explain the experimentally observed emission behavior relies on the knowledge of the wavelength dependence of $P(\lambda, z)$ and $Q(\lambda)$. $P(\lambda, z)$ is available from the transmission spectrum. The wavelength dependence of $Q(\lambda)$ can be obtained from the emission spectrum of individual particles. Theoretical spectra can be deduced from calculations of the scattering characteristics of quartz made by Conel (1969). He calculated absorption and scattering cross sections. The absorption cross sections for quartz are shown in Fig. 3. These curves

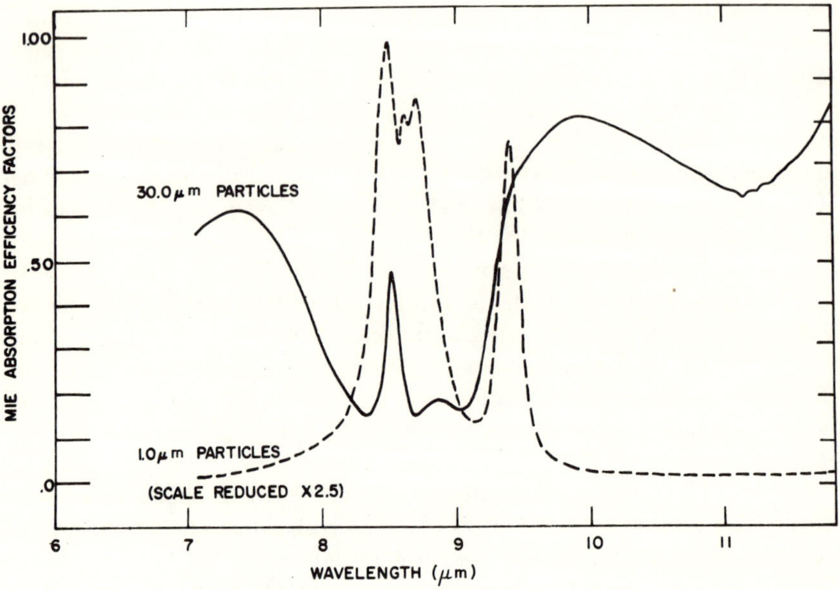

FIG. 3. Theoretical single-particle emission spectra derived from the Mie absorption efficiency factors as calculated by Conel (1969).

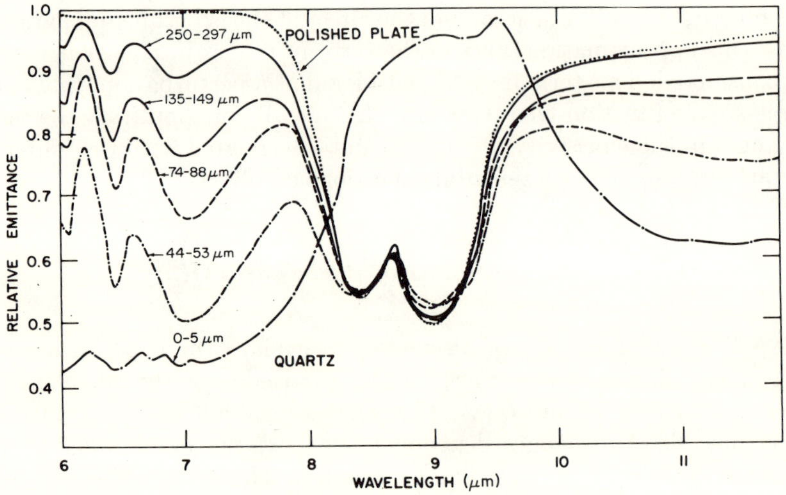

FIG. 4. Relative emittance spectra of quartz separated into narrow size ranges. The spectra displaying well-defined minima were normalized at 8.5 μm (1176 cm^{-1}).

correspond to both the single-particle true absorption and single-particle emission spectra. For the 1.0-μm particles, the maximum cross section occurs in the absorption band region, whereas the 30-μm particles have a minimum in this region and a maximum in the vicinity of the Christiansen frequency.

This type of behavior has been observed experimentally (see quartz data in Fig. 4). The quantity of interest here is the "source function," defined as the wavelength profile on which energy is radiated by particles having the size range of the samples used for the emission measurements. This is most easily obtained by recording the emission spectrum of an optically thin layer of the particles suspended on a mirror. Such spectra are presented in Fig. 5. All the 0–5 μm samples display an emission maximum in the absorption band region. The source functions for the 0–74 μm samples are less well-defined because of the large number of particles smaller than 5 μm in the 0–74 μm samples. It is evident that the source functions for the 0–5 μm samples have pronounced maximums in the absorption band region, while the source functions for the 0–74 μm samples radiate more uniformly in the 6–12 μm (833–1666 cm^{-1}) wavelength range.

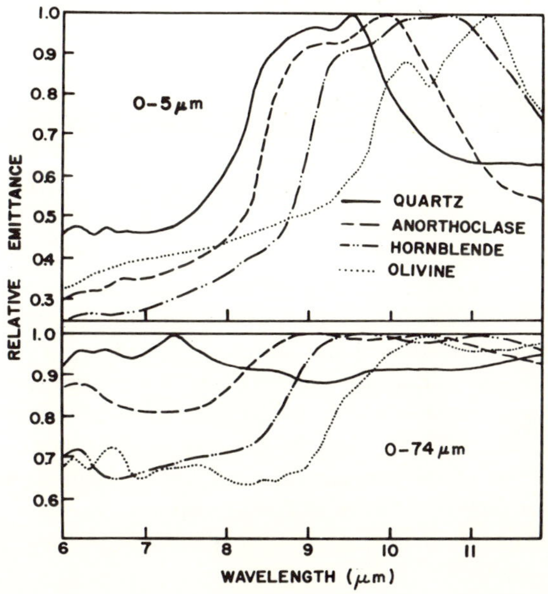

FIG. 5. Source function emittance spectra for 0–5 μm samples (top) and 0–74 μm samples (bottom). The spectra were recorded from a dispersed monolayer of material on a mirror, under vacuum, heated with visible energy with a 77°K background shield.

V. Analysis of Effect of Experimental Parameters on Emission Behavior

The effects of the emissivity spectrum of particle size, background temperature, packing, pressure, and incidence angle can now be considered in terms of the above theory. The basic instrumentation and techniques used have been described in detail (Logan and Hunt, 1970) as has the precise data reduction procedure (Logan *et al.*, 1973). Experimental data were collected using common rock-forming minerals, because they provide better characterized samples than do rocks. The spectral properties of these minerals are collected in Table I below. The effect of each of the parameters will be discussed separately.

A. Particle Size

Samples of both 0–5 μm and 0–74 μm size ranges display well-defined emissivity maxima that occur at different wavelengths. The emissivity maximum of the 0–5 μm sample is generally at longer wavelengths than that of the 0–74 μm sample of the same material. There are also significant changes in the average emissivity with particle size. The spectra of felsic minerals tend to show greater contrast (i.e., lower average emissivity) in the 0–74 μm sample than in the 0–5 μm sample, while the spectra of mafic minerals show more contrast in the 0–5 μm sample than in the 0–74 μm sample.

The effect of particle size on the wavelength of the emissivity maximum is the largest of any of the parameters studied. Such large shifts as a function only of particle size are of considerable concern, because it would appear that the usefulness of the position of the emissivity maximum as a diagnostic tool is drastically reduced unless information is available about the particle size distribution of the sample or target.

Analysis of the 0–5 μm samples shows that they contain very few particles larger than 1 μm, most being considerably smaller. Consequently, such samples represent a limiting case for particle size distribution and are unrealistic as analogs of remote targets with particulate surfaces. They are, however, of considerable interest for considerations of the behavior of circumstellar dusts and of particles suspended for long periods in the Martian atmosphere.

For remote sensing of particulate surfaces, it becomes important to evaluate the magnitude of the particle size effect for reasonable variations of particle size distribution. The 0–74 μm samples used here are adequate lunar analogs, having size distributions that follow the grinding law, but

TABLE I

SPECTRAL PROPERTIES OF MINERALS

SAMPLE	TRANSMISSION MAXIMUM	Emissivity data for 0–5 μ particle size: 295°K shield, Fairy castle, $\lambda_{\epsilon\,max}$	295°K, Fairy castle, ϵ_{av}	295°K, Packed, $\lambda_{\epsilon\,max}$	295°K, Packed, ϵ_{av}	77°K shield, Fairy castle, $\lambda_{\epsilon\,max}$	77°K, Fairy castle, ϵ_{av}	77°K, Packed, $\lambda_{\epsilon\,max}$	77°K, Packed, ϵ_{av}
Quartz	7.35	7.45	0.884	7.16	0.904	7.44	0.838	7.18	—
Albite	7.63	7.83	0.885	7.49	0.946	7.80	0.855	7.47	0.937
Anorthoclase	7.78	8.30	—	—	—	8.22	0.906	7.93	0.961
Labradorite	7.91	8.12	0.882	7.97	0.957	8.03	0.873	7.88	0.944
Hornblende	8.47	8.84	0.885	8.80	0.949	8.75	0.858	8.72	0.938
Olivine	8.70	9.40	0.848	9.23	0.952	9.45	0.793	9.15	—

SAMPLE	TRANSMISSION MAXIMUM	Emissivity data for 0–74 μ particle size: 295°K shield, Fairy castle, $\lambda_{\epsilon\,max}$	295°K, Fairy castle, ϵ_{av}	295°K, Packed, $\lambda_{\epsilon\,max}$	295°K, Packed, ϵ_{av}	77°K shield, Fairy castle, $\lambda_{\epsilon\,max}$	77°K, Fairy castle, ϵ_{av}	77°K, Packed, $\lambda_{\epsilon\,max}$	77°K, Packed, ϵ_{av}
Quartz	7.35	7.16	0.821	7.16	0.867	7.15	0.789	7.15	0.899
Albite	7.63	7.55	0.884	7.53	0.914	7.55	0.841	7.53	0.886
Anorthoclase	7.78	7.68	0.897	7.76	0.917	7.65	0.931	7.61	0.936
Labradorite	7.91	7.71	0.887	7.75	0.951	7.71	0.894	7.75	0.937
Hornblende	8.47	8.50	0.952	8.60	0.961	8.43	0.922	8.55	0.939
Olivine	8.70	8.73	0.900	8.73	0.936	8.88	0.853	—	—

there is some variation from material to material because of the responses of different hardnesses and fracture habits to grinding.

The observed sensitivity of the emission spectrum to particle size is to be expected, because particle size not only affects the radiative transfer properties of the sample as seen in the transmission spectrum, but also has a large effect on the source function. It is, in fact, the effect on the source function that makes the emission spectra of these materials much more sensitive to particle size than are their transmission spectra.

As is discussed above, samples whose temperature increases with depth have an emission spectrum dominated by the probability of photon escape from the sample, with only some small perturbation from the source function. Probability-of-escape spectra (i.e., transmission spectra) are relatively insensitive to particle size; 0–5 μm samples have more contrast and better resolved transmission minima than do 0–74 μm samples, but the wavelength of the transmission peak is only slightly affected by particle size (see Fig. 2).

Both the probability-of-escape spectrum and the source functions have more contrast in the 0–5 μm samples, and this gives an emission spectrum with reduced contrast on the long-wavelength side of the peak, because of the source function contribution. On the other hand, the 0–74 μm samples have less contrast in both the probability of escape spectra and source function, and so the emissivity maximum is generally shifted to shorter wavelength.

There is also a correlation between the change in spectral contrast as a function of particle size and composition. This can be explained by the variations with composition of the influence of particle size on the source function, i.e., the difference between the source functions for 0–5 and 0–74 μm samples becomes progressively smaller as the composition of the sample material changes from felsic to more mafic. Accordingly, the magnitude of the source function effect is progressively reduced, and the change in contrast with particle size follows more closely than predicted by the transmission behavior.

B. Background Temperature

The shift in wavelength of the emissivity maximum caused by changing the background temperature from 295° to 77°K is always less than 0.1 μm and is greater for mafic minerals than for felsic ones. There is no good relationship between the direction of the shift and type of mineral. In all cases, the colder background temperature produces a decrease in average emissivity that is almost entirely due to a change in emissivity on the long-wavelength side of the peak.

Lowering the background temperature from 295° to 77°K changes the amount of radiation from outside the sample, removes the sample surface reflection component, and alters the thermal gradient within the sample. The increase in contrast cannot simply be explained as a loss of the background radiation reflected from the sample. When the level of the background radiation above the surface is reduced, the temperature gradient into the sample becomes steeper and there is a greater difference between the energy radiated at wavelengths where the transmissivity is low and where it is high. Also, increasing the sharpness of the temperature gradient preferentially removes the emission component arising near the surface which contains most source function character. This accounts for the shift of the emissivity peak of the 0–5 μm samples to shorter wavelength for the low background temperature, because the emission component removed is peaked in the absorption band.

C. Packing

The packing condition has a large effect on both the wavelength of the maximum and the average emissivity. Varying this parameter results in the largest change in contrast observed, especially for minerals characteristic of basic rocks. The contrast is always decreased as the packing is increased. For the 0–5 μm samples, the shift in the maximum is always to shorter wavelengths as the packing is increased, and there are indications that the shift is greater for felsic minerals than for others. The shift is generally smaller for the 0–74 μm samples, and it appears that the peaks of the mafic minerals shift to longer wavelengths, whereas peaks of felsic minerals shift to shorter wavelengths as the packing density is increased.

Changing the packing in the samples alters (1) the scattering properties, as evidenced by the change in the transmission spectrum with packing for the 0–5 μm samples (Logan *et al.*, 1973); (2) the conductivity and, therefore, the thermal gradient in the sample; and (3) the surface area of particles per unit volume (S) and therefore the importance of the contribution from the source function on the steady-state photon density [see Eq. (4)].

The change in contrast can be related to the shallower temperature gradient in the packed samples and to a change in scattering properties with packing.

Two effects can account for the shift in the emissivity peak. First, increasing the packing causes the expression in parentheses in Eq. (4) to be closer to unity, which gives a loss in source function character in the emissivity spectrum. Second, the change in scattering properties with packing could account for the shift in the emissivity maximum, just as it accounts for the shift in the transmission maximum. Thus, a simple change in the scattering properties as evidenced by the change in the transmission spectrum with packing can account for both a change in contrast and a shift in the emissivity maximum. However, the magnitude of the effect indicates that this is not the only factor and that the other effects listed are also operative.

D. Pressure

Rock samples were chosen to study the effect of pressure because of the applicability to remote sensing of Martian surface materials. Results for a 0–74 μm granite sample are illustrated in Fig. 6. The spectrum recorded at a pressure of 8 mm Hg and a background temperature of 77°K simulates Martian surface conditions. All other spectra were recorded with a 295°K background temperature.

As the pressure is increased above 10^{-4} mm Hg, the position of the maximum shifts progressively to longer wavelengths until, at pressures of ~1 mm Hg, a total shift of 0.3 or 0.4 μm can occur, particularly in low-contrast spectra.

In general, the significant loss in contrast with increasing pressure occurs on the long-wavelength side of the peak, while the short-wavelength side remains relatively unaltered. An accompanying change takes place in the average emissivity, being greatest for acidic rocks (for granite the average

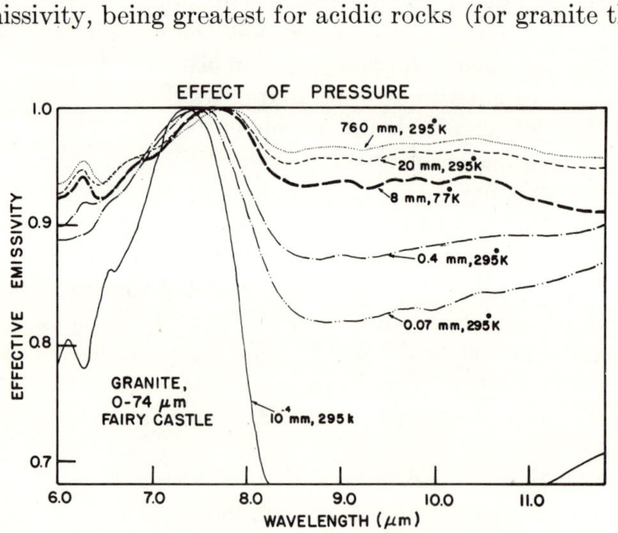

FIG. 6. Effect of pressure. The effective emissivity spectra of a fairy castle 0–74 μm sample of granite were recorded at the pressures indicated on the curves. Only the 8-mm curve was recorded with the background shield at 77°K. For all the others, shield temperature was 295°K.

emissivity increases by about 0.2) and least for basic rocks (for basalt the average emissivity increases by 0.04). Increasing the pressure from 1 to 760 mm Hg produces little additional change in either the contrast or the position of the emission maximum.

The effect of pressure on the emissivity spectrum is analogous to that of background temperature discussed above, because both influence the temperature gradient within the sample. The effect of pressure can be explained as a change in the conductivity between the particles in the sample, which reduces the sharpness of the temperature gradient below that for samples in vacuum. This produces less contrast in the spectrum, particularly on the long-wavelength side of the peak where the influence from source function is greatest.

E. Insolation Angle

Variation of the insolation angle alters the depth scale (normal to the surface) on which visible energy is absorbed. This, in turn, provides shallower temperature gradients as the angle of incidence increases, which accounts for the reduced contrast in the spectrum. This effect was investigated by varying the angle of the plane of the sample surface inside the shield while the phase angle between the illumination and viewing orientations remained unchanged. This experiment then simulates variations in insolation angle due to sloping terrain. The plane of the sample surface was varied through 50°, which gave rise to a small change in contrast, but no discernible shift occurred in the wavelength of the emissivity maximum.

VI. Spectra of Rocks: A Diagnostic Test

Having demonstrated above that emission spectra of silicates are significantly affected by changing experimental conditions, we have attempted to determine the extent to which such spectra are diagnostic of composition when experimental parameters are held constant (Logan *et al.*, 1973). Such a test is a realistic approach to remote sensing conditions on our moon, the moons of other planets, Mercury, and asteroids, where the lack of significant atmosphere and meteoroid bombardment should result in a surface regolith of relatively constant average particle size in a uniform thermal environment. Consequently, we have selected a set of lunarlike experimental conditions for our diagnostic test, using our 0–74 μm particle size range and underdense "fairy castle" samples, in a vacuum with a cold (77°K) background shield to simulate space conditions. In a subsequent section we will consider the infrared remote sensing problems of a planet with an atmosphere, and describe preliminary results of such an experiment for Mars.

A suite of igneous rocks was selected for our reference spectra, each of the 72 samples being characterized petrologically, and chemically analyzed for percent SiO_2 content by weight. Sedimentary and metamorphic rocks were not included, the former for obvious reasons and the latter for less obvious but geologically defensible ones.

Spectra typical of the major igneous rock divisions are shown in Fig. 7. These illustrate the fact that lunar-type experimental conditions tend to maximize spectral contrast, yielding well-defined emissivity maxima and less well-defined minima. Thus, under such experimental conditions, it is the emissivity peak, related to the Christiansen frequency, that is most easily used to determine composition. The spectra in Fig. 7 also illustrate the systematic way in which both peak position and spectral contrast vary with rock type. The more basic the rock, the longer the wavelength at which its emissivity peak falls. Spectral contrast, however, decreases from the acidic into the basic range, and then increases again in the ultrabasic range. There is a close correlation between visible albedo and spectral contrast in emission, due to the effect of the optical depth in the visible on the thermal profile within the sample.

Emissivity peak position and spectral contrast data for our entire suite of reference samples are summarized in Fig. 8. The symbols representing the members of each of the four major rock divisions are joined at a common point. It is evident from Fig. 8 that there is only limited overlap in the emission properties between the acidic, intermediate, basic, and ultrabasic divisions. It is important to note, in addition, that the distribution of

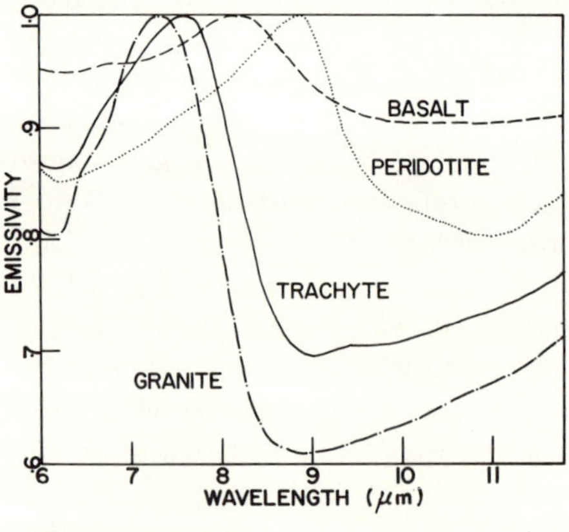

FIG. 7. Typical effective emissivity spectra of four 0–74 μm rock samples. The spectra were recorded under vacuum with a 77°K background shield.

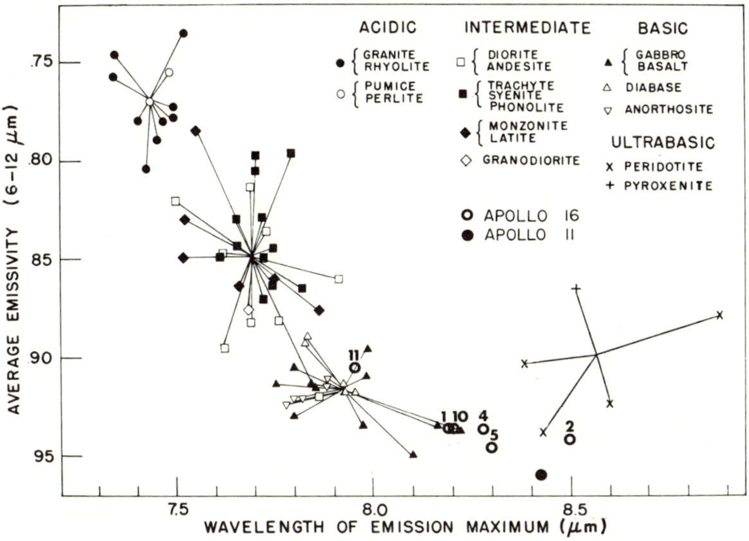

FIG. 8. Wavelength of the effective emissivity maximum versus the effective emissivity averaged over the 6–12 μm wavelength range for fairy castle 0–74 μm particle size rock samples and some Apollo 16 and 11 soil samples. The data were recorded under vacuum with a 77°K background shield.

symbols within each rock division is not random but is strongly influenced by mineralogical differences among kindred rocks. For example, the wavelength of the emissivity peak is very sensitive to the presence of quartz. All of the intermediate rocks that fall close to the acidic range are rich in quartz, and the effect of quartz is also evident in the basic rocks. Some minerals, like quartz, have emissivity maxima at relatively short wavelengths, and when present in a rock serve to substantially reduce the wavelength of its emissivity peak.

All terrestrial igneous rocks are altered to some extent, but if weathering or alteration effects are kept at a minimum, we find that within the basic rock division those rocks whose symbols are to the left in Fig. 8 contain quartz as well as plentiful plagioclase, those in the central region contain predominantly plagioclases, while those to the right contain progressively more abundant pyroxene and olivine.

In the ultrabasic range, the wavelength of the emissivity peak shifts over a large range. The principal variable here is olivine content. In general, rocks to the right in the ultrabasic range contain progressively more olivine, particularly magnesian olivine. The extreme long-wavelength point arises, in fact, from a peridotite that might more specifically be called a dunite.

VII. Lunar Data: Comparison with Laboratory Standards

A. Remote Observational Data

The best available observations of lunar spectral emissivity are those made by Murcray *et al.* (1970) with a balloon-borne telescope. The spot size in these early observations was large (180 arc sec), and the target areas were therefore regional in size. However, definite emissivity peaks were obtained, the wavelength of which differed from region to region, as illustrated in Fig. 9. It is also interesting that, despite the large spot size, these spectra correlate well with laboratory spectra of lunar soil samples, as is discussed below.

B. Laboratory Spectra of Lunar Soils

Laboratory measurement of the emission spectra of Apollo soils in our simulated lunar environment (Fig. 10) show that they display emission peaks similar to those of lithologically equivalent terrestrial rocks recorded under the same experimental conditions, and more importantly, similar to those recorded from the balloon system. For example, the emissivity peak displayed by the center of Mare Imbrium (Fig. 9) at 8.37 μm (1,195 cm^{-1}) is very close to that of 15601 sample [8.42 μm (1187 cm^{-1})] taken from the

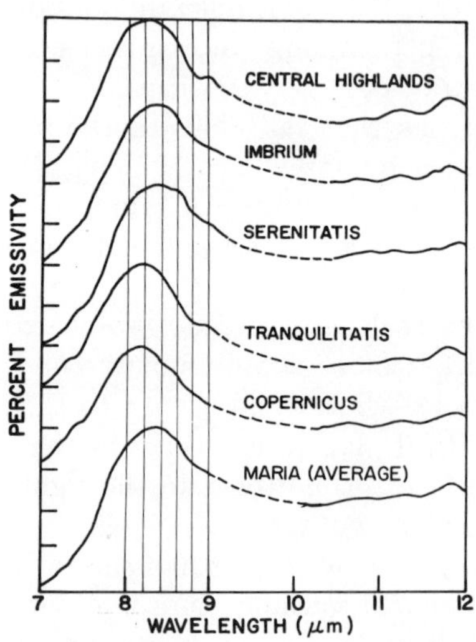

FIG. 9. The spectral emissivity of various regions on the moon calculated from the balloon flight data of Murcray *et al.* (1970). The dotted lines connect the data collected each side of the 9.64 μm ozone absorption band.

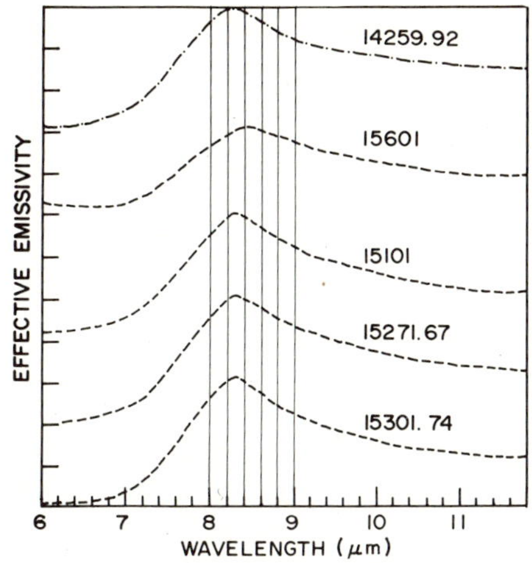

FIG. 10. Effective emissivity spectra of one Apollo 14 (top) and four Apollo 15 soil samples. The spectra were recorded under vacuum with a 77°K background shield.

Mare Inbrium margin. The maximum obtained in the Central Highlands spectrum at 8.24 μm (1213 cm^{-1}) is virtually identical with that displayed by Apollo 14 soil at 8.22 μm (1216 cm^{-1}), and this result is reasonable if the regolith in the Central Highlands and the debris ejected from the Imbrium Basin to form the Fra Mauro formation are both derived from a primaeval lunar crust.

Although these results suggest that there may be regional spectral characteristics, laboratory study of soil samples from different Apollo landing sites has shown that there are significant (and remotely mappable) differences in the emission spectra of soils from different lithologic units in individual localities, as well as from locality to locality (Logan *et al.*, 1973; Salisbury *et al.*, 1973).

VIII. Remote Sensing of a Planet with an Atmosphere: Martian Application

Planets with atmospheres are highly individual, making it difficult to generalize on planetary remote sensing in the infrared. However, infrared spectra are available for Mars, and we have made a preliminary analysis of these data (Hunt *et al.*, 1973). Consequently, we will use Mars as an example, and the reader may extrapolate from this example to other planetary applications, bearing in mind the significant spectral effects of experimental conditions.

Any remotely sensed infrared spectrum of Mars can be affected by the presence of fine silicate particles in the atmosphere. The spectrum will be a composite of surface emission, attenuation of surface emission by particle haze transmission and cloud emission.

A. Surface Emission

The effect on the appearance of a spectrum caused by increasing the pressure from the near-vacuum of the moon to the 3.8–7.6 mm of Hg appropriate for Mars is to degrade the contrast and shift the emission maximum to longer wavelengths as shown in Fig. 6. In general, acidic rocks retain more spectral contrast and suffer less peak shift than do basic rocks.

B. Transmission Effect of Haze Particles

Figure 11 illustrates the appearance of transmission spectra of suspended 0–5 μm particles. The transmission measurements record the effect of scattering out of the beam, as well as true absorption by the particles. Two features in these spectra are of particular interest. First, well-defined transmission minima appear (for quartz and olivine) in the molecular vibration region. As Lyon (1963), for instance, pointed out, the positions and shapes

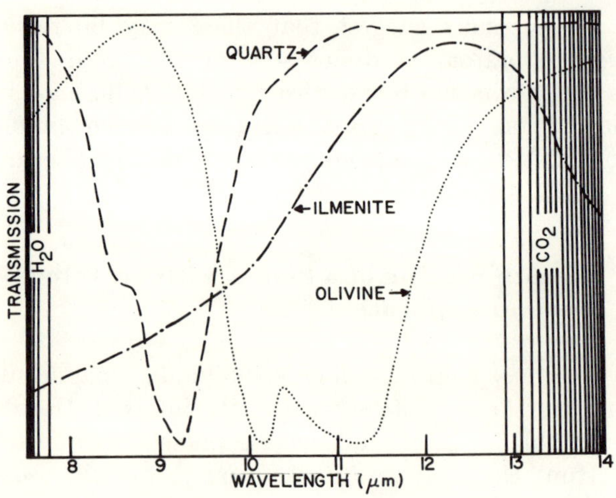

FIG. 11. Transmission spectra of a cloud of 0–5 μm particles of quartz, olivine, and ilmenite. Vertical lines indicate onset of H_2O and CO_2 absorption. Ordinate is percentage of transmission increasing upward in arbitrary units.

of features in this region are compositionally diagnostic. Second, transmission maxima occur (for olivine and ilmenite) to shorter wavelength, which are related in position to the Christiansen frequency.

Obviously, any source behind the Martian atmosphere will be attenuated by this type of spectrum, the extent of such attenuation will depend upon both the concentration of particles in the atmosphere and the path length through it.

C. Emission from Cloud Particles

Referring first to emission from an optically thin haze, an emission peak occurs in the vibrational band region near 10 μm (1000 cm^{-1}) which is essentially the inverse of the absorption band. It becomes less defined as the physical (and optical) thickness of the layer is increased and tends to fill in the transmission feature. When the particles of such a layer are cold and dispersed relative to the surface particles, their emission contribution is not significant. A contribution from this type of emission should be apparent only when significant emission from the surface can be excluded, such as when the polar regions form the background.

Given an optically thick cloud of particles, on the other hand, both the character and importance of their emission changes dramatically. Their radiative interaction produces a spectrum with a minimum in the vibration band region and an emission peak near the Christiansen frequency. Obviously, the appearance of the emission spectrum recorded is governed by many parameters. Certainly the emission spectrum from an optically thick cloud of particles suspended in the atmosphere will show higher contrast than the spectrum of a surface layer in contact with the same atmosphere.

When the Mariner 1971 reached Mars, the planet's surface was overlain not by a thin haze, but by a dust cloud thick enough to obscure most surface features. For a small slant path the cloud was not optically thick in the infrared, having an absorbance estimated to be about 0.5 (Chase *et al.*, 1972). Infrared spectra of Mars obtained under these conditions have been published by Hanel *et al.* (1972), and are shown in Fig. 12. They attribute the diffuse emission features near 9.3 (1075 cm^{-1}) and 21.3 μm (470 cm^{-1}) to dust in the atmosphere and compare them with absorption spectra of fine dusts recorded in KBr pellets by Lyon (1964).

The preliminary conclusion of Hanel *et al.* from the polar spectra that the cloud particles are intermediate in composition is consistent with the indeterminacy of the transmission minima near 9.3 and 21.3 μm (1075 and 470 cm^{-1}). Typically, however, intermediate rocks display multiple bands in the 20-μm (500 cm^{-1}) region rather than a single band (see Fig. 13).

Also it is difficult to rationalize the Martian reflection bands near 0.95 μm (10,526 cm^{-1}) and between 1.4 and 2.2 μm (7143 and 4545 cm^{-1}) (McCord *et al.*, 1971), and near 3.0 μm (3333 cm^{-1}) (Sinton, 1967; Houck *et al.*, 1973) with intermediate composition for the surface dust layer, because these bands are not typical of the spectra of intermediate rocks. At first glance, basalt appears to be a good candidate among the primary igneous rocks because of its single band at 21.3 μm (470 cm^{-1}). Indeed, basaltic rocks may also display near-infrared bands near 0.95 μm (10,526 cm^{-1}) as a result of which Adams and McCord (1969) have suggested a basaltic composition for the Martian surface. However, the spectral fit of the band envelopes in the mid-infrared is poor near 9.3 μm (1075 cm^{-1}) and the poorly defined bands between 1.4 and 2.2 μm (7,143 and 4,545 cm^{-1}) and the band at 3 μm (3,333 cm^{-1}) do not occur in basaltic rocks.

The clay mineral montmorillonite exhibits spectroscopic properties which closely match the available Martian data (see Fig. 13). Not only does montmorillonite display two major bands in the appropriate positions in the mid-infrared, while lacking any intense subsidiary, but it also displays bands near 0.95, 1.4, and 1.9 μm (10,526, 7,143, and 5,260 cm^{-1}) (see Chapter 4, Fig. 5) in the near-infrared (Hunt and Salisbury, 1970), which do not conflict with the bright area spectral features found by McCord *et al.* (1971) from ground-based observations. Certainly, the Martian 3.0-μm (3,333 cm^{-1}) band first observed by Sinton (1967) is consistent with the abundant presence of such a hydrated mineral as montmorillonite.

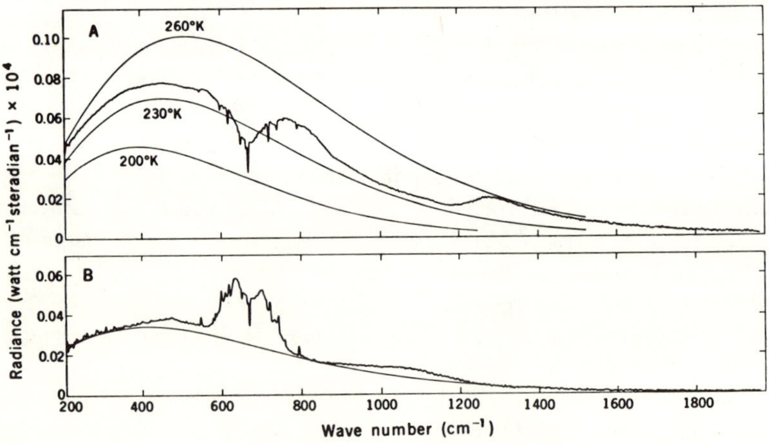

FIG. 12. A. Example of Martian nonpolar thermal emission compared to three blackbody curves. B. Example of polar thermal emission. The smooth curve is the composite of two blackbody spectra. (From Hanel *et al.*, 1972.)

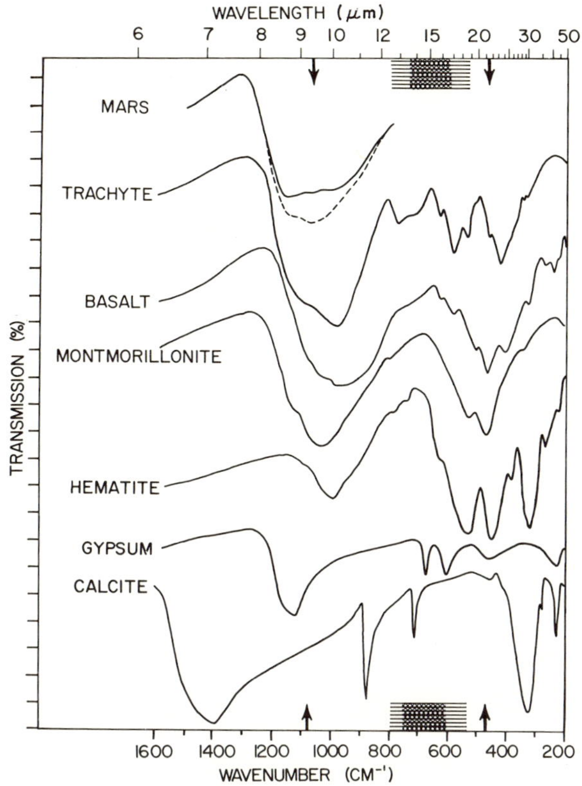

FIG. 13. Transmission spectra of rocks and minerals ground to 0–5 μm and suspended on a mirror. Curves have been separated vertically for clarity. Peak transmission near 8 μm (1250 cm^{-1}) is typically about 90% for the rock spectra, and ordinate divisions are 10%. Shaded region of the spectrum is obscured by Martian CO_2 lines in Mariner 1971 spectra. Recalculated Mariner 1971 nonpolar spectra are shown at top. Vertical arrows indicate positions of polar radiance maxima, near which we also expect the nonpolar transmission minima.

It is of interest to note that the present data exclude the possibility of either hematite or carbonates as major components of the Martian dust cloud, as indicated by their spectra in Fig. 13. Other salts, such as sulfates, on the other hand, could be present in large amount with montmorillonite, without substantially distorting the observable portion of its mid-infrared spectrum. A thin ferric oxide stain on the soil particles, which is commonly associated with montmorillonite, could account for the reddish color of the Martian regolith, as suggested by Van Tassel and Salisbury (1964) and Binder and Cruikshank (1964).

References

Adams, J. B., and McCord, T. B. (1969). *J. Geophys. Res.* **74,** 4851–4856.

Adams, J. B., and McCord, J. B. (1972). *Proc. 3rd Lunar Sci. Conf., Geochim. Cosmochim. Acta,* **3,** Suppl. 3, 3021–3034.

Aronson, J. R., Emslie, A. G., Allen, R. V., and McLinden, H. G. (1967). *J. Geophys. Res.* **72,** 687.

Binder, A. B., and Cruikshank, D. P. (1964). *Commun. Lunar Planet. Lab.* **2,** 193–196.

Chase, S. C., Hatzenbeler, H., Kieffer, H. H., Miner, E., Munich, G., and Neugebauer, G. (1972). *Science* **175,** 308–309.

Conel, J. E. (1969). *J. Geophys. Res.* **74,** 1614–1643.

Emslie, A. G. (1966). *Progr. Astronaut. Aeronaut.* **18,** 281–290.

Hanel, R. A., Conrath, B. J., Hovis, W. A., Kunde, V. G., Lowman, P. D., Pearl, J. C., Prabhakara, C., Schlachman, B., and Levin, G. V. (1972). *Science* **17,** 305–308.

Henry, R. L. (1948). *J. Opt. Soc. Amer.* **38,** 775–789.

Houck, J., Pollack, J. B., Sagan, C., Schaack, D., and Dekker, J. (1973). *Icarus* **18,** 470–480.

Hunt, G. R., and Salisbury, J. W. (1970). *Mod. Geol.* **1,** 283–300.

Hunt, G. R., and Logan, L. M. (1972). *Appl. Opt.* **11,** 142–147.

Hunt, G. R., Logan, L. M., and Salisbury, J. W. (1973). *Icarus* **18,** 459–469.

Logan, L. M., and Hunt, G. R. (1970). *J. Geophys. Res.* **75,** 6539–6548.

Logan, L. M., Hunt, G. R., Balsamo, S. R., and Salisbury, J. W. (1972). *Geochim. Cosmochim. Acta* **3,** 3069–3076.

Logan, L. M., Hunt, G. R., Salisbury, J. W., and Balsamo, S. R. (1973). *J. Geophys. Res.* **78,** 4983–5003.

Lyon, R. J. P. (1964). *NASA Tech. Note.* **NASA TN D-1871.**

McCord, T. B., Elias, J. H., and Westphal, J. A. (1971). *Icarus* **14,** 245–251.

Murcray, F. H., Murcray, D. G., and Williams, W. J. (1970). *J. Geophys. Res.* **75,** 2662–2669.

Passell, T. O. (1964). *In* "Evaluation of Infrared Spectrophotometry for Compositional Analysis of Lunar and Planetary Soils: Rough and Powdered Surfaces," Stanford Res. Inst. Proj. No. PSU-3843, Final Rep., Part II, Contract No. NASr-49(04), pp. 104–121.

Salisbury, J. W., and Glazer, P. E. (1964). "Studies of the Characteristics of Probable Lunar Surface Materials," AFCRL-64-70, Spec. Rep. No. 20. Air Force Cambridge Res. Lab., Cambridge, Massachusetts.

Salisbury, J. W., Hunt, G. R., and Logan, L. M. (1973). *Geochim. Cosmochim. Acta* **3,** Suppl 4, 3191–3196.

Samuelson, R. E. (1967). Ph.D. Dissertation, Georgetown University, Washington, D.C.

Sinton, W. M. (1967). *Icarus* **6,** 222–228.

Spitzer, W. G., and Kleinman, D. A. (1961). *Phys. Rev.* **121,** 1324–1335.

Straton, J. A. (1941). "Electromagnetic Theory." McGraw-Hill, New York.

Van de Hulst (1957). "Light Scattering by Small Particles." Wiley, New York.

Van Tassel, R. A., and Salisbury, J. W. (1964). *Icarus* **3,** 264–269.

Vincent, R. K., and Hunt, G. R. (1968). *Appl. Opt.* **7,** 53.

6

APPLICATIONS OF INFRARED SPECTROSCOPY AND RADIATIVE TRANSFER TO EARTH SCIENCES

James R. Aronson and Alfred G. Emslie

I. Introduction

The subject matter of this chapter is concerned with the infrared spectra of particulate media, radiative transfer, and their application to earth sciences. The topics range from dusty planetary atmospheres through particulate surface layers and into the deep interiors of planets.

We may define infrared spectroscopy as the study of the interaction of electromagnetic radiation in the infrared spectral region (~0.7 μm–1 mm wavelengths) with matter. This interaction involves considerations of the radiative transfer within diverse media. Somewhat more narrowly, spectroscopy implies selective effects that in our case are dependent on the radiative cross sections of individual particles which, in turn, depend on the optical constants and are thus strong functions of the precise wavelength involved. In the infrared spectral region, the absorption of radiation is caused principally by the internal vibrations of the species. At the short-wavelength end of the infrared region some electronic transitions do occur so that these must be considered as well. The selective absorption of infrared radiation is caused by transitions between different energy levels in the material involved such that the measured frequency of absorption or emission corresponds to an energy change in the species.

Infrared spectroscopy has been extensively used in detailed investigations of molecular structure and as a diagnostic tool since every species for which the molecular motion causes a change in dipole moment has a spectrum characteristic of its own structure. Although infrared techniques are most commonly developed for qualitative studies, careful examination of intensities leads to a most useful tool for quantitative analysis as well. While organic chemists have made the most use of infrared spectroscopy, it has long been recognized that the same principles apply to inorganic species, and, in particular, to minerals. In fact, vibrations in minerals tend to have large absorption coefficients and so tend to produce strong bands.

Most infrared studies have been carried out using transmission techniques, but the present intense interest in remote sensing usually requires that reflection or emission spectra be measured. These techniques provide essentially the same information as do transmission spectra but are somewhat more difficult to interpret and have had much less development. Spectra can be observed from materials which give specular or diffuse reflection or, in the usual case, an intermediate result. If the sample is suitable for specular measurements (highly polished homogeneous slabs), Fresnel's equations, which give the reflection coefficient as a function of the complex index of refraction $m = n - ik$ and the angle of incidence, are rigorous. Here n is the ordinary index of refraction and k is an absorption index related to the Lambert absorption coefficient α by $k = \alpha\lambda/4\pi$ where λ is the radiation wavelength. The Fresnel equations vary with angle of incidence as given by Simon (1951), but their effects can be indicated by the normal incidence equation

$$R = \frac{(n - 1)^2 + k^2}{(n + 1)^2 + k^2} \tag{1}$$

It is apparent why even Fresnel reflection spectra differ from transmission spectra in that the real part of the refractive index has a major effect on the reflection spectrum. The real part, n, in turn, depends on k in such a way that a given absorption produces an "anomalous" drop in n on the short-wavelength side of the band (reaching values as low as 0.1 for many minerals) which then rises to a peak on the long-wavelength side of the absorption peak.

Diffuse reflection spectra are much more difficult to characterize simply, being made up as they are of Fresnel effects from particle surfaces, transmission through particles (which is affected by refraction), diffraction effects and interference phenomena. Scattering includes all these effects and refers to any direction of view rather than simply the backward directions (see Chapter 5, Section III).

II. Remote Sensing

The reflectance spectra of many minerals were measured near the turn of the century by Coblentz (1905). However, it was the advent of the space program that provided the impetus for detailed investigations of the potential of middle-infrared spectroscopy for remote sensing of the composition of planetary surfaces. Lyon (1965), in particular, as well as many others, recognized the great potential in the method. Practically alone among remote sensing techniques, infrared spectroscopy can provide direct unambiguous petrological information. The techniques used are similar to those applicable in the visible and near-infrared regions, but the information obtained is much more characteristic of particular species as it arises directly from the molecular motions rather than from electronic transitions which are modified by the crystal field at the site of the emitting ion (see Chapter 2). The prospect of making observations from satellites opened up infrared spectroscopy as a fruitful tool for planetary astronomy. However, the first and by far the principal use of remote measurements of infrared radiation has been to infer many properties of various surfaces from the apparent or brightness temperatures. With regard to earth sciences, such diverse topics as water pollution, crop health, geological fault location, oceanic current boundaries (and hence fishing locations), and climatological radiation balance are among applications of infrared radiometry. These applications usually depend on the assumption of black or at least gray body behavior for the targets in the wavelength range investigated. As such, they do not involve spectroscopy. Improved interpretations would undoubtedly result from spectroscopic measurements in these areas (see Chapter 7, Figs. 11 and 12).

The purpose of spectroscopic remote sensing was first demonstrated by Bell and Eisner (1956). They identified the gypsum sands at the White

Sands National Monument in New Mexico by their *reststrahlen** bands from a mobile laboratory. However, as more workers (Lyon, 1965; Hovis and Callahan, 1966; Aronson *et al.*, 1966; Hunt and Vincent, 1968) began to examine the problem, it soon became apparent that the spectral signature was a complicated function of various physical factors, such as particle size, opacity, volume fraction, surface roughness, and temperature gradient. In particular, some bands were observed to increase in intensity with particle size and some to decrease in the same particle size range. In considering this problem, we realized that the reflectance or emittance spectrum could only provide a suitable signature for remote analysis if various physical parameters of the target are known. This, however, would require an untenably large reference catalog unless some way were found to model the physical effects, given only the chemical information that is inherent in m, the complex index of refraction. We therefore set out to develop a theory of the reflectance or emittance spectra of minerals. The theory involves detailed considerations of the interaction of the radiation with individual particles and the way in which the radiation traverses layers composed of those particles. The details have been reported in two publications (Emslie and Aronson, 1973; Aronson and Emslie, 1973). We shall refer to these as AO-I and AO-II. During the time in which we have been developing our theory, infrared minerological data have been obtained remotely from spacecraft for both the earth and Mars (Hanel *et al.*, 1972a, b) (see Chapter 5, Fig. 12).

III. Radiative Transfer

A. Application of Radiative Transfer to Infrared Spectra

In most infrared work involving remote sensing, one deals with a medium containing particles, and/or gases, and the medium emits, scatters, and absorbs the infrared radiation under the influence of a nonuniform temperature distribution. The earth's atmosphere, for example, contains water droplets, ice crystals, and dust or aerosol particles immersed in the absorbing gases CO_2 and H_2O, and has a temperature that varies strongly with height. The earth's land surface, and that of the moon and planets, usually has a layer of dust, snow, or vegetation which has a temperature gradient. The interior of the earth and planets, although not accessible to remote sensing, is composed of compacted particles of various kinds which emit, absorb, and scatter, again with a radial temperature gradient (For the causes of temperature gradients, see Chapter 5, Section IV, B).

* *Reststrahlen* (residual rays) are described in chapter 7, Section II, A.

One must analyze these different situations by a radiative transfer theory that takes account of the emission at various temperatures, the reabsorption, and the multiple scattering of the radiation, in order to be able to interpret remotely acquired spectral information in terms of the nature of the medium, or to model heat flow in inaccessible regions such as planetary interiors.

An exact theory of radiative transfer has been given by Chandrasekhar (1950). In this theory the diffuse radiation in the medium is regarded as an infinite number of streams of radiation which interact with each other owing to the presence of the scattering, absorbing, and reemitting particles. The theory requires that the nature of the scattering and absorption per unit volume of the medium be known in detail. However, since this is seldom possible, as in the usual situation of irregularly shaped particles of unknown size distribution and perhaps unknown composition, a theory based on a relatively small number of discrete radiation streams is generally adequate.

In earth science problems where the infrared radiation is almost always symmetrical about the vertical (z-axis), a six-stream model reduces effectively to only two streams. Sagan and Pollack (1967) and Lyzenga (1973) have shown that such a two-stream model gives an excellent agreement with the exact theory of Chandrasekhar for a wide variety of conditions.

B. The Six-Stream Model

A six-stream model can be set up in two ways: (*a*) with two streams along the positive and negative z directions and the remaining four streams horizontal, or (*b*) with three upward-directed streams and three downward-directed streams all inclined at an angle $\cos^{-1}(1/\sqrt{3})$ to the vertical. In our previously reported work on powders (AO-I and AO-II) we have used configuration (*a*). Here we shall use configuration (*b*), which has been used by Conel (1969), since it allows one to determine both the absorption and scattering coefficients of a remotely observed atmosphere from vertical radiance measurements for two different temperature profiles.

From symmetry the three upward-slanting streams have the same spectral intensity I at any point, measured in units of watt/cm^2 per wave number across planes perpendicular to the streams. One can combine the three identical streams into an effective resultant stream which, owing to the $\cos^{-1}(1/\sqrt{3})$ inclination, also has intensity I, now relative to a horizontal plane. Similarly, the three equal downward-slanting streams, each of intensity J, combine into a resultant downward flux density J. In terms of the resultant intensities I and J, the radiative transfer equations take the very simple form

$$dI/dz = -(K+S)I + SJ + KB \tag{2}$$

$$-dJ/dz = -(K+S)J + SI + KB \tag{3}$$

where K is the total effective absorption cross section of all particles in a unit of volume at a given point in the medium, S is the effective backscattering cross section per unit volume, and B is the Planck function corresponding to the temperature T at the given point. The quantities K, S, and B are all functions of the radiation frequency ν.

K and S are given in terms of the average absorption, backscattering, and transverse-scattering cross sections σ_a, σ_b and σ_t of the individual particles by the expressions

$$K = \sqrt{3}N\sigma_a \tag{4}$$

$$S = \sqrt{3}N(\sigma_b + \tfrac{1}{2}\sigma_t) \tag{5}$$

where N is the number of particles per unit volume. The factor $\sqrt{3}$ arises from the slant distance along a discrete stream relative to distance along the z-axis. The factor $\frac{1}{2}$ is the fraction of the transversely scattered radiation that contributes to scattering into the backward hemisphere.

We may calculate the cross sections σ_b and σ_t from the average single particle scattering distribution $p(\phi)$ by integrating with a weighting factor $\cos^2 \phi$ to get the backward and forward cross sections σ_b and σ_f and with a factor $\sin^2 \phi$ to get the transverse cross section σ_t. We consider that this procedure is physically more reasonable than the usual method of approximating the phase function $p(\cos \phi)$ as a sharply truncated expansion in Legendre functions. The forward scattering is regarded as part of the undeflected radiation and does not contribute to S.

The phase function p is usually calculated by Mie theory (1908) on the assumption that the particles are spherical. Since most real particles, however, have edges and other asperities, we prefer to determine p separately by a coarse-particle theory based mainly on geometrical optics and including the effects of edges and asperities, and by a fine-particle theory based on wave optics applied to ellipsoids of various shapes. It should be noted that experimental results for coarse corundum powders having chunky and platelet geometries indicate that the precise particle shape is relatively unimportant compared to the effect of edges. The values of K and S derived by the two theories are averaged with suitable weighting factors. The method is described in AO-I and summarized later along with an account of some extensions of the theory.

C. *Applications of the Transfer Equations*

1. *Determination of K and S for a Uniform Atmospheric Layer*

The simplest application of the radiative transfer equations is to the case of an approximately isothermal atmospheric layer which contains a uniform distribution of scattering and absorbing particles such as ice crystals, aerosol, or dust, and absorbing gases such as carbon dioxide and water vapor. In such a layer the general solution of the transfer equations is

$$I = Ae^{-\gamma z} + Ce^{\gamma z} + B \tag{6}$$

$$J = RAe^{-\gamma z} + \frac{1}{R} Ce^{\gamma z} + B \tag{7}$$

where A and C are arbitrary constants, B is the Planck function corresponding to the temperature T of the layer, γ is the extinction coefficient, and R is the reflectance of a semiinfinite layer of the same composition. The parameters γ and R are related to K and S by the expressions

$$\gamma = (K^2 + 2\,KS)^{1/2} \tag{8}$$

$$R = 1 + \frac{K}{S} - \left(\frac{K^2}{S^2} + 2\,\frac{K}{S}\right)^{1/2} \tag{9}$$

where K includes both gas and particle absorption. All quantities are functions of the infrared frequency ν.

If the spectral fluxes I_1, J_1 at the bottom of the layer ($z = 0$) and I_2, J_2 at the top of the layer ($z = h$) and the temperature T are measured from an aircraft, as has been done by Kuhn (1970), one can determine γ and R for the layer by solving the four simultaneous equations

$$I_1 = A + C + B \tag{10}$$

$$J_1 = RA + C/R + B \tag{11}$$

$$I_2 = Ae^{-\gamma h} + Ce^{\gamma h} + B \tag{12}$$

$$J_2 = RAe^{-\gamma h} + Ce^{\gamma h}/R + B \tag{13}$$

for the four unknowns γ, R, A, and C. From γ and R it is easy to obtain K and S from Eqs. (8) and (9), from which it follows that

$$K = \gamma(1 - R)/(1 + R) \tag{14}$$

$$S = 2\,\gamma R/(1 - R^2) \tag{15}$$

Spectral data on I_1, J_1, I_2 and J_2 can therefore provide, in addition to direct information on the earth's energy balance, valuable information on the nature of the particles in the atmosphere and how their concentration is changing with time.

2. *Interpretation of Remotely Measured Spectral Radiance*

A more difficult theoretical problem is that of interpreting remote spectral radiance data obtained by a satellite above a planet. One may assume that the temperature and pressure distribution in the atmosphere can be found by the Kaplan (1959) method applied to a strong gaseous absorption band of, for example, CO_2. The problem then is to interpret the observed nongaseous infrared radiance spectrum in terms of the particles present in the atmosphere.

One method is to assume a particle type, size, concentration versus height, and ground emissivity; calculate $K(z, \nu)$ and $S(z, \nu)$; calculate spectral radiance by Eqs. (2) and (3); compare with measured radiance; iterate with respect to all the assumed parameters.

A second method, somewhat analogous to (1.) above, is to determine $K(z, \nu)$ and $S(z, \nu)$ so as to give the best fit to the measured spectral radiance and then to search for particles that have this dependence of K and S on ν.

Both methods depend on the calculation of the radiance by Eqs. (2) and (3) for any frequency ν, for given $K(z)$, $S(z)$, $B(z)$ and ground emissivity ϵ. One can carry out this calculation on a computer by dividing the atmosphere into n layers in each of which K, S, and B are taken as approximately uniform. Then the fluxes I and J in each layer are given by the solution [Eqs. (6) and (7)], and characterized by a pair of arbitrary constants A and C. The total number of constants for the n layers is therefore $2n$. At the top of the atmosphere $J = 0$, and at the ground $I = (1 - \epsilon)J$, which provide two more linear equations in the A's and C's. Therefore, there are $2n$ linear equations in all, which can be solved for the A's and C's by Gaussian elimination. The radiance at the top of the atmosphere can then be determined.

For both methods a measure of fit between the calculated and observed spectral radiance is needed that pays particular attention to essential spectral shape rather than the absolute radiance level. Such a measure of fit is described later.

In the second method, where one first determines $K(\nu)$ and $S(\nu)$ before searching for a candidate particle material, it is necessary to measure spectral radiance for two different distributions of temperature in the atmosphere since at each frequency one needs two experimental quantities to determine the unknowns K and S. The accuracy of the determination depends strongly, of course, on how different the two temperature profiles are. If reasonable accuracy can be attained this method should be superior to the first method since the measure of fit, applied separately to the spectral distributions of both K and S, provides a more critical test of a proposed particle species than when applied to spectral radiance only. We plan to apply both methods to the Mariner 9 spectra of the 1971 dust storm on Mars.

3. *Radiative Transfer in a Planetary Interior*

At the high temperatures existing in the deep interior of the earth, moon, or planets, the radiative component of outward heat flow may greatly exceed the conductive component and is, therefore, of considerable geophysical interest.

One can derive an expression for the radiative conductivity by subtracting Eqs. (2) and (3) to obtain

$$d(I + J)/dz = -(K + 2S)(I - J) \tag{16}$$

Now, since the temperature changes very little in a distance of the order $1/(K + 2S)$, I and J closely approximate the uniform-temperature values

$$I = J = n^2(\nu)B(\nu, T) \tag{17}$$

where $n(\nu)$ is the average real part of the refractive index of the rock particles.

The net radiative flux density is therefore given by

$$Q_r = \int_0^\infty (I - J)\, d\nu = -\frac{dT}{dz}\int_0^\infty \frac{2n^2(\nu)\dfrac{\partial B(\nu, T)}{\partial T}\, d\nu}{K + 2S} \tag{18}$$

The radiative conductivity is then

$$C_r = -\frac{Q_r}{\dfrac{dT}{dz}} = 2\int_0^\infty \frac{n^2(\nu)\dfrac{\partial B(\nu, T)}{\partial T}\, d\nu}{K + 2S} \tag{19}$$

Expressions of this form, based on the slightly different radiative transfer model of Schuster (1905) have been used by Clark (1957a, b) and Aronson *et al.* (1967, 1970) to evaluate C_r. It is found that providing the grain sizes in the rocks at great depths are of the order of a millimeter or larger, S is small compared with K. Therefore, the effect of scattering of the radiation is negligible.

Thus it is possible to investigate the likely radiative transfer processes in the lunar interior simply by studying the effects of high temperature (Aronson *et al.*, 1967, 1970; Fukao *et al.*, 1968) and pressure (Shankland, 1970) on the absorption coefficients of the expected minerals. The results of these studies indicate that previous studies which had relied on room temperature spectra of minerals made a serious error in that high temperatures significantly modify the spectral absorptions. The modifications take the form of increases in the band widths for both the vibrational transitions and an electronic transition due to ferrous iron located near 1 μm. The band broadenings effectively close a window in the near infrared spectrum through which radiation could be passed and thus significantly reduce the magnitude of the heat flow by radiative transfer. This is an important result as radiative transfer had been postulated as a mechanism for reducing the temperature in planetary interiors and so reducing the likelihood of melting and subsequent volcanic activity. The pressure effect is negligible compared to the temperature effect.

4. *Radiative Transfer in a Powder*

In a dense powder the infrared absorption and scattering is often so strong, especially in reststrahlen bands, that the intensities I and J undergo large fractional changes as the radiation traverses a single layer of particles. Under these conditions the differential Eqs. (2) and (3) which, by their nature, describe the properties of a continuum with continuously distributed parameters K and S, must be replaced by difference equations involving ΔI and ΔJ instead of dI/dz and dJ/dz; K and S are then replaced by the absorption α and reflectance ρ of a horizontal monolayer of particles.

For a powder composed of particles of average volume v and with a volume packing fraction f we can think of an average particle as located at the center of a cube of edge a where

$$a^3 = v/f \tag{20}$$

A monolayer may then be assumed to have a thickness a and an average transverse area a^2 per particle. The absorption and reflectance of the monolayer are therefore given approximately by the expressions

$$\alpha = \frac{\sqrt{3}\sigma_a}{a^2} = \sqrt{3}\sigma_a \left(\frac{f}{v}\right)^{2/3} \tag{21}$$

$$\rho = \frac{\sqrt{3}(\sigma_b + \sigma_t/2)}{a^2} = \sqrt{3}(\sigma_b + \sigma_t/2)\left(\frac{f}{v}\right)^{2/3} \tag{22}$$

The factor $\sqrt{3}$ allows for the fact that the projected area per particle for a slant stream inclined at $\cos^{-1}(1/\sqrt{3})$ to the vertical is $a^2/\sqrt{3}$ rather than a^2.

For the case of spherical particles of diameter d the factor $(f/v)^{2/3}$ becomes $(\pi f/6\, d^3)^{2/3}$.

It is to be noted that Eqs. (21) and (22) do not include any allowance for shadowing of the particles in the monolayer by each other.

Expressions for α and ρ for the other type of six-stream model are given in AO-I.

The reflectance R of a semiinfinite powder regarded as a stack of monolayers is

$$R = \frac{1 + \rho^2 - \tau^2}{2\,\rho} - \left\{\left(\frac{1 + \rho^2 - \tau^2}{2\,\rho}\right)^2 - 1\right\}^{1/2} \tag{23}$$

where τ, the monolayer transmittance, is given by

$$\tau = 1 - \alpha - \rho \tag{24}$$

One can show that Eq. (23) reduces, as it should, to Eq. (9) when f tends to zero.

It is worth remarking that the monolayer method is valid only when the particles are so far apart that coherent scattering is unimportant. In the case of very closely spaced small particles, one must use the Lorentz-Lorenz theory of cooperative scattering as in AO-I.

When a temperature gradient is present in a powder, the difference equations for the nth discrete monolayer take the form

$$I_{n+1} = \tau_n I_n + \rho_n J_{n+1} + B_n \tag{25}$$

$$J_n = \tau_n J_{n+1} + \rho_n I_n + B_n \tag{26}$$

where I_n, J_n are the ingoing and outgoing fluxes, respectively, on the lower side of the monolayer and I_{n+1}, J_{n+1} are the outgoing and ingoing fluxes on the upper side. For any chosen number of monolayers, there are exactly enough boundary conditions to determine all the fluxes.

IV. Calculation of Particle Absorption and Scattering Cross-Sections

A. Coarse Particles

Our method of calculating the scattering and absorption cross sections σ_b, σ_t, and σ_a for nonspherical particles is explained in detail in paper AO-I. We summarize here the basic approach used.

For particles larger than a wavelength in size we regard edges and other asperities as perturbations of the surface of a sphere of volume equal to that of the actual particle. We first calculate the scattering and absorption of the unperturbed sphere by geometrical-optics ray tracing and then correct the results by a wave-optics treatment of the effects of the edges and asperities. The most important and unexpected effect of the edges and asperities is that they cause considerable additional absorption by the particles, especially in *reststrahlen* bands.

The enhanced absorption may be considered as arising from a greater exposure of the absorbing interior of the particle to the incident wave. The edges and asperities would be expected to act in the same way as smaller particles clinging to a main particle. A number of experiments described in AO-II give evidence for the additional absorption by edges and asperities and clinging fine particles. The evidence may be summarized as follows:

1. The theoretical and experimental spectra of smooth glass spheres are in excellent agreement without any additional hypotheses.
2. An inordinately large reduction in the strength of the *reststrahlen* bands in quartz is observed when a small volume of clinging fine particles is present.
3. Scanning electron micrography shows the presence of steplike surface features on samples of large particles of crushed corundum. The samples show a qualitatively similar spectral feature to the smaller size particles in a region of a very strong *reststrahlen* band.
4. The feature in question can be produced in a single-crystal emittance spectrum by suitably roughening the surface with diamond paste.
5. Changes in the surface character of corundum particles modify the shape and level of the spectrum in the range of the anomolous feature in ways similar to those found theoretically when the distribution of the shapes of the asperities is altered.

To treat the edges and asperities theoretically we regard them as elliptical dipoles, having a wide variety of shapes, distributed over the surface of the unperturbed spherical particle, and calculate the absorption by a generalization of Rayleigh's analysis for small spherical dipoles

The combined ray-tracing and distributed dipole calculations yield both the angular distribution of scattered radiation and the total radiation absorbed by a single particle for an incident plane wave of given intensity. The absorption cross section σ_a is calculated in the usual way as the ratio of the power absorbed to the incident power density. We calculate the power effectively scattered into four discrete streams in the transverse plane by multiplying the scattered intensity distribution by $\sin^2 \phi$ and integrating over all angles, where ϕ is the angle of scattering. The transverse cross section σ_t is this power divided by the incident power density. Likewise, we find σ_b by multiplying the scattered intensity by $\cos^2 \phi$ and integrating over the backward hemisphere only. One could calculate the forward-scattering cross section σ_f by multiplying by $\cos^2 \phi$ and integrating over the forward hemisphere, but this discrete component of the scattered radiation is regarded merely as part of the undeflected incident beam.

A further contribution to the scattering, not included in AO-I, is that caused by the edges and asperities themselves. To calculate this scattering we assume that the radiation reflected externally by an element of area of the unperturbed sphere leaves the sphere perfectly diffusely instead of perfectly specularly. Under these conditions the scattering cross sections for external reflection become

$$\sigma_{\mathrm{refl},t} = \frac{\pi d^2}{4} \int_0^{\pi/2} R_0(\tfrac{1}{4} + \tfrac{1}{4}\cos^2\theta + \tfrac{1}{2}\sin^2\theta)\sin 2\theta \, d\theta \tag{27}$$

$$\sigma_{\mathrm{refl},b} = \frac{\pi d^2}{4} \int_0^{\pi/2} R_0(\tfrac{1}{8}\sin^2\theta + \tfrac{1}{2}\cos^2\theta)\sin 2\theta \, d\theta \tag{28}$$

where θ is the nominal angle of incidence at a point on the particle and R_0 is the total reflectance at the point including the absorption by edges and asperities, as calculated in AO-I.

In the case of a powder, a correction to σ_b and σ_t, not included in AO-I and AO-II, is required to allow for the loss of scattering due to wave interference in the regions around the points of contact. Such a region has very low reflection and refraction, and corresponds to the central dark area in a Newton's ring interference pattern. We assume for simplicity that optical bonding is complete wherever the gap between two particles is less than $\lambda/10$, where λ is the free-space wavelength. For two spheres of diameter d the optically bonded area is $\pi d\lambda/20$ and the number of contact points per sphere is shown in AO-I to be of the order of $12f$, where f is the volume fraction occupied by the particles. The fraction of the surface area πd^2 of the particle that remains for reflection and refraction is, therefore, $1 - 6f\lambda T/10d$, where T is the usual particle transmission factor. Since this expression becomes negative for certain combinations of f, λ, and d, we prefer to regard it as the first two terms in the expansion of an exponential. The "contact factor" is therefore taken as

$$C = e^{-6f\lambda T/10d} \tag{29}$$

We multiply σ_b and σ_t by C to allow for this optical bonding.

B. Fine Particles

For particles smaller than the wavelength the ray tracing method breaks down. Also, individual edges and asperities are too small to interact strongly with the incident wave. However, the overall shape and orientation of a particle do affect the electric dipole moment induced in the particle. To take the gross shape into account we assume that the particles are ellipsoids with a wide range of the depolarization factors L, M, and N. For a particle of volume v the power scattered and absorbed can readily be calculated and averaged over all orientations of the ellipsoid and both states of polarization of the incident plane wave. One can then separately average the scattered and absorbed powers, for constant v, over all values of L, M, and N subject to the condition that the sum of these parameters is 4π, and hence derive the following expressions for the total scattering cross section σ and the absorption cross section σ_a

$$\sigma = \frac{4\pi^5\nu^4 d^6}{27} \mid m^2 - 1 \mid^2 \frac{\left\{Im(1n\, m^2) + Im\left(\dfrac{1n\, m^2}{m^2 - 1}\right)\right\}}{Im(m^2)} \tag{30}$$

$$\sigma_a = \frac{2\pi^2\nu d^3}{3} Im\left(\frac{-m^2 1n\, m^2}{m^2 - 1}\right) \tag{31}$$

Since the averaged scattering is approximately isotropic, we have $\sigma_b = \sigma/6$, and $\sigma_t = 4\sigma/6$.

In the case of a powder, where the spacing of the fine particles is less than a wavelength, coherence in the scattering from different particles must be taken into account. As described in detail in AO-I, we do this by first finding the average complex refractive index $\bar{m}$ for the medium by means of a Lorentz-Lorenz type of calculation, and then calculate the average scattering in the same way as before but using the relative index $m/\bar{m}$ instead of the absolute index m of the ellipsoidal particles.

Owing to the coherence between particles and because the changes in I and J are small for a monolayer of fine particles, we use the continuum form of the radiative transfer equations rather than the discrete monolayer model. The discontinuity between the Lorentz–Lorenz index $\bar{m}$ and the vacuum index of unity must be allowed for in the boundary conditions at the surface of the medium, which, in general, gives rise to both external and internal reflections.

C. *Bridging between Coarse and Fine Particle Theories*

The cross sections σ_b and σ_t vary approximately as d^2 for the coarse particle theory and d^6/λ^4 for the fine particle theory. Therefore, in the region $d \ll \lambda$ where the fine particle theory is valid, the coarse particle theory gives scattering cross sections that are too large. On the other hand, for $d \gg \lambda$ where the coarse particle theory applies, the fine particle theory greatly overestimates the cross sections. This suggests that in the transition range of particle size, where neither theory is clearly valid, one should use a bridging formula for the scattering cross sections that weights more heavily the theory that gives the smaller cross section.

A suitable bridging formula is

$$1/\sigma_{br}^2 = 1/\sigma_c^2 + 1/\sigma_f^2 \tag{32}$$

where, σ_c, σ_f are the coarse and fine particle theory values of $\sigma_b + \sigma_t/2$, respectively, and σ_{br} is the bridged value. We also calculate a weighting factor σ_{br}/σ_0, where σ_0 is the smaller of σ_c and σ_f, and apply this factor to the absorption cross section given by the theory that produces σ_0.

In the case of powders where the fine particle theory has a significant contribution from the surface reflection by the Lorentz-Lorenz medium, one must bridge the coarse and fine particle theories with respect to reflectance as in AO-I and AO-II.

V. Comparison of the Theory with Experimental Results for the Case of Powders

Figures 1, 2, and 3 show experimental and theoretical results for the reflectance of powders of corundum, quartz, and mixtures of the two. As in AO-II, the experimental results pertain to emittance measurements on approximately isothermal powders of large depth ($\epsilon = 1 - R$). The theoretical curves are based on the theory given in AO-I together with the three significant modifications described in the present publication: discrete monolayer model; diffuse reflection by the surfaces of the coarse particles; optical bonding of contacting particles.

A considerable improvement in the reflectance level in the reststrahlen bands for the coarser particles can be seen when compared with AO-II. Some increase in the level resulting from the discrete-monolayer model and the diffuse reflection by the particle surfaces, for those regions where the particles are semitransparent, is effectively nullified by the contact factor. The comparison between these three figures and their counterparts

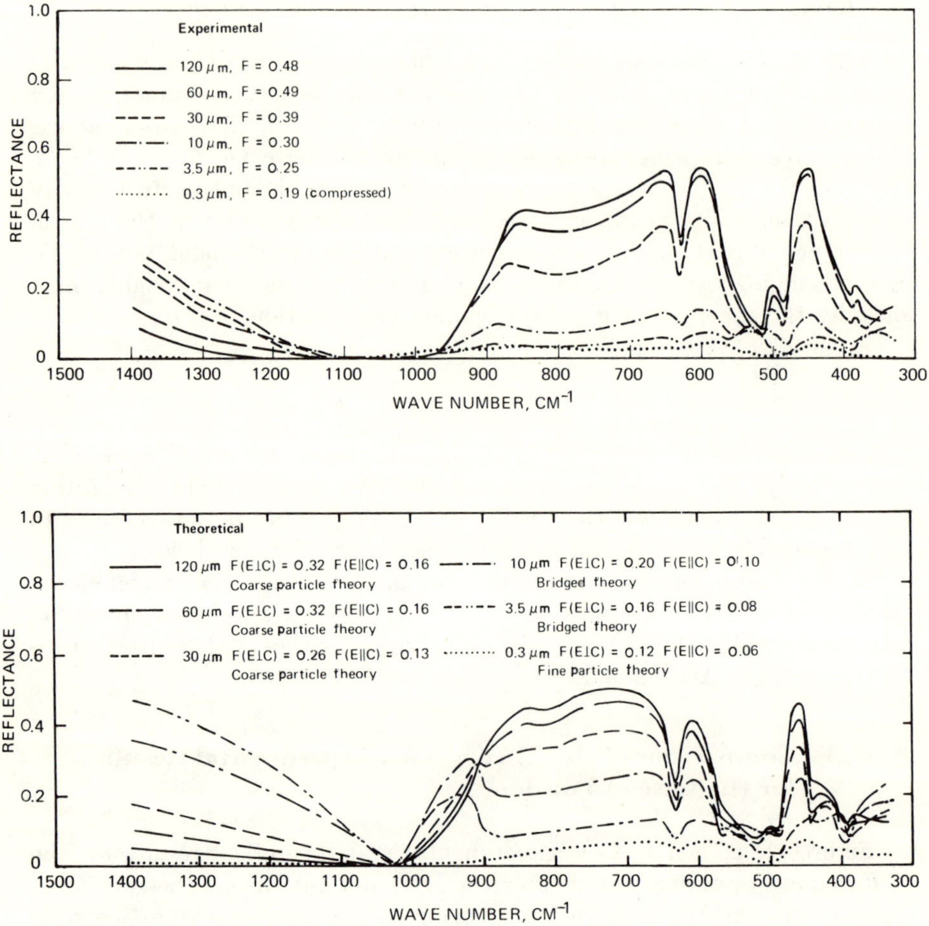

FIG. 1. Comparison of theoretical and experimental reflectance of corundum powders. (Experimental spectra from Aronson and Emslie, 1973).

in AO-II demonstrates the essential validity of the discussion of AO-I concerning the inadequacy of a continuum model for this case.

VI. The Inverse Problem

A. Powder Reflectance

The inverse problem consists of finding a suitable way to use the theory of reflectance to analyze the experimental spectrum of an unknown mixture of powdered minerals in order to derive quantitative information about volume fractions f_j and particle sizes d_j of the minerals j. For the sake of clarity, the appropriate summations over the mineral species have been neglected in this chapter but are given in AO-I. It is assumed, of course, that the mixture contains a reasonably small number of candidate minerals of known optical constants.

Our theory indicates quite strongly that a simple linear mixing rule for reflectances is not a good basis for the inverse problem. For example, Fig. 4 shows the variation of reflectance predicted by the continuum model,

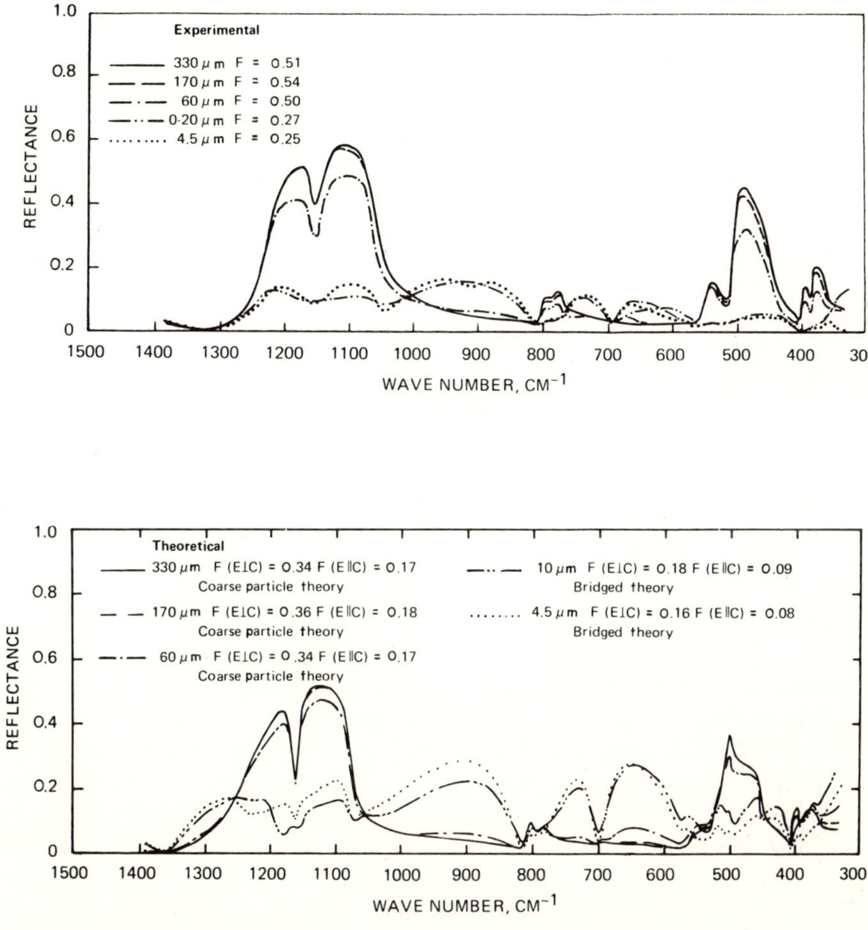

FIG. 2. Comparison of theoretical and experimental reflectance of quartz powders. (Experimental spectra from Aronson and Emslie, 1973.)

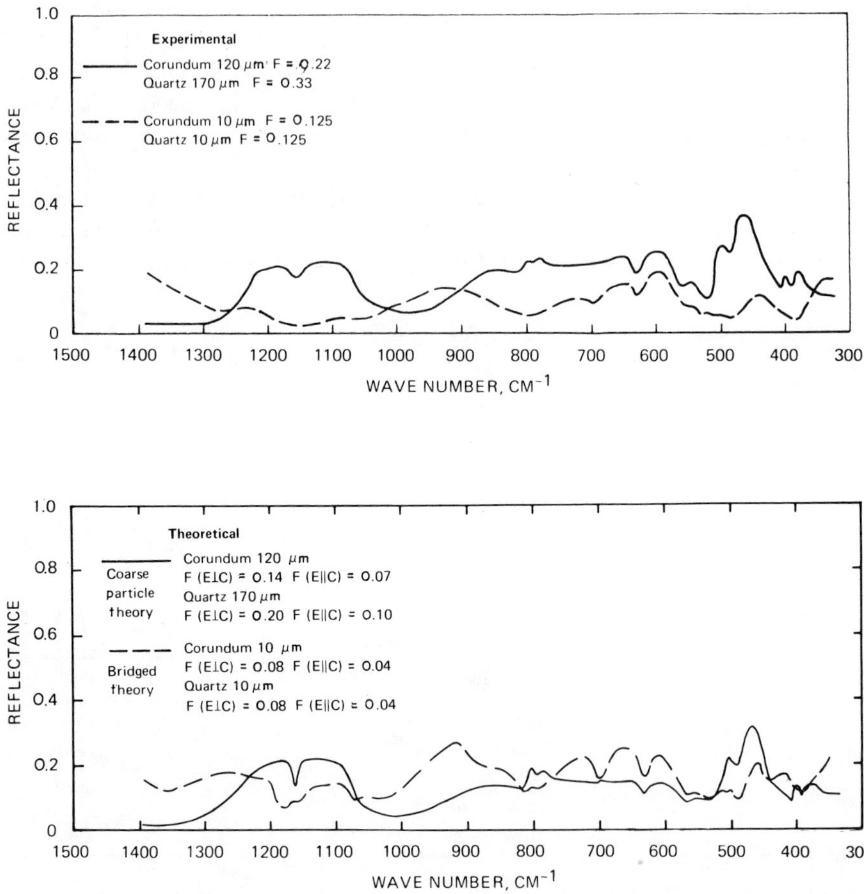

FIG. 3. Comparison of the theoretical and experimental reflectance of mixtures of quartz and corundum. (Experimental spectra from Aronson and Emslie, 1973.)

described in AO-I as a function of the relative proportions of the two components in a 10 μm quartz–corundum mixture. The five frequencies chosen illustrate examples of rather different behavior. The important point is the significant nonlinearity of the curves, the extreme example being that for 630 cm^{-1}.

A further consideration is that no theory can possibly include all of the parameters needed to describe a real mixture, such as the particle size distributions of all component minerals, the extent of surface asperities on each type of particle, the degree of aggregation, etc. This means that the theoretical spectrum will always differ quantitatively from the experimental spectrum even with the optimum choice of the adjustable theoretical parameters. Such differences will occur in spectral level and in the detailed shape of the spectral bands. It appears, however, that the locations of the maxima and minima of the experimental and theoretical spectra agree quite well, for optimum values of the theoretical parameters, and that the ordering of the peaks and valleys with respect to amplitude is about the same for both spectra.

On the basis of these considerations, we have developed a new quantitative method for comparing a theoretical spectrum with an experimental one. The method pays attention only to the ordering of the amplitudes in each spectrum at a number of frequencies selected from the experimental spectrum on the basis of features of interest such as peaks, valleys, abrupt changes of slope, etc. Some twenty to thirty frequencies are often sufficient to characterize an infrared spectrum.

After selection of N frequencies, we compare the amplitude at the first frequency with the amplitudes at the other $N - 1$ frequencies, and record a $+$, 0, or $-$ at each of these frequencies depending on whether the ampli-

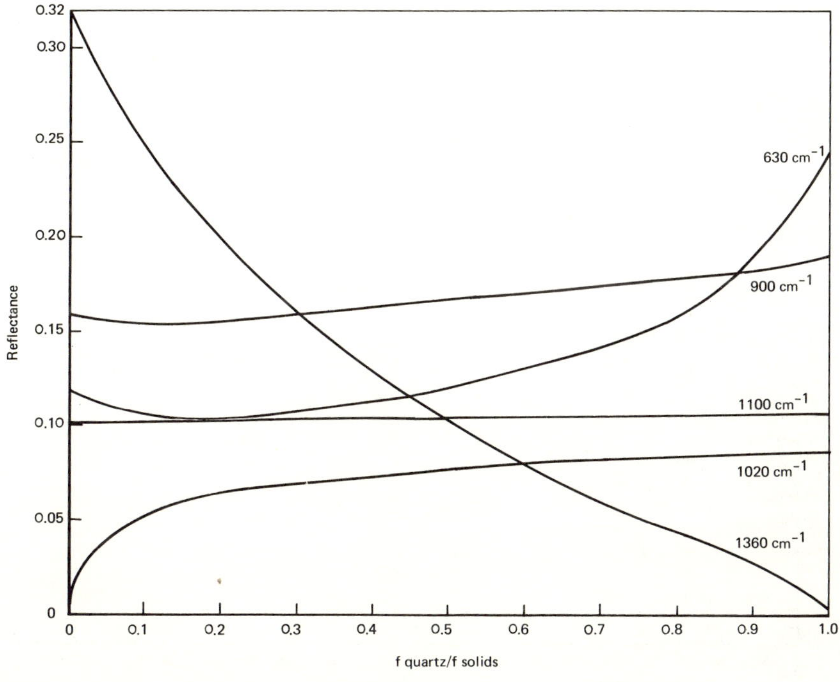

FIG. 4. Theoretical variations in reflectance with the mixing ratio in mixtures of corundum and quartz (10 μm powders).

tude is greater than, equal to, or less than that at the first frequency. We then repeat the comparison starting with the second frequency and comparing its amplitude with those of the succeeding $N - 2$ frequencies. On continuing the process we obtain a triangular matrix of $N(N - 1)/2$ elements which have the symbols +, 0, or −. This matrix characterizes the essential shape of the experimental spectrum. The symbol 0 is assigned when the amplitudes at two frequencies are equal within some noise level.

We now compute a theoretical spectrum, for a set of assumed values of the f_j and the d_j, and characterize it in the same way as for the experimental spectrum, using the same set of N selected frequencies.

Next we compare the theoretical and experimental matrices and count the number of coincidences in the two sets of $N(N - 1)/2$ elements. The final score, which lies between 0 and $N(N - 1)/2$, is a quantitative "measure of fit" between the theoretical and experimental spectra and can be used as the criterion for determining the optimum theoretical parameters.

A computer program was constructed to carry out such an analysis of our data. It was used for two mixtures of quartz and corundum. The coarse-particle theory results were used for a 170 μm quartz–120 μm corundum mixture, and the bridged theory for a 10 μm mixture. The procedure used the continuum model theory (AO-I) and so its facility for ignoring the spurious spectral level was more severely tested than would presently be the case. The analysis was carried out in each example after choosing 20 spectral points from the experimental data. Figure 5 shows the measure of fit for each of the chosen ratios of the components. The arrows indicate the measured ratios in the real mixtures, but it must be recognized that this ratio is the average for the entire sample, and may not precisely

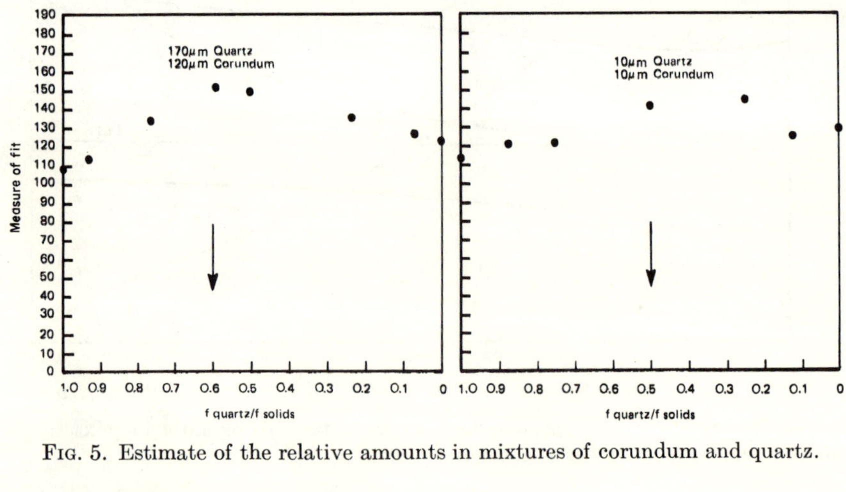

FIG. 5. Estimate of the relative amounts in mixtures of corundum and quartz.

describe the top layer where the measured radiation originates. No attempt has yet been made to interpolate between the individual theoretical results although this could clearly be done. An increase in the number of spectral points would also almost certainly improve the technique as 20 points did not sufficiently describe all features in the spectrum. We consider that this initial success should be readily extendable to mixtures having more components.

B. Atmospheric Radiance

As pointed out in Section III, C, 2, identification of the particle species present in an atmosphere may be made less ambiguous by the simultaneous optimization of two parameters such as K and S. The "measure of fit" just described can be used equally well for this purpose.

ACKNOWLEDGMENT

This work was supported in part by NASA under contracts NAS9-10875 and NAS5-23257.

References

Aronson, J. R., and Emslie, A. G. (1973). *Appl. Opt.* **12,** 2573–2584.

Aronson, J. R., Emslie, A. G., and McLinden, H. G. (1966). *Science* **152,** 345–346.

Aronson, J. R., Eckroad, S. W., Emslie, A. G., McConnell, R. K., Jr., and von Thuna, P. C. (1967). *Nature (London)* **216,** 1096–1087.

Aronson, J. R., Bellotti, L. H., Eckroad, S. W., Emslie, A. G., McConnell, R. K., and von Thuna, P. C. (1970). *J. Geophys. Res.* **75,** 3443–3456.

Bell, E. E., and Eisner, L. (1956). *J. Opt. Soc. Amer.* **46,** 303–304.

Chadrasekhar, S. (1950). "Radiative Transfer." Oxford Univ. Press (Clarendon), London and New York (reprinted by Dover, New York, 1970).

Clark, S. P., Jr. (1957a). *Trans. Amer. Geophys. Union* **38,** 931–938.

Clark, S. P., Jr. (1957b). *Amer. Mineral.* **42,** 732–742.

Coblentz, W. W. (1905). "Investigations of Infra-Red Spectra." Carnegie Inst., Washington, D. C. (reprinted by the Coblentz Society, 1962).

Conel, J. E. (1969). *J. Geophys. Res.* **74,** 1614–1634.

Emslie, A. G., and Aronson, J. R. (1973). *Appl. Opt.* **12,** 2563–2572.

Fukao, Y., Mitzutani, H., and Uyeda, S. (1968). *Phys. Earth Planet. Interiors* **1,** 157–162.

Hanel, R. A., Conrath, B. J., Kunde, V. G., Prabhakara, C., Revah, I., Salomonson, V. V., and Wolford, G. (1972a). *J. Geophys. Res.* **77,** 2629–2641.

Hanel, R. A., Conrath, B. J., Hovis, W., Kunde, V. G., Lowman, P., MacGuire, W., Pearl, J., Pirraglia, J., Prabhakara, C., Schlachman, B., Levin, G., and Straat, P. (1972b). *Icarus* **17,** 423–442.

Hovis, W. A., Jr., and Callahan, W. R. (1966). *J. Opt. Soc. Amer.* **56,** 639–643.

Hunt, G. R., and Vincent, R. K. (1968). *J. Geophys. Res.* **73,** 6039–6046.

Kaplan, L. D. (1959). *J. Opt. Soc. Amer.* **49,** 1004–1007.
Kuhn, P. M. (1970). *J. Atmos. Sci.* **27,** 937–942.
Lyon, R. J. P. (1965). "Evaluation of Infrared Spectrophotometry for Compositional Analysis of Lunar and Planetary Soils," NASA CR100, Nat. Aeron. Space Admin., Washington, D. C.
Lyzenga, D. R. (1973). *Icarus* **19,** 240–243.
Mie, G. (1908). *Ann. Phys. (Leipzig)* [4] **25,** 1377–1445.
Sagan, C., and Pollack, J. (1967). *J. Geophys. Res.* **72,** 469–477.
Schuster, A. (1905). *Astrophys. J.* **21,** 1–22.
Shankland, T. J. (1970). *J. Geophys. Res.* **75,** 409–413.
Simon, I. (1951). *J. Opt. Soc. Amer.* **41,** 336–345.

7

REFLECTANCE AND EMITTANCE OF TERRAIN IN THE MID-INFRARED (6–25μm) REGION

R. J. P. Lyon and A. A. Green

I. Introduction

Nichols and Rubens, in 1897, studied the metallike reflectance of inorganic mineral crystals. Today we have used their *reststrahlen* phenomenen (residual ray) to observe the earth's surface from the 270 mile (500 km) orbit of SKYLAB in the S-191 experiment. Coblentz's painstaking point-by-point digital measurements of one spectrum per four hours, working in the basement of Cornell University, and at the National Bureau of Standards from 1903 through 1920, have now been repeated with an aircraft-mounted spectrometer which takes six spectra per second. These new measurements were made in the 6.5–13.0 μm region (1538–770 cm^{-1}) of the electromagnetic spectrum to monitor the fundamental vibrations of the silicate minerals which make up the surface rock of most planets and especially the earth. Yet today many of those who study scattering properties to derive the optical constants of powdered material using attenuated total reflectance techniques, little realize that the dust clouds which hover over Mars also have been determined as silicate powders from spectra taken at least 3000 miles (4800 km) away from the planet's surface using the spectrometers on Mariner IV. Although many thousands of man-years of effort have been devoted to work in the photographic regions few still are aware of the power of the *reststrahlen* measurements to reveal uniquely the composition of surface materials.

This chapter will serve as a short review of these developments but will also show how very significant are these residual ray measurements in geological remote sensing, in the chemical study of surficial rocks and soils, and for the rapid preparation of geological maps. In our considered opinion, by far the most significant geological information residing in remotely sensed data is in the *reststrahlen* bands (in silicates at 8–12 μm; 1250–850 cm^{-1}).

A step of major importance to geologists, made recently, is in the use of the multiband approach to geological mapping, with the production of image (photographs) of the terrain materials in several wavelength bands. Yet perhaps more significant than each initial image is the ratioed image wherein algebraic manipulation of the several bands of imaged data can be designed to emphasize the spectral information content, while still retaining the pictorial aspects of the presentation, for geologists have always worked with pictured data. Recently we have been the excellent images from the Earth Resources Technology Satellite (ERTS) series. With ERTS the four bands of imaged data which make up the spectra are available in prints, or in digital tapes, each showing the radiant energy distribution of reflected sunlight in the 0.5–0.6 (Band 4), 0.6–0.7 (Band 5), 0.7–0.8 (Band 6), 0.8–1.1 μm (Band 7) regions. Experimenters today are making ratioed images from Band 7/Band 4 and Band 5/Band 4 information to emphasize the ferrous– and ferric–iron distribution in surface materials and soils in the terrain (Billingsley and Goetz, 1973; Vincent, 1972, 1973). These developments have proceeded from the basic spectral measurements, which before were performed with high-resolution instruments under strictly laboratory conditions. In a similar way basic laboratory infrared measurements have culminated in spacecraft experiments, designed specifically to map the composition of terrain (Mineralogical Satellite, Burns and Lyon, 1963) or flown as high resolution surface composition mapping radiometers (W. Hovis, private communication, 1972) in the NIMBUS series of spaceflights.

II. Basic Research

A. Early Absorption and Emission Work

Many fundamental vibrations of crystalline minerals can be observed in the general region of 7–12 μm (1430–830 cm^{-1}). Naturally occurring minerals on the surface of the earth are primarily silicates, but locally carbonates, sulfates and even nitrates and oxides may be significant. Table I shows the wavelength regions associated with their lattice vibrations. Subdivisions of the important silicates are indicated (detailed spectral data on all these minerals are available in Chapter 8).

Absorption maximums occur at the frequency of these fundamental vibrations. These are typically observed in the laboratory in the transmission mode, with the powdered mineral sample embedded at a very low concentration in a matrix, like KBr, CsI, or Nujol oil. Because of the rapid change in the values of the optical constants [refractive index (n), and absorptivity (k)] at these maximums, reflectivity of the solid sample also changes markedly, becoming almost metallike (40–80% reflective) at the fundamental vibration frequencies. This effect was first examined by Nichols and Rubens in 1897, who found that if a beam of polychromatic infrared radiation was reflected off a crystalline material (NaCl, KCl, quartz, etc.) the reflected beam was enriched in a narrow band of wavelengths characteristic of that crystal. Multiple reflectance off the same

TABLE I

VIBRATION WAVELENGTH (FREQUENCIES) OF MINERALS OF THE EARTH'S SURFACE[a]

MINERAL	WAVELENGTH (μM)	FREQUENCY (CM^{-1})
Silicates		
Tektosilicates	9.1–12.5	800–1100
Inosilicates–pyroxenes	9.3–11.6	860–1080
Inosilicates–amphiboles	9.3–10.5	950–1080
Phyllosilicates	9.1–11.1	900–1100
Carbonates	6.9– 7.0	1429–1450
Nitrates	7.1– 7.7	1300–1400
Sulfates	8.33– 9.1	1100–1200
Oxides		
Iron oxides	17.4–18.2	550–575

[a] After Lyon (1967).

material left only the narrow bands, which he termed residual rays (*reststrahlen*). In those early days before gratings were developed the use of series of these crystals was the only way to secure long wavelength infrared radiations of known values.

From consideration of the general principals of conservation of energy (Kirchhoff's radiation law) we can say that if a body is in thermal equilibrium and optically thick then the relationship,

$$\rho_\lambda + \epsilon_\lambda = 1 \tag{1}$$

or

$$\epsilon_\lambda = 1 - \rho_\lambda \tag{2}$$

where ρ is the reflectance and ϵ the emittance (exitance).

Experimentally, therefore, one could show that the spectrum of emitted radiation from optically thick mineral surfaces would not be blackbody in character, but would depart markedly at these *reststrahlen* wavelengths because of relationship [Eq. (2)] between reflectance and emittance.

B. Laboratory Infrared Measurements

1. Absorption

Mineral and rock samples are prepared in a standardized way (Tuddenham and Lyon, 1960) being ground under alcohol to below 1 μm particle size, and embedded at 0.25% concentration in KBr pellets. The pellets are dried at 110°C overnight and studied in a double-beam transmission with a blank KBr pellet in the reference beam. In general, spectra of single mineral species are extremely informative, and the structural state and crystal chemistry of the crystals can be derived from the wavelength shifts in the transmission spectrum (Launer, 1952; Tuddenham and Lyon, 1960). Rocks (which can be considered as physical mixtures of minerals) show composite spectra, but if the constituent minerals are well ordered structurally showing good diagnostic peaks, modal analyses can be derived from their infrared spectra (Lyon, *et al.*, 1959, for G-1 granite, modal analysis).

2. Reflection

Solid minerals and rocks are studied by reflecting one of the beams of the spectrometer at a high angle off a polished surface. Information of the wavelength position and shape of the *reststrahlen* features can be obtained and diagnostic use can be made for crystal/rock-type determination. Weakly absorptive features (e.g., OH bondings) are not shown.

The composition of powdered materials (where surface scattering is at a maximum) can be determined, at least to 100 μm particle size, by using a hemispherical irradiance technique, and then introducing the reflected ray bundle into a spectrometer (Lyon, 1964, Figs. 11–13). The retention of finer powder on the surface causes excessive scattering and a loss of the reflectance spectrum (Aronson and Emslie, 1973; see Chapter 6).

3. Emission

Techniques used today are adaptations of those used by Coblentz in the 1910–1915 period (Coblentz, 1905, 1906, 1908). Samples are heated up to 100°–200°C above ambient and allowed to emit radiation directly into the optical train of the spectrometer, via some form of collective optics, suitably baffled and apertured to avoid receiving radiation from the heating system (Lyon, 1964).

Again suitable determination of specimen composition can be made, although these are degraded by scattering from all rough surfaces. Most laboratory emission spectrometers are of the conventional type, but modified with receiving optics. Sometimes advanced solid-state detectors are used, but usually it is more simple to heat the sample than to adapt faster, more sensitive detectors. The sample must be hotter than the surroundings else it will show only blackbody radiation—the *hohlraum* effect.

C. The Significance and the Measurements of ϵ_λ and ρ_λ Curves

A major aim in remotely sensing a terrain surface is to determine its composition as an aid to field mapping on foot, the classic manner of securing the geological information. It was proposed that the *reststrahlen* effect could be used for this purpose, with spectra observations from aircraft and spacecraft (Burns and Lyon, 1963; Lyon, 1964). Such observations must be made rapidly with a narrow field of view of the ground, to minimize the ground swath being integrated into a single spectrum.

We now see how it is possible to extend the knowledge of absorption spectra of minerals, to the reflectance, or to emittance observations, and to retain the information on their crystal chemistries, while observing them from a distance. In addition, because of the fortuitous situation that the gases in our atmosphere only absorb strongly at the 6.5 and 15.0 μm (1540 and 667 cm^{-1}) regions, leaving a "window" open from 8 to 14 μm (1250–715 cm^{-1}) for remote observation of these terrain emittances. Thus one can observe the crystal chemistry of silicates in rocks and soil from a distance, initially from only a few feet (Bell and Eisner, 1956; Lyon, 1964) then at longer distances (Lyon and Patterson, 1966), and finally at aircraft altitudes and speeds (Lyon, 1972), leading up to spacecraft observations in the SKYLAB S-191 experiments, where distances of 270 nautical miles (500 km) and speeds of 4 miles/sec/(7.4 km/sec) were involved.

All remote methods used to date have been based on measurements of the target's emitted radiation. Studies with reflected infrared radiation require the larger power levels which have only recently become available from laser systems. In this respect a high power CO_2 laser tuned across its emission lines from 9.5 (1052 cm^{-1}) to 11 μm (910 cm^{-1}) seems as though it may be a promising line of investigation.

III. Field Studies

A. Design of Rapid Scan Spectrometers for Field Studies

Our early field work (Lyon and Patterson, 1966) with the Perkin-Elmer, Model SG-4 Fastie-Ebert, oscillating-grating type spectrometer, had convinced us of two points: (1) that the nonlinear rocking motion of the grating caused serious uncertainties in the wavelength-versus-time plots, and (2) that the high spectral resolution of the grating was unnecessary with the broadened spectra we were obtaining (see later discussion of the causes of this degradation). In addition the need for order-sorting filter made the operation complex optically.

R. S. Vickers, who worked with us from 1967–1969, made a significant development by coupling the grating spindle directly to an optical encoder of 1:1000 resolution. But instead of using this digital readout of angle as data, he further conceived the idea of using the least significant bit (the outside most rapidly changing square-wave pulse) as an external "clock" for the analog–digital conversion in the data system. Not only did this remove the ambiguities of wavelength determination but it gave a sharp leading-edge pulse at a highly reproducible position on the grating excursion. We now did not record the wavelength "voltage" output, but used it solely to trigger the A-D conversion of the detector voltage output, thereby halving our data-recording problem. This added materially to our progress. We were able now to bypass this first problem, in our new design of the spectrometer which carried digital wavelength encoders, and in the more simple data systems. Both the NASA aircraft spectrometer and the Exotech Model 10 and 20 units now carry optical encoders for this purpose.

However, the design goals for the experiment were to fly the spectrometer over terrain, necessitating completing a spectrum in something less than 1 sec. When compared with the 30–90 sec for the SG-4 grating unit (or the 10–30 min for conventional laboratory units) these rapid scan spectrometers needed a dispersive system of a simple mechanical design. Both interferometric systems and fixed grating multiple-detector units were examined in detail. In 1965 interferometers had severe drawbacks in design, and also produced unnecessarily high-resolution spectra. In 1973 the multiple detector approach was simple, either using single amplifiers with multiplexed output, or multichannel amplifiers. Jet Propulsion Laboratory has flown a 29-detector, fixed-grating system in high-altitude balloon and space-craft experiments, and SKYLAB S-192 multichannel scanner used only a single detector Hg-Cd-Te crystal sensitized for several wavelengths from 0.4 to 13.0 μm (2500–770 cm^{-1}).

About 1965 the circular variable filter (CVF) was developed by Optical Coating Labs under the auspices of Warren Hovis of NASA. These wheels are made of a substrate material suitably transparent for the wavelengths being studied (germanium for the 8–14 μm (1250–714 cm^{-1}) areas). On this is coated the interference filter coating of such a design that it possesses a narrow bandpass, usually 1.5% in width of the central wavelength. This coating has a constant bandpass in a radial direction, but this varies progressively around the circumference so as to cause the narrow pass filter to change its central wavelength proportionately with rotation. Usually a wheel is composed of two 180° coatings such that from 0° to 180° the filter traverses λ_1 to λ_2 and from 181° to 359° it traverses λ_2 to λ_1. Often wheels are made from multiple segments. The Beckman Microspec unit has a 3-piece wheel to cover 2.5–15 μm (4000–667 cm^{-1}), while the S-191 spectrometer in SKYLAB used 2 wheels mounted concentrically, with 2 stacked detectors, to give short wavelength (0.4 –2.4 μm; 2500–4160 cm^{-1}) and long wavelength (6.2–15.5 μm; 1615–645 cm^{-1}) data concurrently with each rotation, of a 1-sec period.

B. *Use of Rapid Scan Spectrometers. Ground-, Air-, and Space-Based Measurements*

Two main units have been used in the measurements described in this chapter. The first, a one-of-a-kind made in 1965 for NASA, by Lockheed Missiles and Space Co., was built to spacecraft-type specifications and was tested for the adverse environments of outer space before delivery (Lyon and Patterson, 1969). The same unit is being flown in the RB57 aircraft up to 65,000 ft, in unpressurized conditions. The filter wheel has two 6.5–13 μm (1540–770 cm^{-1}) segments (in the 0°–180°, and the 181°–359° positions) taking 6 spectra per second with a Hg-doped germanium detector system. Resolution is about 1.5% with a field of view 0.4° (7 mrad), that is, 7 m at 1-km distance.

A wideband-pass radiometer (10–12 μm; 1000–850 cm^{-1}) and a 35-mm instrumentation camera with a 6-inch (153 mm) lens are all boresighted with the spectrometer (to several minutes of arc) so that the visual location of the IR data may be made (Lyon and Patterson, 1969). Our latest use of this system has been in the RB57 aircraft (Mission 248) flown August 11, 1973, at the time of the SKYLAB SL-3 overflight along the same target line. The RB57 was at 63,000 ft (19.2 km) while SKYLAB S-191 spectrometer was at 200 nautical miles (370 km) above Walker Lake, Nevada. In each unit boresight film, showing the line of sight for the IR units, were exposed to secure the ground locations for the spectra.

Another CVF unit, with which we have had ground field experience, is the Exotech Model 10 Spectrometer, differing only slightly from the others in design. The optical system uses a narrow beam (1° field of view) Pfund collector with the Hg–Cd–Te detector requiring only liquid nitrogen (instead of liquid helium) for a coolant—a major breakthrough for field logistics. The CVF wheel also has a 1.5% resolution, is single-sided (0°–180° only) with a variable spectral period from 1 sec to about 60 sec with correspondingly matched electronic bandpasses. The time-equivalent of the blank half of the wheel is used for digital recording of housekeeping data—temperatures, times, etc., the sequencing of events being controlled by the digital encoder on the CVF (Lyon and Green, 1973). As in most radiometer-type instruments, the Exotech 10 has a mechanical chopper to alternate the infrared radiance from the target and from an internal blackbody reference at a known temperature, onto the detector. The demodulated detector signal is then proportional to the difference between the target and the reference radiances. A parallel analog-recording system, using an X–Y recorder, with subsequent hand A–D conversion (using a chart digitizer) has proved a useful field method in providing instantaneous chart output for observation of the spectra during the experiment.

C. Acquisition of Field Spectra

One of the most attractive possibilities for this *reststrahlen* technique lies in modal analysis (mineral volume percentages) of rock targets. The targets for observation should be sawed (or smooth) slabs, of suitable areal extent so as to contain several hundred crystal grains, thus avoiding sampling errors. In addition, if the rocks show preferred orientations of the crystals, two or more slabs cut normal to, and parallel with, the *a*, *b*, and *c* fabric directions should be used.

The specimens may be studied in the laboratory, using heating mounts or placing the slabs in an oven to receive an even temperature which should remain relatively constant over the 10–60 sec required for a spectral scan.

Our mode of observation for specimens is to find a convenient location in the open under sunlit, clear sky conditions, usually on the patio in front of our building, or, if in the field, at some central base location. Using spectral periods of 30 sec, for about 5 consecutive spectra, will allow a large (30–50) number of samples to be viewed in a working day. The specimens are placed horizontally in a shallow box or tray lined with aluminum foil (low emissivity, high sky reflectivity) at about 15–30 ft (0.46–9.2 m) from the spectrometer, whose horizontal, 1°-wide, line of sight is bent downward on to the target by a 45° front-surfaced mirror.

D. *Data Analysis*

The voltage outputs of these types of spectrometers are proportional to the difference between the radiance reaching the spectrometer from the target (L_{sp}) and the radiance of the internal blackbody reference (L_{ref}).

$$V = At(L_{sp} - L_{ref}) \qquad \text{(for any given } \lambda) \tag{3}$$

The proportionality factors A and t can be regarded as an electronic gain term and an optical transmission term for the spectrometer system, respectively. In more exact work corrections should be made for the energy emitted from the chopper blade and the different optical paths followed by the reference and target signals.

The value of At must be determined by calibration against blackbody targets of known temperature. Generally the calibration temperatures range should at least span the expected range of target temperatures in order to determine any nonlinearity in the system. The instrument must also be calibrated so that each angular position of the CVF can be correlated with its correct wavelength. This is usually accomplished by observing a blackbody through a thin polystyrene sheet which has a well-known absorption spectrum.

In order to be further processed the data (volts) must be recorded in digital form. The wavelength span of the CVF is divided into small intervals during which the voltage is sampled and recorded (100–300 points/spectrum of 30 or 60 sec period). At this stage averaging and digital least squares smoothing techniques can be implemented (Savitzky and Golay, 1964) to remove any remaining high-frequency noise components. Using the values of At determined at each sampling point (λ_i) it is then possible to calculate each $L_{sp}(\lambda_i)$.

If we are close enough to the target and in a wavelength region of low atmospheric absorption, then $L_{sp}(\lambda_i)$ can be considered to be the radiance emitted by the target $L_t(\lambda_i)$. Its magnitude will depend upon both its emittance and the target temperature T.

$$L_t(\lambda_i) = \epsilon(\lambda_i)L_{BB}(\lambda_i, T) + [1 - \epsilon(\lambda_i)]L_{Sky}(\lambda_i) \tag{4}$$

Where L_{BB} represents the radiance calculated with the Planck function and L_{Sky} is the radiance incident on the sample from the background.

As the chemical information of interest is contained in the emittance $\epsilon(\lambda)$ curve we need to know the rock temperature and the incident sky radiance. The latter can be easily measured with the spectrometer but the value of T is very difficult to determine. If we assume that $\epsilon(\lambda_i) = 1$ at one of the measured points we can calculate a curve which will approximate the true $\epsilon(\lambda)$ curve. This can be done by selecting the value of $L_{\text{sp}}(\lambda_i)$ which corresponds to the highest blackbody temperature and taking this as the target temperature T then substituting into the above equation and solving for $\epsilon(\lambda_i)$ at each sampling point. The curve generated will not be the true $\epsilon(\lambda)$ curve, but if $L_{BB} \gg L_{\text{Sky}}$ and $\epsilon \sim 0.9$ then the only real error will be a small slew in the whole curve because of a low estimate for the temperature.

E. Examples of Field Emittance Spectra of Natural Materials

Several sets of emittance spectra generated in the field in the manner described above have been selected for brief discussion in this chapter, with a more thorough mineralogically oriented coverage reserved for a later paper. These curves illustrate typical spectra for freshly broken rock samples or crushed mineral sands. Although the materials now do not represent the surface as seen by a remote sensor we can observe the information content potentially present before the effects of environmental degradation. However the spectra do represent what can be achieved in short-path length observations, by using a technique readily adaptable to observation of specimens in most field locations.

1. Nonsilicates

Figure 1 shows the emittance spectra taken in the field of specimens of a sulfate, gypsum (A), and three carbonates, calcite (B), dolomite (C), and malachite (D). Their absorption data are as follows:

Material (refer to Fig. 1)	*Peak absorptions* (major italicized)	
Sulfate		
Gypsum crystal (A)	*8.70* μm	(1149 cm^{-1})
Carbonate		
Calcite crystal (B)	*11.40* μm	(877 cm^{-1})
Dolomite (C)	*11.20* μm	(893 cm^{-1})
Malachite (D)	9.70, *12.10* μm	(1031; 826 cm^{-1})

Fig. 1 clearly shows the difference in peak position for the sulfate and carbonate anion groups, as well as the shift to longer wavelengths between the calcium, calcium–magnesium, and copper carbonates with the increasing atomic weight of the cation. Such data have been available in many papers before, but not in the form of emittance spectra, taken under field conditions (Hunt *et. al.*, 1950).

For purposes of comparison the curves of all seven figures (Figs. 1–7) have been displaced vertically, but their absolute values may be compared by the following method: at the vertical arrow on each spectrum ϵ_λ = 1.00, and at the horizontal arrow ϵ_λ = 0.90. Spectral information was plotted every 0.1 μm (10 cm^{-1} at 1000 cm^{-1}) and this coarse grid now results in some lack of precision in peak positions.

2. Silicate Mineral Specimens

Figure 2 shows the emittance spectra for several silicates either in solid crystalline masses or as sand (garnet and actinolite).

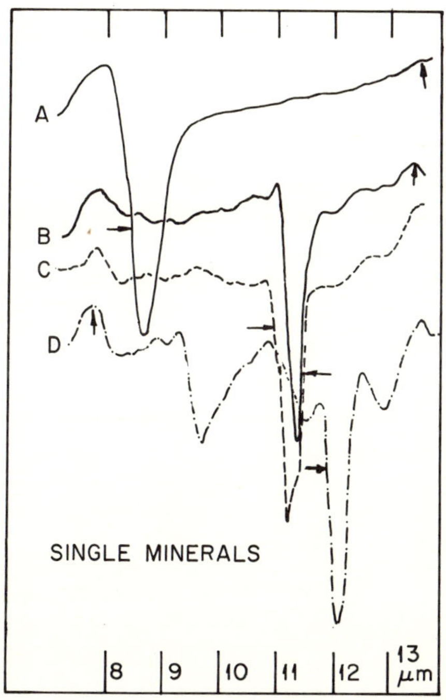

FIG. 1. Field emittance spectra of samples of single, nonsilicate minerals. All spectra were prepared on the same scale and displaced vertically for ease of viewing. Vertical arrows indicate where ϵ_λ is assumed to be 1.0, horizontal arrows where ϵ_λ = 0.90. Gypsum (A), calcite (B), dolomite (C), and malachite (D) curves are shown. Wavelengths range from 7.20 to 13.50 μm (1380–741 μm^{-1}), as shown on lower scale figures.

Materials (refer to Fig. 2)	*Peak position* (major italicized)
Tektosilicates	
Quartz (A)	8.30, *9.20*, 12.75 μm (1205, 1087, 784 cm^{-1})
Microline-perthite (B)	8.70, *9.90* μm (1149, 1010 cm^{-1})
Nesosilicates	
Garnet (grossularite) (C)	*10.40*, 11.60 μm (962, 861 cm^{-1})
Inosilicates	
Actinolite (D)	8.80, 9.40, *10.10* μm (1136, 1064, 990 cm^{-1})
Cyclosilicates	
Beryl (E)	8.20, 8.90, *10.30*, 12.20 μm (1220, 1124, 971, 820 cm^{-1})

Similarities in the structural groups can be seen even in these emittance spectra. Their absorption data are given in the above tabulation.

3. *Coarse-Grained Granites*

Figure 3 illustrates a group of similar, coarse-grained granites from some of the classic localities of New England, including several color variants of each. No effect of the color differences can be seen, but the bulk mineralogical (modal) differences between the quartz-rich (Conway) and feldspar-rich (Quam) types are observable. Their absorption data are given in the following tabulation.

Type (refer to Fig. 3)	*Peak positions* (major italicized)[a]
Quartz-rich	
Conway, gray (F6-91) (D)	8.40, *9.30*, 12.80 μm (1190, 1075, 781 cm^{-1})
Conway, tan-pink (F6-80) (E)	8.40, *9.25*, 12.80 μm (1190, 1081, 781 cm^{-1})
Barre, dark gray, (F-105) (G)	8.45, *9.35*, 12.90 μm (1183, 1070, 775 cm^{-1})
Westerley, blue (F6-28) (C) very fine grained	8.55, *9.30*, *9.80*, 12.90 μm (1170, 1075, 1020, 775 cm^{-1})

Continued

Type (refer to Fig. 3)	*Peak positions* (major italicized)[a]
Westerley, gray (F6-34) (B)	8.40, 9.35, *9.80*, 12.85 μm (1190, 1070, 1020, 778 cm^{-1})
Westerley, pink (F6-26) (A)	8.50, *9.30*, 9.80, 12.80 μm (1176, 1075, 1020, 781 cm^{-1})
Feldspar-rich	
Quam, dark gray (F6-112) (F) fine grained	8.60B,[a] 9.30, *9.70*, 12.90 μm (1163, 1075, 1031, 775 cm^{-1})

[a] B, Broad peak.

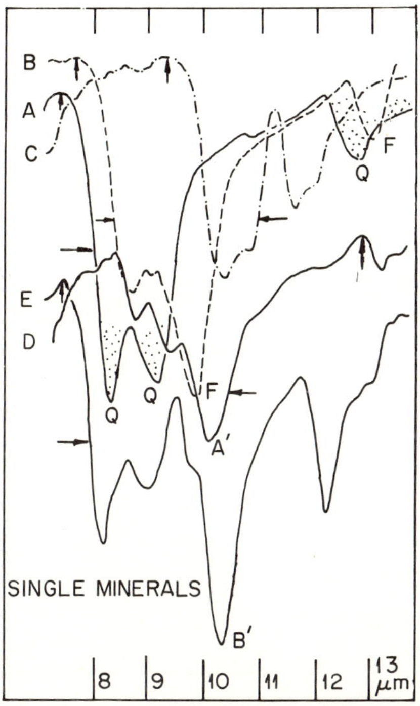

FIG. 2. Field emittance spectra for single mineral silicate samples. All curves prepared on same scale and displaced vertically. Note arrows for emittance scale. Quartz (A) is also indicated by (Q) at the peak absorbancy. Microcline–perthite feldspar (B) is indicated at peak values by (F). Garnet (grossularite) (C) sand spectrum also appears on Fig. 6 indicated by (G″). Actinolite (amphibole) (D) and a large beryl crystal (E) show markedly different emittance spectra. Peak for actinolite is shown as (A).

4. *Sierran Crystalline Rocks*

Figure 4 covers emittance spectra of a group of six crystalline rock suites from Sonora Pass in the Sierra Nevada, California. The Brown Bear basalts (A and B) differ markedly from the others in the absence of quartz, and having a higher plagioclase content. The porphyritic quartz monzonites (F, G, and H) show more equal peaks for quartz and feldspar than the quartz monzonites (C, D, and E) which resembles the granodiorite from Cinko Lake much more closely. Their absorption data are summarized in the following tabulation.

Type (refer to Fig. 4)	Modal analysis[a]				Major peak positions (major italicized)
	Q	*K-Spar*	*Plag*	*Femic*	
Basalt, Brown Bear (A, B) NASA #621	—	5	70	5	*9.70*, 12.90 μm 1031, 775 cm^{-1}
Quartz monzonite, porphyritic					
Topaz Lake (F) NASA #331	25	30	30	4	*8.40*, *9.15*, 12.75 μm 1190, 1095, 784 cm^{-1}
Cathedral Peak (G) fresh	30	30	30	5	*8.45*, *9.25*, 12.95 μm 1183, 1081, 772 cm^{-1}
Mill Creek (H) NASA #308	30	30	20	15	8.40, *9.25*, 12.80 μm 1190, 1081, 781 cm^{-1}
Crow Springs (C) Q-8	25	30	30	10-R	8.80, *9.30*, 12.80 μm 1136, 1075, 781 cm^{-1}
—(D) Q-18	25	30	35	5-S	8.50, *9.30*, 9.70, 12.90 μm 1176, 1075, 1031, 775 cm^{-1}
—(E) Q-18	25	30	35	5-R	8.40, *9.40*, 12.80 μm 1190, 1064, 781 cm^{-1}
Granodiorite,					
Cinko Lake (I)	30	—	20	45-P	8.50, 9.40, *9.90*, 12.90 μm 1176, 1064, 1010, 775 cm^{-1}

[a] Rocks collected and modal analyses made by J. Quade, University of Nevada. R. rough; S, sawed; P, polished.

5. *Bushveldt Ore Complex Suite*

Examples are shown in Fig. 5 from a suite of specimens from the Bushveldt complex. All specimens were fresh broken surfaces of 2–5 small

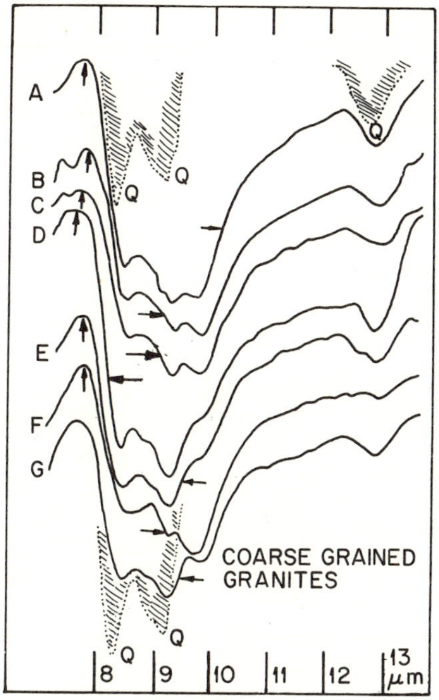

FIG. 3. Field emittance spectra from samples of 7 New England granites of note are shown, vertically displaced. Arrows have same meaning as in Fig. 1. Although these granites are well known for their general similarity, subtle mode-related differences in their mineralogies may be noted in their spectra. Q is quartz and shaded for comparison. Conway (gray) and Conway (tan–pink) are D and E, the most quartz-rich. Barre, is dark gray (G). Westerley blue (C), grey (B) and pink (A) are all very similar to each other, but differ from Quam, dark gray (F).

specimens, 1–3 inches in size, from the Stanford Ore Deposits Collection (OD#). Their absorption data are given in the following tabulation.

Type (refer to Fig. 5)	*OD #*		*Peak position* (major italicized)[a]
Spotted anorthosite	(A)	45461	8.80, *10.65* μm 1136, 939 cm^{-1}
Anorthosite	(B)	45525	8.65, *9.90* μm 1156, 1010 cm^{-1}
Kimberlite, brown (Dutoitspan)	(C)	45720	*9.90* μm 1010 cm^{-1}

Continued

Type (refer to Fig. 5)	*OD #*	*Peak position* (major italicized)[a]
Dunite	(D) 45574	*9.50* μm 1053 cm^{-1}
Pyroxenite	(E) 45453	8.40, *9.25* μm 1190, 1081 cm^{-1}
Pyroxenite and plagioclase	(F) 45551	9.15, *10.40,* 11.50 μm 1093, 962, 870 cm^{-1}
Pyroxenite and olivine	(G) 45611	*9.30, 10.40,* 11.45 μm 1075, 962, 873 cm^{-1}
Spotted norite	(H) 45554	9.80, *10.40*B μm 1020, 962 cm^{-1}
Migmatite	(I) 45589	8.40, 9.80, *10.35* μm 1190, 1020, 986 cm^{-1}
Hornfels	(J) 45571	8.40, *9.20,* 12.90 μm 1190, 1087, 775 cm^{-1}
Granite dike	(K) 45992	*8.40,* 8.90B, 12.80 μm 1190, 1124, 781 cm^{-1}

[a] B, broad peak.

6. Porphyry Copper Suites

Figure 6 shows emittance spectra from the Mission and Twin Butte mines (porphyry and feldspathized argillite) and specimens from the tactite ore with garnet concentrations (G′) and some residual limestone (C′). Portions of the spectra for pure quartz (Q) and grossular garnet sand (G″) are shaded in the upper parts of the figures. Their absorption data are summarized in the following tabulation.

Type (refer to Fig. 6)	*Peak positions* (major italicized)
Mission	
Hornfels (A)	8.40, *9.30,* 11.20 μm 1190, 1075, 893 cm^{-1}
Mission	
Tactite (B and C)	8.30, 9.20, *11.00,* 11.95 μm 1205, 1087, 909, 837 cm^{-1}
Twin Buttes	
Limestone ore with garnet (E, F, and G)	10.10, *11.30*B, 12.20, 13.00 μm 990, 885, 820, 769 cm^{-1}

Figure 7 illustrates the emittance spectra for several other mineralized porphyritic rocks from Bingham pit, Utah, the porphyry from underground at Magma [still retaining some carbonate wall rock (C′)] and the diabase from Ray, Arizona pit. For comparison a polished surface of blue gabbro and a rough specimen of anorthosite from Lake Superior are also included. The absorption data are given in the following tabulation.

Type (refer to Fig. 7)	*Peak positions* (major italicized)
Bingham, Utah	
Pyritic alterations (A)	8.7, *9.30* μm 1149, 1075 cm^{-1}
Actinolite syenite (B)	8.30, *9.20*, 12.75 μm 1205, 1087, 784 cm^{-1}
Magma, Arizona	
Porphyry from 2800 ft. level (C)	8.40, *9.50*, 11.30 (calcite) μm 1190, 1053, 885 cm^{-1}
Ray, Arizona	
Diabase (D) (Pearl Handle pit)	8.50, *9.30*, 9.85 μm 1176, 1075, 1015 cm^{-1}
Duluth, Minnesota	
Blue gabbro (E) (plag, An_{60}, augite, biotite)	9.10, *10.50* μm 1099, 952 cm^{-1}
Ajo, Arizona	
Quartz monzonite Cornelia (F)	8.40, *9.25*, 9.80, 12.80 μm 1190, 1081, 1020, 781 cm^{-1}
Pancake Island, Lake Superior	
Anorthosite (P14-13) (G)	8.80, *10.80*, 13.00 μm 1136, 926, 769 cm^{-1}

F. Factors Causing Degradation of the Spectra

The spectral information illustrated above was obtained from fresh rock surfaces in close proximity to the spectrometer. Many factors combine to degrade the information content when *in situ* targets are viewed from remote platforms. We shall consider some of them in the following sections.

1. Vegetation Cover

This is the primary factor limiting the usefulness of any geological exploration program using spectral techniques with the wavelengths between

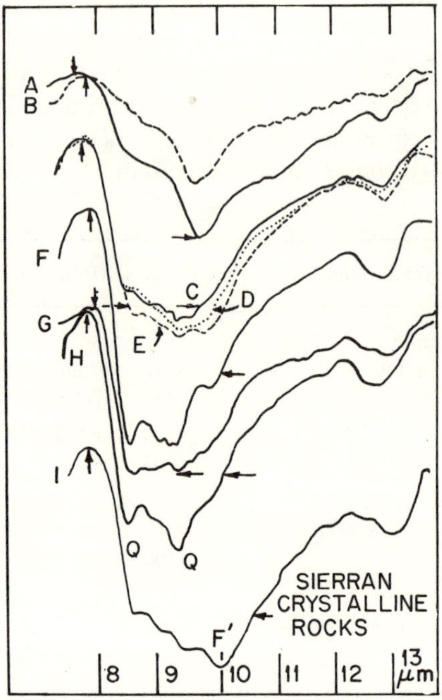

FIG. 4. Field emittance spectra from samples of 9 Sierran crystalline rocks, ranging from quartz monzonite to basalt. The two Brown Bear basalt spectra (A and B) contain no quartz (Q) and hence differ markedly from the rest. The seven quartz monzonites are similar but noticeably different from the granodiorite in its ferromagnesian minerals and in its feldspar (F′) content around 10.0 μm. Modal analyses by J. Quade, University of Nevada, are shown in the main text.

0.3 and 1000 μm (33000–10 cm^{-1}). Any vegetation from lichen cover to tropical rain forest will mask the information in rocks.

In the emissive region ($\lambda > 3.5$ μm; 2850 cm^{-1}) most vegetation is a blackbody radiator generally unrelated spectrally to the soil and rock composition. Thus in areas of high vegetative cover the percentage of the signal which contains geological information is considerably lowered. However, any deviation from blackbody behavior can still be ascribed to the *reststrahlen* effect of that part of the surface which is exposed. If the percentage ground cover and the mean vegetation temperature can be estimated (say from aerial photographs and air temperature data) then it should be possible to correct the data to get $\epsilon(\lambda)$ curves.

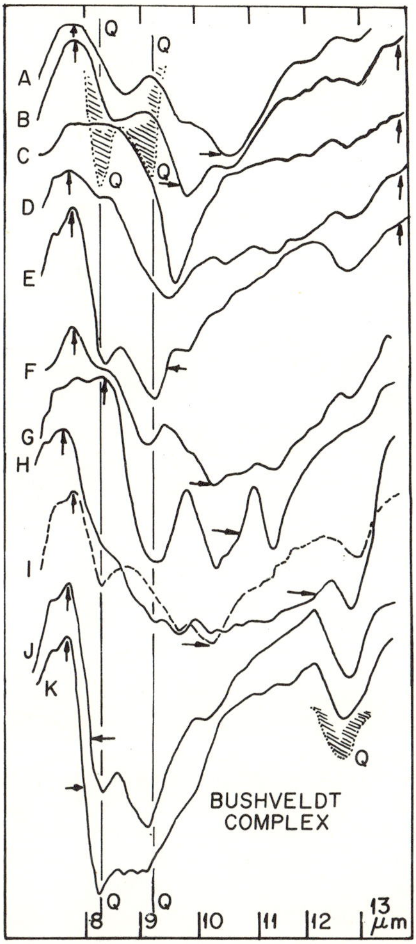

FIG. 5. Field emittance spectra from 11 samples selected from the Bushveldt Complex collection at Stanford University are shown again vertically displaced. Arrows have same meaning as in Fig. 1. Quartz (Q) peaks are shaded for reference. Two anorthosites (A and B) contrast strongly with the brown kimberlite (C) from Dutoitspan pipe and the dunite (D). The pyroxenite (E) spectrum shows a high quartz content (?). The pyroxenite (F) which contains plagioclase and the pyroxenite (G) with considerable olivine produce different spectra. The spotted norite (H) and migmatite (I) show broadly similar spectra which differ from all the rest of the suite. The hornfels (J) and granite dike (K) spectra are again typical of quartz–feldspar-rich felsic rocks.

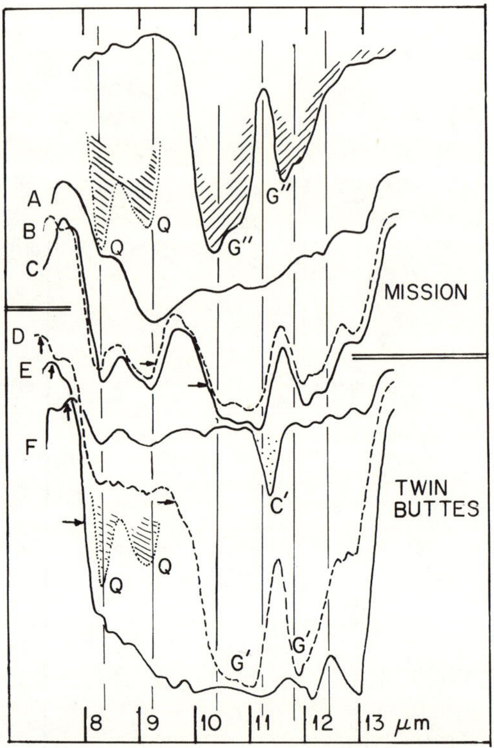

FIG. 6. Field emittance spectra from specimens of mineralized "porphyry coppers" from Tuscon, Arizona area are shown vertically displaced (with arrows). The feldspathized argillite (A) contrasts with the ore tactites (B and C). These show quartz (Q) and garnet (G′) introduced into the remanent limestone host (C′) for both ore bodies—Mission and Twin Buttes. Both ores contain a similar garnet, which differs in peak position from pure grossularite (G″).

2. Terrain Variability

As a spectrometer system takes a finite time to scan through a spectrum, the forward motion of the vehicle will smear each spectrum out over a considerable area on the ground, and a close correlation must be maintained between the boresight camera data and the spectrometer data. In the case of the NASA spectrometer system in the NP3 aircraft this smear is approximately 40 ft (12.2 m) long (Lyon and Marshall, 1971). If the surface composition changes in this distance the spectrum will be unreliable. For this reason we usually show the standard deviation (σ_λ) of the spectra as well as the mean spectrum. A high σ everywhere in the spectrum occurs when the target temperature is variable, but a high σ value only in the *reststrahlen* wavelengths indicates a variable surface composition (see Figs. 8–10).

Even when the rock surface is well exposed the structure of the surface layer determines the amount of information which can be extracted from emission measurements. As discussed in Chapters 5 and 6, as the particle size decreases the amount of contrast in the *reststrahlen* emission minimum decreases toward that of blackbody behavior (Aronson and Emslie, 1973).

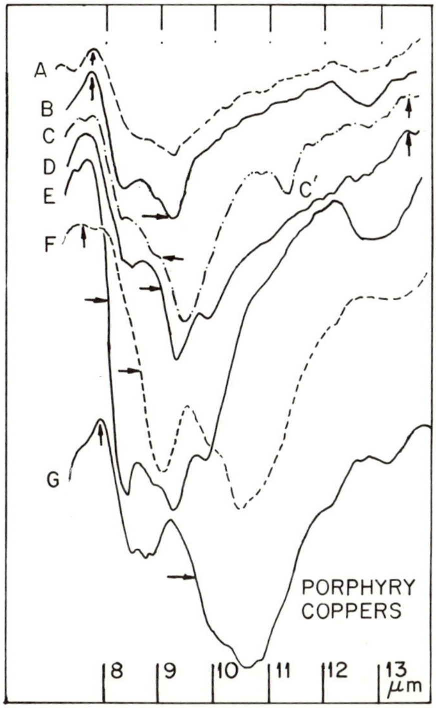

FIG. 7. Field emittance spectra from specimens of "porphyry coppers" are shown in this figure, vertically displaced, again with arrows for scale. Two rock types from Bingham, Utah mine (A and B) principally show the presence of their quartz content. The feldspar-rich porphyry from underground at Magma, Arizona (C) also shows a carbonate (C′) content from the enclosing limestones. The mineralized diabase from the Pearl Handle pit (D) now shows some quartz content and contrasts with the unmineralized blue gabbro from Duluth, Minnesota. The gabbro (E) resembles the anorthosite (G) from Lake Superior [c/f spectrum (A) in Fig. 5, Bushveldt suite]. The quartz monzonite from Ajo, Arizona (Cornelia, N slope of Campbell back, Mt. Ajo) (F) is very typical of similar feldspar and quartz rich rocks (from Figs. 3 and 4).

3. Atmospheric Attenuation and Emission

If we consider a nonscattering atmosphere between the surface and the spectrometer, the radiance reaching the spectrometer at any given wavelength will be given by

$$L_{\text{sp}}(\lambda) = L_t(\lambda)\tau_a(\lambda) + L_a(\lambda, T_a) \tag{5}$$

where $L_t(\lambda)$ is the radiance leaving the target, and L_a is the radiance emitted by the atmospheric path of transmission τ_a.

In general L_a and τ_a are very complex functions of the atmospheric temperature profile and the atmospheric water vapor distribution (Haughton and Smith, 1970).

If a detailed knowledge of the atmospheric profile is available these two parameters should be calculable. However, it is generally easier to calculate them by obtaining the temperature of two known blackbody targets on the ground, then solving two simultaneous equations. When no ground temperature data is available a similar approximation to that made in calculating the emittance curves can be made. Thus the atmosphere is assumed to have $\tau_a = 1$ in one wavelength interval and an approximate surface temperature obtained assuming $L_{\text{sp}} = L_t$. In this way most of the spectral fine structure superimposed by the gaseous emission and absorption processes can be determined.

When only one known blackbody target is available we must make the approximation,

$$L_{\text{sp}} = L_t(\lambda)\tau_a(\lambda) + (1 - \tau_a)L_{BB}(\lambda, T_a) \tag{6}$$

and determine T_a from a wavelength value where it is expected that $\tau_a = 0$. The above equation can then be solved for $L_t(\lambda)$ by assuming $\tau_a(\lambda) = 1$ at some other suitable wavelength.

For short paths 1000–10000 ft (300 m–3 km), we have successfully calibrated-out the atmospheric effects (or at least minimized them) by initially flying spectra over a body of water, using the fact that in the 8–14 μm (1250–715 cm^{-1}) region water behaves almost as a blackbody (i.e., nonspectral) emitter, ($\epsilon = 0.98$). One must then presume that within the area being studied this atmospheric correction remains constant, a fact which we check as often as practical on a mission by reflying any water body.

Figures 8–10 show the emittance spectra of three different rock units obtained flying over Pisgah Crater in Southern California. These rock units were selected on the basis of the known geology of the area by using the data from the boresight camera to split the aircraft flight lines into seg-

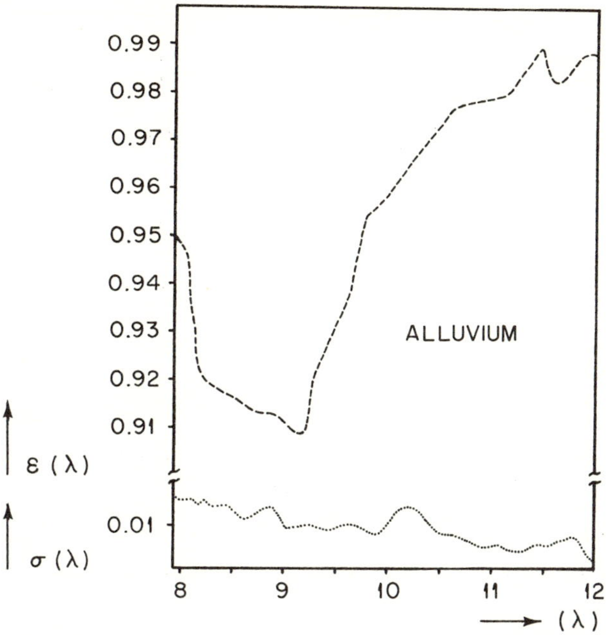

FIG. 8. Airborne emittance spectra from Alluvium-A terrain, Pisgah Crater, California (NASA flight MX108). Top curve is mean spectral emittance, the lower the standard deviation of each measurement, $N = 27$ samples, or from approximately ($27 \times 40 \times 2 = 2160$ ft; 658 m) of flight line, over the same rock type. Minimum emittance is at 9.13 μm (1095 cm^{-1}). Area is No. 1 of Fig. 11. Note the low standard deviations.

ments. The spectra from each segment were averaged and then processed in the manner described above.

These figures are instructive in showing two types of information: (*a*) that their wavelengths of minimum emittance are different and vary in proportion with the ratio of felsic and ferric minerals in the rocks (roughly equivalent to SiO_2 percentage, as used by Vincent and Thomson, 1972a), and (*b*) that over the alluvium and basaltic lava areas the standard deviations of the emittance data are evenly low, while over the dry lake sediments a high standard deviation in the *reststrahlen* area indicates changing surface compositions (and probably particle size effects as well).

Significantly these characteristics are also brought out in the emittance-ratio imagery flown by Vincent over the same flight area (Fig. 11). In that figure Vincent used the ratio,

$$R = \frac{\text{brightness of 8.2–10.9 } \mu\text{m image (A)}}{\text{brightness of 9.1–12.1 } \mu\text{m image (B)}} \tag{7}$$

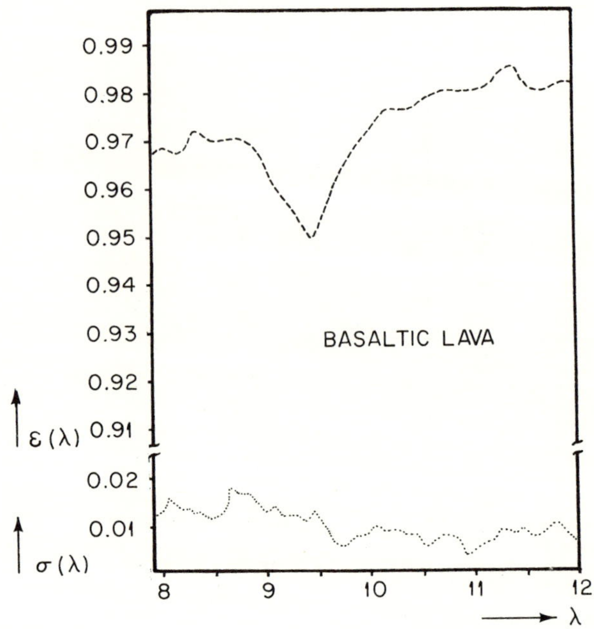

FIG. 9. Airborne emittance spectra from older, Type II, olivine basalt lava (lava flowage, II-C) terrain, Pisgah Crater, California (MX108), $N = 10$ samples, or from 800 ft (243 m) of flight line. Area is No. 5 of Fig. 11. Note low standard deviation. Lowest emittance is at 9.53 μm (1049 cm^{-1}), closely resembling that of the dry lake sediments (clays) of Fig. 10. This similarity has been observed and checked in many of our spectral observations of these two materials.

to produce a new image (A/B) designed to emphasize spectral emittance differences. A low R ratio is shown as white on this image and is indicative of felsic materials (high silica). A high R-ratio is shown as black characterising high femic mineral contents (low silica). Intermediate grays have intermediate felsic/femic mineral ratios. The images cover a flight length of about 2.3 miles and show spectral information for about 1 square mile of terrain.

In the Crater Area (top 3 images of Fig. 11) Areas 1 and 3 are windblown, quartz-rich sand, and Area 2 is a mixed-rock, older alluvium; Area 1 spectrum is shown in Fig. 8. Areas 4 (and 5 in the Lake Area image) are the olivine basalt flows whose spectrum (for Area 5) is shown in Fig. 9. In the Lake Area ratio image (for which original images A and B are not shown) the dry lake sediments occupy Areas 6, 7, 8, and 9 showing by their variance in gray tone their variability in felsic/femic contents. Figure 10 shows the spectrum of those sediments. Area 10 is a repeat occurrence of the older alluvium like that of Area 2, but appears richer in felsic content.

Thus two separate methods for indicating compositional variability by emittance measurements agree very closely in their results. The point can be simply made, however, that the ratioed imagery is by far the easier for the average viewer to interpret.

Two other experimental confirmations of this concept were conducted. Using the SKYLAB S-191 spectrometer, from its altitude of 270 nautical miles (500 km) we have observed lakes in Navada and California on the SL-2 and SL-3 missions. The astronaut, using the gimballed mounting for the telescope, acquired the lake, at some forward angle of view ($< 45°$), and tracked it until the spacecraft passes near-vertically over the lake,

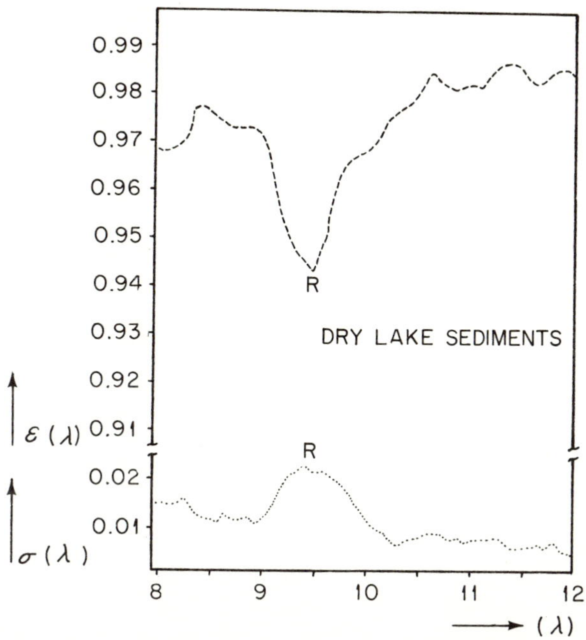

FIG. 10. Airborne emittance spectra from the Dry Lake Sediments (a) of Lavic Lake, Pisgah Crater area, California (MX 108), N = 53, or from 4240 ft (1292 m) of flight line. Area is from a second flight line, at right angles to the lower (a/b) image of Fig. 11, roughly through area No. 7 but containing portions of Areas 6, 7, and 8. The standard deviation curve clearly shows a high chemical variability (Fig. 10, lower curve) and occurs at the *reststrahlen* point (R) for these clay–sand sediments. The same effect is seen as the gray-tone variability in Fig. 11 (lower image between Areas 6, 7, 8, and 9) which to the eye are all the same type of playa sediment. Thus the two methods check each other (Lyon, 1972; Vincent and Thomson, 1972a). Lowest emittance value is also at 9.53 μm (1049 cm^{-1}).

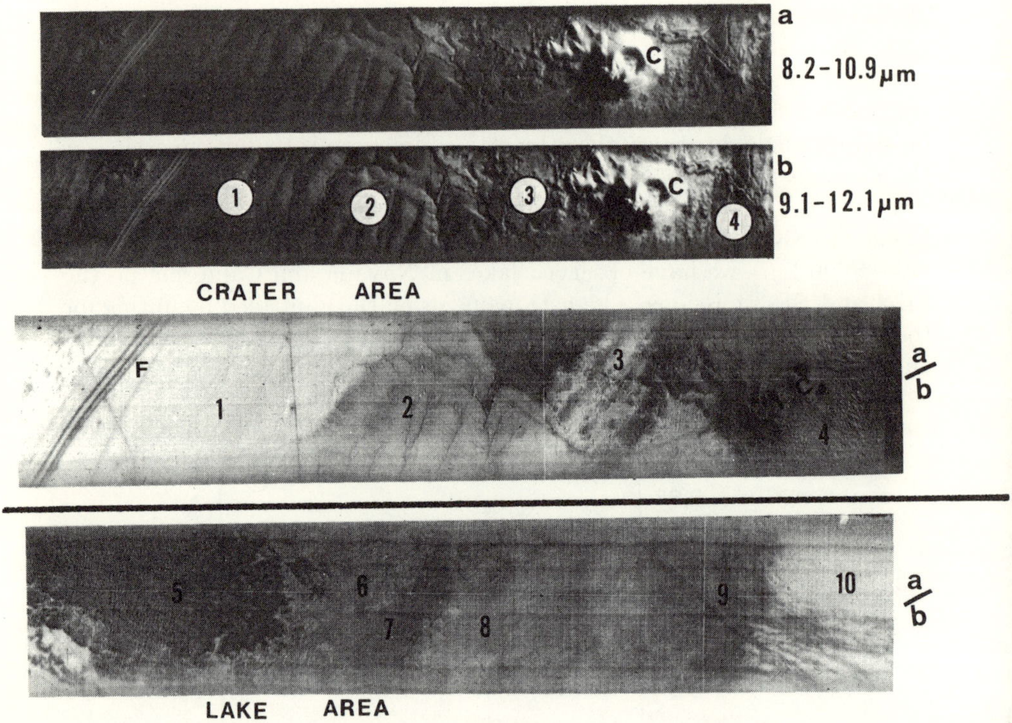

FIG. 11. Airborne spectral images, (a) in the 8.2–10.9 μm (1219–917 cm^{-1}) bandpass and (b) in the 9.1 to 12.1 μm (1099–286 cm^{-1}) bandpass, of Pisgah Crater (C), California area. Areas 1, 2, 3, and 4 (and C) appear on the top three images. The lower two images are "spectral ratio" images (Vincent and Thomson, 1972a) wherein the ratio of channels (a) and (b) is displayed to emphasize their chemical information (spectral emittance or "brightness temperature") contents (see Section III, F). The lowermost image ratio (a/b) is for the adjoining Lavic Lake area.

thus giving us several spectra [1/sec, spacecraft motion, 4 miles/sec (7.4 km/sec)] with decreasing effective air paths. Then the ground targets nearby were acquired and held in the 1 mrad field of view (0.04 nautical miles^2; 0.14 km^2) for as long as they remained in the field of view, usually above 1–10–sec.

At approximately the same time that day the Lockheed Airborne Rapid Scan Spectrometer was flown at 63,000 ft (10 nautical miles; 19.2 km) along the same line in a high altitude aircraft (RB57) taking spectra (6/sec) of the terrain traverses. Radiant temperatures of the lakes were monitored using a PRT-5 unit in a light plane, and checked by similar temperatures taken from a boat.

SKYLAB data have not yet been received from NASA so their interpretation is premature. It is hoped that from these experiments bounds may be established for the path lengths over which the *reststrahlen* method for compositional analysis may be used.

The above discussion has involved the analysis of spectrometer data taken in many wavelength intervals. However, it can be seen from the shape of typical $\epsilon(\lambda)$ curves that high resolution is not needed if we only wish to characterize the wavelength position of the SiO_4 emission minimum. The relative signals between two channels centered at 9 and 10 μm (1110–1000 cm^{-1}) would indicate whether the emission minimum had shifted to longer or shorter wavelengths. These channels can be made wide enough to have the very fast response times and good signal-to-noise characteristics required for line-scan operation. The two images produced (one from each detector) can then be compared and the relative magnitudes of the signal in each channel then give an indication of the position of the emittance minimum. Several methods of redisplaying the image to emphasize these emissivity differences have been used. W. Hovis (private communication, 1972) displays the difference in apparent temperatures indicated between the two channels while Vincent *et al* (1972) display the ratio of the voltage outputs of the two detectors.

Both these methods produce parameters that are proportional to the wavelength position of the emission minimum. However, they are not completely independent of the temperature of the target. Another parameter can be obtained by taking the ratio of the signal from one channel to the signal that same channel would have if observing a blackbody at the apparent temperature indicated by the second channel. This last parameter does not show the same dependence on the target temperature (Lyon and Green, 1973).

Two figures show the results of these techniques. Figure 11 (previously described) shows the images (A and B) formed by two channels of the University of Michigan scanner flying over the Pisgah Crater test site in Southern California. Figure 11 (lower half) shows the ratio image (A/B) of the same area emphasizing the quartz-rich felsic materials (white) and the more femic basaltic flows (black). The lowermost ratio (A/B) image is further south along the flight line showing the end of the olivine basalts (5, black) and the varying composition of the dry lake sediments (6, 7, 8, and 9) and alluvium (light gray, 10).

Figure 12 shows NIMBUS 5 imagery from the Surface Composition Mapping Radiometer (SCMR) whose two channels were specifically

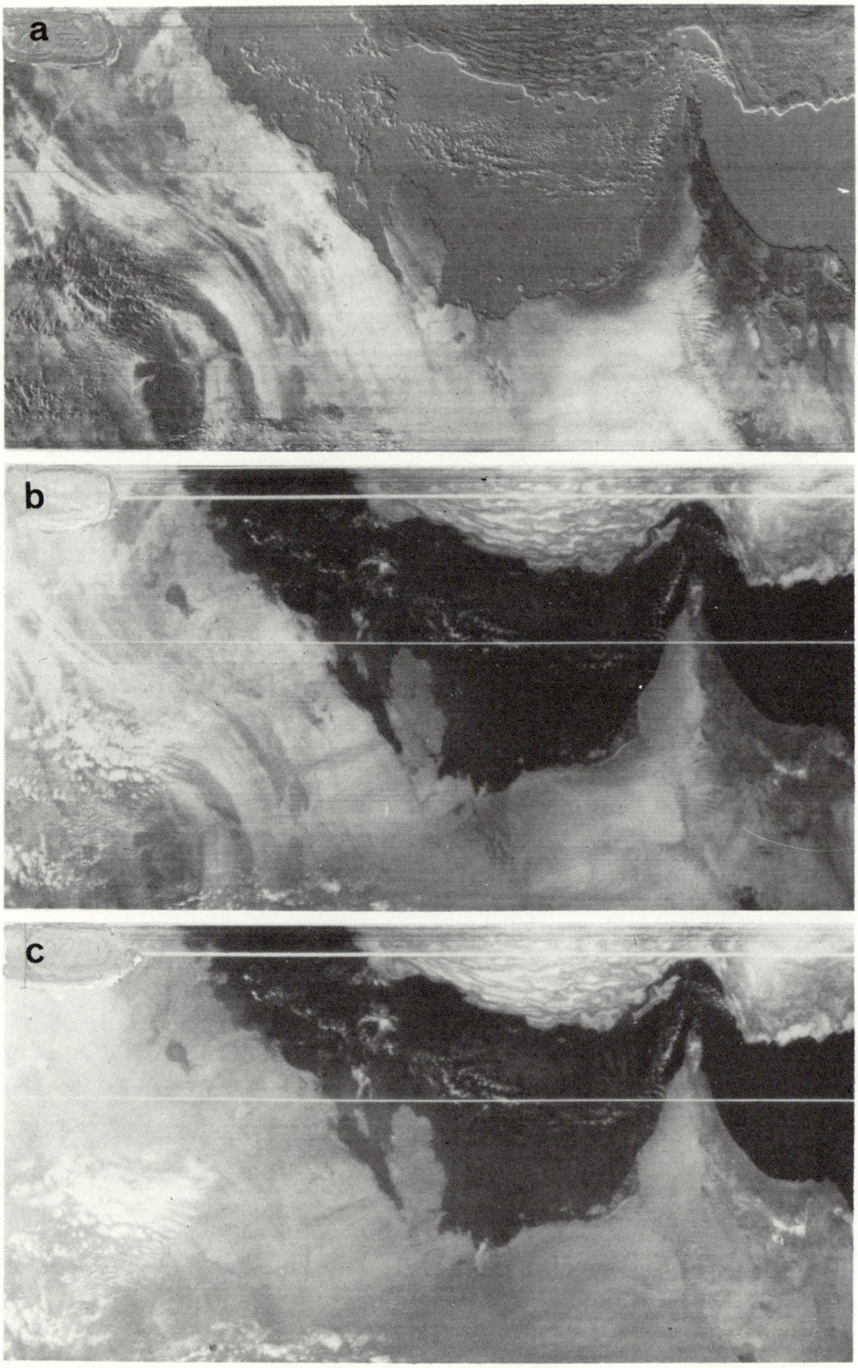
a
b
c

designed to observe *reststrahlen* effects in terrestrial materials from space. The channels are centered at 8.8 μm (1136 cm^{-1}) for Image B and 10.7 μm (935 cm^{-1}) for Image C. Silicic materials image colder (white) in band B and warmer (black) in band C. The difference image A (delta-mode) emphasizes the felsic materials while the femic materials appear darker. In addition, vegetation which appears equally warm in either band (blackbody emission) is medium gray. Carbonates also appear medium gray as their *reststrahlen* band is shorter in wavelength (7.0 μm, 1430 cn^{-1}) than band B and they show graybody behavior equally across both bands B and C.

Similar ratio images are expected from the 13-channel, spectrally imaging scanner in SKYLAB in the S-192 experiment, although for some inexplicable reason only one channel was chosen within the thermal band (8–14 μm; 1250–715 cm^{-1}).

IV. Conclusions

Reststrahlen methods are useful for the derminatioten of surficial materials even in the presence of varying path lengths of atmosphere. Measurements with targets only 5–20 ft (1.5–6.1 m) away minimize the atmospheric effects, and can be made easily and repeatably. Spectral details can be correlated with mineralogical characteristics in the targets, so that modal analyses of rough, sawed, or polished rock slabs under exterior, sunlit conditions make use of the method of considerable importance to geologists.

More remote observations, over path lengths of several hundred to thousand of feet, are always affected- by the atmosphere which causes varying repeatibility, i.e., higher noise in the spectral signatures used and lower success in deriving modal analysis. The low humidity of desert terrains and surfaces free of vegetation markedly assist the determin-

---

FIG. 12. NIMBUS *Spacecraft* imagery of the Persian Gulf (black) and surrounding land masses (gray). North is toward the top and the images cover about 900 × 400 miles (1450 × 650 km). Images were made using the two channels of the Surface Composition Mapping Radiometer (SCMR) which measure the thermal emission in 8.3–9.3 μm (1205–1075 cm^{-1}) (Image b), and in the 10.2–11.2 μm (980–893 cm^{-1}) Image c bands. Quartz-rich materials (sand, granites, etc.) appear colder (white) on Image b, and warmer (dark) on Image c. Image a represents the difference between the voltages of the two channels, emphasizing the quartz-rich areas as brighter tones. Vegetation, carbonates, and femic rocks appear darker in these "delta-mode" images. (from W. Hovis, private communication, 1972.)

ations. Trees and grass invariably show blackbody radiances, with brightness temperatures not far removed from that of the ambient air. One advantage of this blackbody character is that vegetation can be removed from spectra ratio imagery, by dropping out all ratios close to unity in the final image.

Spectral images, ratioed to produce a final single image, offer the most significant data to geologists, as their maplike format permits ready identification of anomalous spectral emittances keyed to identifiable land forms for ground location. Maps of silica percentages (Vincent and Thomson, 1972*a*, *b*) then may be field checked and image anomalies due to varying rock compositions rapidly located. Thin dikes and small outcrops of rock are not missed, if their unusual chemistry is flagged at an early stage in mapping. Such field correlation and checking is a vital part of the operation and anyone who believes the maps made remotely are the final product of the study is very naive.

Clearly, short air paths and higher spectral resolution aid in the classification of the surficial materials. Spacecraft units really are not suitable, because of their fast speed, high altitudes, and of course their high cost. Low- and slow- is the best operational mode for these observations, and light aircraft operative 1000–5000 ft (305 m–1.5 km) above the terrain make the best platforms for sensing.

In summary, the method is of use to the field geologist, provided the spectra are taken from small terrain elements. Preferably if they are reproduced in image (or image-ratio) format then this eases the ground location. Light aircraft rather than spacecraft would make the method usable for persons with relatively small budgets.

References

Aronson, J. R., and Emslie, A. G., (1973). *Appl. Opt.* **12,** 2573–2584.

Bell, E. E., and Eisner, I. C. (1956). *J. Opt. Soc. Amer.* **46,** 303–304.

Billingsley, F. C., and Goetz, A. F. H. (1973). *NASA Spec. Publ.* ***NASA SP*-327.**

Burns, E. A., and Lyon, R. J. P. (1963). *Proc. Int. Astronaut. Congr., 14th, 1963* Vol. 2, pp. 237–250.

Coblentz, W. W. (1905, 1906, 1908). *Carnegie Inst. Wash. Publ.* **35, 65, 97** (reproduced as a single volume under joint sponsorship of the Coblentz Society and Perkins-Elmer Corporation, 1962).

Haughton, J. T., and Smith, S. D. (1970). *Proc. Roy. Soc., Ser. A* **320,** 23.

Hunt, J. M., Wisherd, P., and Bonham, L. C. (1950). *Anal. Chem.* **22,** 1478–1497.

Launer, P. J. (1952). *Amer. Mineral.* **37**, 764–784.

Lyon, R. J. P. (1964). *NASA Contract. Rep.* ***NASA CR*-100,** 208.

Lyon, R. J. P. (1967). "Physical Determinative Methods in Mineralogy" (J. Zussman, Chapter 8, pp. 371–403. Academic Press, New York.

Lyon, R. J. P. (1972). *Science* **175**, 983–986.

Lyon, R. J. P., and Green, A. A. (1973). *Stanford Remote Sens. Lab., Tech. Rep.* **73–1**.

Lyon, R. J. P., and Marshall, A. A. (1971). *IEEE Trans. Geosci. Electron.* 131–138.

Lyon, R. J. P., and Patterson, J. (1966). *Proc. Symp. Remote Sens. Environ., 4th, 1966* pp. 215–230.

Lyon, R. J. P., and Patterson, J. (1969). *Proc. Int. Symp. Remote Sens. Environ., 6th, 1969* pp. 527–552.

Lyon, R. J. P., Tuddenham, W. M., and Thompson, C. S. (1959). *Econ. Geol.* 1047–1057.

Savitzky, A., and Golay, M. J. E. (1964). *Anal. Chem.* **36**, 1627–1639.

Tuddenham, W. M., and Lyon, R. J. P. (1960). *Anal. Chem.* **32**, 1630–1634.

Vincent, R. K. (1972). *Proc. Int. Symp. Remote Sens. Environ. 8th, 1972* pp. 1239–1247.

Vincent, R. K. (1973). *Symp. Proc. Manage. Util. Remote Sens. Data, 1973* pp. 377–397.

Vincent, R. K., and Thomson, F. (1972a). *Science* **175**, 968–988.

Vincent, R. K., and Thomson, F. (1972b). *J. Geophys. Res.* **77**, 2465-2471.

Vincent, R. K., Thomson, F., and Watson, K. (1972). *J. Geophys. Res.* **77**, 2473–2477.

8

SELECTED TERRESTRIAL MINERALS AND THEIR INFRARED ABSORPTION SPECTRAL DATA (4000–300 cm^{-1})

Homer C. Liese

I. Introduction

The primary purpose of this chapter will be to present infrared absorption spectral data for as many terrestrial minerals as is practicable. In addition, selected recent research studies are discussed very briefly emphasizing their geological application. The use of infrared absorption analyses with emphasis on special advantages and limitations also is outlined.

No attempt will be made to deal with the theoretical aspects of infrared absorption analysis nor the empirical assignment of absorption bands to types of vibrations, bonds and/or atomic or molecular combinations. Information regarding these aspects may be found in Chapter 13, as well as in Lyon (1967), Rao (1964), and Stubican and Roy (1961a,b,c). Chapter 9 concerning far-infrared spectroscopy of minerals should also be consulted.

Of special interest is the article by Lyon (1967), and it is highly recommended that the researcher become familiar with this work. In addition to outlining the basic theory of infrared absorption analysis, Dr. Lyon discusses (1) the various experimental absorption–analysis methods; (2) such general analytical results as the spectral data from synthetic versus natural minerals, quantitative analysis, and the distinction between absorbed water, hydroxyl, and water of crystallization; and (3) actual spectra and limited spectral data for selected mineral groups. Finally, the article lists more than 100 references.

The experimental procedure followed in the infrared analysis was the KBr technique, and the instrument employed was the Perkin-Elmer Model 621 grating spectrophotometer. Information regarding sample preparation, instrument settings, and precision and accuracy evalution can be found in Liese (1974).

II. Uses of Infrared Absorption Analysis

Lyon (1967) lists four major mineral groups that can be distinguished by infrared absorption analytical techniques. (*a*) Minerals of relatively constant composition (e.g., quartz); (*b*) minerals that exhibit ranges of composition (e.g., olivine); the spectra clearly show small differences in the amounts of major elements; (*c*) minerals of constant composition but with different crystal structures (e.g., pyrite and marcasite); (*d*) minerals that vary continuously both in structure and chemical composition (e.g., Na-rich plagioclase with high- and low- temperature forms which may range from An_0 to An_{20} in composition).

A. Advantages

There exist specific advantages in employing this analytical technique. (1) Qualitative analysis is truly relatively quick and relatively easy for many minerals, especially if the investigator has in his possession a good set of known spectra or some good spectral data. (2) It is possible to investigate glasses that yield absorption spectra, but show very little in the way of an x-ray diffraction pattern. (3) x-Ray analysis is sensitive to long-range order, whereas infrared analysis is much more sensitive to short-range ordering, or ordering on a nearest-atom basis. (4) An especially small amount of sample is required (0.3 mg), thus enabling a greater number and variety of geological problems to be investigated. (5) Essential cations of similar size and charge, but of substantially different weights, would prove to be more accurately measured than by x-ray diffraction techniques. (6) Certain mineral impurities are clearly detectable in the spectra of some minerals to be determined.

B. Limitations

While there exist a few examples of truly quantitative analysis in the literature (Lyon, 1967), the majority of the investigations deal, at best, with semiquantitative data. This is due simply to the fact that too many factors influence the frequencies of absorption and/or the intensities of the absorption bands. For example, intramolecular interactions (such as mass effects, electrical effects, symmetry, bond-angle strain, hydrogen bonding, field effects, etc.) do not affect the normal vibrational frequencies greatly, but may influence the detailed structure and intensities of the absorption bands (Rao, 1964). Added to these, of course, are the ever-variable geological environmental factors.

Another limitation would be the failure of the absorption spectra to reveal any spectral variation owing to variation of minor or trace element concentration.

III. Results and Discussion

Infrared absorption spectral data for more than 150 selected rock-forming and economic minerals are presented.* Each table presents the following data for each mineral: chemical formula; crystal system and class; selected frequency ranges for a few absorption bands are given in order to present a more realistic view of the infrared data for minerals (such ranges may be interpreted to be caused by a variation in composition, temperature, pressure, and/or operator, etc.); the spectral data for an example mineral are then presented to inform the worker of all the infrared absorption

* Raman spectral data for 56 of these same minerals are listed in Chapter 12, Section IV.

activity within the region scanned (the example mineral also illustrates the various combinations of absorption band frequency and intensity).

In addition, the relative magnitudes of absorption are presented. The total number of infrared spectra that were observed and evaluated from the literature and from this laboratory in order to obtain these absorption ranges also are included. Finally, each table lists its minerals in the approximate order as presented by Dana, i.e., the standard classification, and the mineral names are also included in the Subject Index. A brief discussion concerning the geological applications via recent research is summarized for each group to be discussed.

Some of the literature that was especially rewarding in the compilation of these spectral data were Moenke (1962), Lyon (1962, 1963, 1967), Grim (1968), Wachter and Lyon (1970), and Nyquist and Kagel (1971).

A. Native Elements

Sulfur, antimony, arsenic, and graphite were analyzed. None revealed any absorption troughs except perhaps graphite, and this occurred between 400 and 300 cm^{-1} and was of extra weak magnitude. Anderson and Wong (1969) studied rhombic sulfur and obtained infrared spectra from 500 to 50 cm^{-1} at 95°K and no peaks appeared. DeCamargo *et al.* (1970) obtained spectra of some Brazilian diamonds.

B. Sulfides (Table I)

Earlier studies of selected sulfides are listed in Lyon (1962) and Rao (1964), and those containing actual spectra or spectral data are few in number. More recently, Karr and Kovach (1969) analyzed 6 sulfides of practical and economic interest, for example, the examination of cinnabar in a high-quartz vein. In a study more involved with crystal structure, symmetry, band assignment, and type of vibration, Forneris (1969) analyzed realgar and orpiment in some detail. Yu and Soltai (1972) investigated the crystallography of a high-temperature phase of realgar. Nyquist and Kagel (1971) obtained spectra for 8 sulfide minerals employing Nujol mull in the region of absorption rather than KBt.

Liese (1974) investigated 54 different sulfides and sulfosalts by infrared absorption (435–250 cm^{-1}) and then by ultraviolet emission analyses. Results revealed that many of the absorption bands were diagnostic, and that some could be correlated with compositional variations.

C. Oxides and Hydroxides (Table II)

Boldyrev and Povarennykh (1968a) show the infrared absorption spectra for 28 oxides and discuss the crystal chemistry of each mineral in a correlation attempt.

TABLE I

INFRARED ABSORPTION SPECTRAL DATA (450–250 CM^{-1}) FOR SELECTED SULFIDE MINERALS[a]

MINERAL NAME, FORMULA CRYSTAL SYSTEM (AND CLASS)	SELECTED FREQUENCY RANGES AND ABSORPTION MAGNITUDES (EXAMPLE MINERAL: SHOWS ALL ABSORPTIONS PLUS LOCALITY)	NO. OF SAMPLES
Chalcocite, Cu_2S	415–412s, 345–340m to w	5
O(2/m 2/m 2/m)	(415s, 345mw, Tsumeb, S.W. Africa)	
Bornite, Cu_5FeS_4	351–345s	3
I(or T)(4/m $\bar{3}$ 2/m?)	(345s, Pinal Co., Ariz.)	
Sphalerite[b] (Zn, Fe)S	306–280s	5
I($\bar{4}$3m)	(300s, Ontario, Canada)	
Galena, PbS	No absorption	6
I(4/m $\bar{3}$ 2/m)		
Chalcopyrite, $CuFeS_2$	372–370sh, 359–355s, 316–315m	7
T($\bar{4}$2/m)	(370sh, 358s, 316m, Zacatecas, Mexico)	
Pyrrhotite, $Fe_{1-x}S$	No absorption	5
H(6/m 2/m 2/m)		
Cinnabar,[b] HgS	346–342s	3
Trig (32)	(342s, Pershing Co., Nevada)	
Covellite, CuS	No absorption	3
H(6/m 2/m 2/m)		
Realgar,[b] AsS	365–362m, 358–355m, 339–338s	2
M(2/m)	(365m, 355m, 338s, Humboldt Co., Nev.)	
Orpiment,[b] As_2S_3	390xw, 380–378m, 362–360sh, 345–344m, 300xs	4
M(2/m)	(390w, 379m, 362sh, 350sh, 345m, 300xs, Allchar, Greece)	
Stibnite, Sb_2S_3	335–332s	4
O(2/m 2/m 2/m)	(332s, 302xs, 290sh, 250bs, Bau, Borneo)	
Pyrite, FeS_2	415–410s, 348–345m, 292–285w	10
I(2/m$\bar{3}$)	(411s, 345m, 289w, Logrono, Spain)	
Marcasite, FeS_2	419–418s, 398–396s, 352–350m, 328–325m	2
O(2/m 2/m 2/m)	(418s, 396s, 352m, 325m, Kent, England)	
Arsenopyrite, FeAsS	435–431s, 369–362ms	4
O or M(2/m 2/m 2/m)	(431s, 367ms, 278w, Freiberg, Germany)	
Molybdenite, MoS_2	381–380s	4
H(6/m 2/m 2/m)	(380s, Deepwater, N.S.W., Australia)	
Enargite[c], Cu_3AsS_4	379–377s, 260–255mw	3
O(2mm)	(379s, 350sh, 260m, San Juan Co., Colo.)	

[a] Abbreviations: Crystal systems: I(isometric); T(tetragonal); H(hexagonal); Trig.(trigonal); O(orthorhombic); M(monoclinic); Tri(triclinic). Absorption magnitudes: s(strong); m(medium); w(weak); sh(shoulder); br or b(broad); x(extra)

[b] Raman data in Chapter 12, Section IV,A,1.

[c] Enargite is classified as a sulfosalt.

TABLE II

INFRARED ABSORPTION SPECTRAL DATA (4000–300 CM^{-1}) FOR SELECTED OXIDE AND HYDROXIDE MINERALS[a]

MINERAL NAME, FORMULA CRYSTAL SYSTEM (AND CLASS)	SELECTED FREQUENCY RANGES AND ABSORPTION MAGNITUDES (EXAMPLE MINERAL: SHOWS ALL ABSORPTIONS PLUS LOCALITY)	NO. OF SAMPLES
	Oxides	
Cuprite, Cu_2O	625–615s	2
I(4m $\bar{3}$ 2/m)	(615xs, Butte, Mont.)	
Zincite, ZnO	540–535m, 440–435s	2
H(6mm)	(540sh, 435s, 355sh, 325sh, Franklin, N.J.)	
Spinel, $MgAl_2O_4$	685–677s, 515–505s, 301m	3
I(4/m $\bar{3}$ 2/m)	(766sh, 677s, 570sh, 505s, 301m, Burma)	
Magnetite, Fe_3O_4	575–570s	60
I(4/m $\bar{3}$ 2/m)	(575s, 375sh, Degrasse, N.Y.)	
Franklinite, (Zn, Mn, Fe) $(Fe, Mn)_2O_4$	565–555s, 435–428s	2
I(4/m $\bar{3}$ 2/m)	(555s, 428s, 390sh, 325sh, Franklin, N.J.)	
Chromite $(Mg, Fe)Cr_2O_4$	635s	2
I(4/m $\bar{3}$ 2/m)	[695sh, 635s, 607s(?), 450sh, Phillipine Islands]	
Chrysoberyl, $BeAl_2O_4$	975–945m, 438–433m	2
O(2/m 2/m 2/m)	(1080m, 975m, 762ms, 635m, 610–590sh, 570–530xw, 511w, 438m, 360m, Custer, South Dakota)	
Corundum,[b] Al_2O_3	607–595s, 460–450s	3
Trig ($\bar{3}$ 2/m)	(1078m, 775w, 635w, 595s, 453s, 378w, 355sh, Ceylon)	
Hematite, Fe_2O_3	550–540s, 471–460ms, 333ms	10
Trig($\bar{3}$ 2/m)	(630sh, 540s, 460ms, 378w, 333ms, Minas Gerais, Brazil)	
Ilmenite, $FeTiO_3$	545–540s, 461–440ms, 325s	7
Trig($\bar{3}$)	(540s, 460s, 355sh, 325s, Ishpeming, Mich.)	
Psilomelane $(Ba, H_2O)_2Mn_5O_{10}$	715m, 535–533s	2
M(2/m)	(715xbm, 533s, 460xbrw, 415sh, 320xbsh, near Prescott, Ariz.)	
Rutile,[b] TiO_2	530–512s, 402–395s, 315–305s	6
T(4/m 2/m 2/m)	(525xbrs, 460sh, 402ms, 360sh, 330ms, 305ms, Mexico)	
Pyrolusite, MnO_2	550–532s, 418–405m	2
(T(4/m 2/m 2/m)	(680–650sh, 532s, 405m, 335wm, Lake Valley, New Mex.)	
Cassiterite,[b] SnO_2	620–613s, 540–525s	3
T(4/m 2/m 2/m)	(613s, 530ms, 478w, 330sh, 300sh, Rockford, Ala.)	
Anatase,[b] TiO_2	470–460s	2

TABLE II—(*Continued*)

MINERAL NAME FORMULA CRYSTAL SYSTEM (AND CLASS)	SELECTED FREQUENCY RANGES AND ABSORPTION MAGNITUDES (EXAMPLE MINERAL: SHOWS ALL ABSORPTIONS PLUS LOCALITY)	NO. OF SAMPLES
T(4/m 2/m 2/m)	(460xbrs, Schweiz)	
Brookite, TiO_2	425–410s, 340–335ms	4
O(2/m 2/m 2/m)	(620–560s, 525sh, 490sh, 477s, 425s, 361sh, 340s, Magnet Cove, Ark.)	
	Hydroxides	
Brucite, $Mg(OH)_2$	3698–3695s, 362–358s	2
Trig($\overline{3}$ 2/m)	(3695s, 640sh, 565sh, 430s, 358s, Lodi, Nev.)	
Boehmite, AlO(OH)	3290s, 3100–3090s, 1072–1065s, 525–500s	2
O(2/m 2/m 2/m)	(3290s, 3100s, 1072s, 1040sh, 760s, 605s, 525s, S. France)	
Manganite, MnO(OH)	510–500wm	2
M(2/m)	(680–640sh, 580–540sh, 510wm, 490–450sh, 392w, Tuba City, Ariz.)	
Goethite, $HFeO_2$	3140xbrs, 904–883m, 803–790m, 415–402s	3
O(2/m 2/m 2/m)	(3140xbrs, 893m, 793m, 620m, 495sh, 402s, Biwabik, Minn.)	
Gibbsite, $Al(OH)_3$	3540–3525s, 970–960m, 418–405m	2
M(2/m)	(3620m, 3525s, 3469s, 3410sh, 1013s, 970m, 910m 838sh, 795s, 770sh, 740s, 658s, 553s, 515s, 445s, 418m, 360s, Minas Gerais, Brazil)	

[a] See Table I, footnote [a] for abbreviations.
[b] Raman data in Chapter 12, Section IV,A,2.

White (1967) and White and Keramidas (1971) applied infrared spectroscopy to order–disorder problems in oxides. Preud'homme and Tarte (1971) studied the zinc spinels, chromites, and several other minerals. Differences in the spectra of cassiterite were attributed to variations in temperature and pressure of formation according to Gotman *et al.* (1963). The shapes of the absorption curves for wood tin and cassiterite are similar, but their intensities differ (Dolomanova and Rudnitskaya, 1969).

Differential infrared studies indicate that interlayer OH ions are present in some manganese nodules. However, owing to the poor spectral resolution and band overlap caused by mineral impurities, infrared spectroscopy offers little promise for identifying nodule mineralogy (Elderfield and Glasby, 1973).

There have been several recent studies of the bauxite group minerals. Among these are the infrared study of bauxite and bauxite constituents, i.e., alumina and aluminum (White, 1971); the determination of the genetic type of bauxite from the position of OH stretching and deformation bands (Shcherban and Nurmagambetov, 1970); and finally, Ziborova and Khramtsov (1969) studied the use of infrared spectrometry as applied to the origin and alteration of bauxite.

D. Halides (Table III)

Some of the more common halides do not show any absorption bands at all until they approach low frequencies of infrared radiation. For example, fluorite, halite, and sylvite show a very strong, broad absorption band at 305–280, 250–150, and 200–140 cm^{-1}, respectively. These spectra can be observed in Nyquist and Kagel (1971). Fijal (1970) very briefly discussed the use of infrared absorption spectroscopy in the study of minerals from salt deposits; included in it were CO_3, SO_4, BO_4, as well as the halides.

TABLE III

INFRARED ABSORPTION SPECTRAL DATA (4000–280 CM^{-1}) FOR SELECTED HALIDE MINERALS[a]

MINERAL NAME, FORMULA CRYSTAL SYSTEM (AND CLASS)	SELECTED FREQUENCY RANGES AND ABSORPTION MAGNITUDES (EXAMPLE MINERAL: SHOWS ALL ABSORPTIONS PLUS LOCALITY)	NO. OF SAMPLES
Halite,[b] NaCl	No absorption bands	4
I(4/m $\bar{3}$ 2/m)	[(?) locality]	
Sylvite, KCl	No absorption bands	2
I(4/m $\bar{3}$ 2/m)	[(?) locality]	
Fluorite,[b] CaF_2	305–280s	3
I(4/m $\bar{3}$ 2/m)	(430sh, 350sh, 305–280s, Rosiclaire, Ill.)	
Atacamite, $Cu_2(OH)_3Cl$	3340–3335s, 598–593s, 410–399s	2
O(2/m 2/m 2/m)	[3440s, 3335s, 985m, 948m, 913w, 892m, 863sh, 846m, 818w, 593s, 513s, 477m, 441m, 399s, 318 (?), locality]	
Carnallite, $KMgCl_3 \cdot 6H_2O$	1638–1630s, 300ms	2
O(2/m 2/m 2/m)	(1630s, 300ms, Suria, Spain)	
Cryolite,[b] Na_3AlF_6	585–570s, 402–394ms	3
M(2/m)	(575s, 402ms, 310–300sh, Ivigtut, Greenland)	

[a] See Table I, footnote [a] for abbreviations.

[b] Raman data in Chapter 12, Section IV,A,3.

E. Carbonates, Nitrate, and Borates (*Table IV*)

Adler and Kerr (1963a,b) studied in detail the spectral changes of carbonates and discussed how these were dependent on specific crystallochemical changes. Boldyrev and Povarennykh (1968b), in a similar but less detailed study obtained similar results and also included some nitrate minerals.

Chester and Elderfield (1966, 1967, 1971) have contributed since to the quantitative (or semiquantitative) determination of carbonate mineralogy. For example, they have determined calcite:dolomite ratios in carbonate rocks and calcite:aragonite ratios in shell material. This kind of data can of course be extremely useful in making semiquantitative interpretations regarding environmental conditions of deposition.

Fijal and Klapyta (1971) compare the x-ray and infrared methods of quantitative phase analysis for the calcite–dolomite system. Elderfield (1970) demonstrates the ease of distinguishing calcite from aragonite in a calcareous sediment by infrared absorption analysis.

F. Sulfates, Chromate, Molybdates, and Tungstates (*Table V*)

Adler (1965) studied the sulfates in detail, correlating spectral variations with changes in composition and structure.

Seidl *et al.* (1969) show that spectral features of gypsum, previously explained by distinct H_2O molecules or by proton tunneling, are due to vibrational coupling. Povarennykha and Gevorkyan (1971) present the spectra of gypsum and its isostructural minerals. Wachter and Lyon (1970) show spectra for several sulfate minerals. For example, gypsum dried for 48 hr at 110°C shows a shift in response toward that of anhydrite.

Plyusnina and Zaytseva (1969) have outlined infrared spectra and characteristic frequencies of synthetic and natural tungstates and molybdates. Zorina and Syritso (1972) have found that a systematic relationship exists between the scheelite structure and the infrared spectrum.

G. Phosphates, Arsenate, and Vanadates (*Table VI*)

Earlier work by Adler (1964) was later confirmed and supplemented by Povarennykha and Gevorkyan (1970), Klee and Engel (1970), and Fortsch (1970). Far-infrared absorption spectra of apatite and pyromorphite were obtained by Omori (1969). Engel and Klee (1970) discuss the OH valence variations in apatites. LeGeros *et al.* (1968, 1970) studied the carbonate in carbonate-containing apatites.

TABLE IV

INFRARED ABSORPTION SPECTRAL DATA (4000–300 CM^{-1}) FOR SELECTED CARBONATE, NITRATE, AND BORATE MINERALS[a]

MINERAL NAME, FORMULA CRYSTAL SYSTEM (AND CLASS)	SELECTED FREQUENCY RANGES AND ABSORPTION MAGNITUDES (EXAMPLE MINERAL: SHOWS ALL ABSORPTIONS PLUS LOCALITY)	NO. OF SAMPLES
	Carbonates	
Calcite,[b] $CaCO_3$	1440–1420s, 877–870m, 715–708wm, 315–307ms	8
Trig($\bar{3}$ 2/m)	(1420s, 871m, 709wm, 313ms, N. Vernon, Ind.)	
Magnesite,[b] $MgCO_3$	1450–1435s, 887–885m, 750–740m, 380s	3
Trig($\bar{3}$ 2/m)	(1435s, 885m, 750m, 380s, 302w, Stevens Co., Wash.)	
Siderite,[b] $FeCO_3$	1425–1410s, 866–864m, 740–737m, 335–325s	5
Trig($\bar{3}$ 2/m)	(1410s, 865m, 737m, 325s, Roxbury, Conn.)	
Rhodochrosite,[b] $MnCO_3$	1435–1410s, 867–865m, 730–725m, 307–300s	6
Trig($\bar{3}$ 2/m)	(1410s, 865m, 728m, 307s, Butte, Mont.)	
Smithsonite,[b] $ZnCO_3$	1435–1415s, 871–867m, 744–743m, 300sh	3
Trig($\bar{3}$ 2/m)	(1415s, 867m, 744m, 300sh, Laurian, Greece)	
Dolomite,[b] $CaMg(CO_3)_2$	1450–1435s, 881–875m, 730–722wm, 360–358s	4
Trig($\bar{3}$)	(1435s, 876m, 725wm, 360s, 315m, Monroe Co., N.Y.)	
Ankerite, $CaFe(CO_3)_2$	1450–1440s, 877–875m, 727–726m	2
Trig($\bar{3}$)	(1440brs, 875m, 727m, 355brs, Gilman, Colo.)	
Aragonite,[b] $CaCO_3$	1470–1420s, 866–854m, 712–709w, 699–696w	3
O(2/m 2/m 2/m)	(1420s, 1079wm, 854m, 709w, 696w, Aragon, Spain)	
Witherite,[b] $BaCO_3$	1445–1435s, 860–855m, 695–693m	3
O(2/m 2/m 2/m)	(1435s, 1062w, 859m, 695m, 488w, Northumberland, England)	
Strontianite,[b] $SrCO_3$	1470–1452s, 860–858ms, 708–706m, 695–693m	3
O(2/m 2/m 2/m)	(1452s, 1072wm, 858ms, 708m, 702m, 487xw, Humm, Germany)	
Cerussite,[b] $PbCO_3$	1405–1390s, 841–832m, 677–670m	4
O(2/m 2/m 2/m)	(1390s, 1048m, 832m, 670m, 450xbrsh, Vanadium, New Mex.)	
Malachite, $Cu_2(CO_3)(OH)_2$	3409–3405ms, 3322–3320ms, 1425–1423s, 1395–1392s, 430–428wm	2
M(2/m)	(3409ms, 3320ms, 1510s, 1425s, 1392s, 1100wm, 1050ms, 878m, 820ms, 752wm, 575w, 525wm, 505w, 430wm, 355sh, 325sh, 300m, Washington Co., Utah)	
Azurite, $Cu_3(CO_3)_2(OH)_2$	3430–3424m, 1466–1463s, 1418–1413s, 460–450m	2
M(2/m)	(3430m, 1490s, 1463w, 1413s, 1090w, 950m, 830m, 815w, 770w, 745w, 490m, 450m, 400m, 340m, 300m, Washington Co., Utah)	

TABLE IV—(*Continued*)

MINERAL NAME, FORMULA CRYSTAL SYSTEM (AND CLASS)	SELECTED FREQUENCY RANGES AND ABSORPTION MAGNITUDES (EXAMPLE MINERAL: SHOWS ALL ABSORPTIONS PLUS LOCALITY)	NO. OF SAMPLES
Trona, $Na_2(CO_3)\cdot Na(HCO_3)\cdot 2H_2O$ M(?)	3470s, 1700–1688s, 1472–1462s, 600–595s (3470s, 3040m, 1688s, 1462s, 1190m, 1015m, 850s, 680sh, 650sh, 595s, Westvaco, Wyo.)	2
	Nitrate	
Soda-niter,[b] $NaNO_3$ Trig($\bar{3}$ 2/m)	1390–1382s, 1122–1095ms (3400w, 3200w, 1382s, 1095xbrms, 750xbrxw, 626w, 605w, 450xbrw, 390xw, Tarapaca, Chile)	2
	Borates	
Kernite,[b] $Na_2B_4O_6(OH)\cdot 3H_2O$ M(2/m)	3552–3550ms, 1375–1358s, 1020–1010s, 832–825mw, 553wm (3550ms, 3410sh, 3315brms, 3000w, 1706w, 1358s, 1150sh, 1060sh, 1010s, 970s, 950s, 870w, 825mw, 760w, 740w, 725w, 690m, 650sh, 615w, 553wm, 495w, 456w, 430xw, 390xxw, 340xxxw, Kern Co., Calif.)	2
Colemanite,[b] $CaB_3O_4(OH)_3\cdot H_2O$ M(2/m)	3603–3600ms, 1365–1360s, 1046–1030m, 581–580m (3600ms, 3518wm, 3255s, 3130xw, 1430–1400sh, 1360s, 1325w, 1275xw, 1226w, 1030m, 930m, 885xw, 815sh, 758m, 730m, 694xw, 581m, 514s, 435xbrm, 338xw, Inyo Co., Calif.)	2
Ulexite, $NaCaB_5O_9\cdot 8H_2O$ Tri(?)	3600–3590s, 1325–1318s, 1065–1050m, 868–854s, 450–442m (3590s, 3520m, 3420m, 3120xw, 1625wm, 1388s, 1350xw, 1318s, 1200m, 1050m, 998wm, 854s, 738w, 713wm, 632w, 598xw, 530m, 442m, 372w, Chile?)	2

[a] See Table I, footnote [a] for abbreviations.
[b] Raman data in Chapter 12, Section IV,A,4.

In the studies by Bhatnager (1968) and Kislovskiy and Knubovets (1970) infrared spectra of hydroxyapatite and fluorapatite were obtained, and the nature and extent of the changes in vibration frequency with the substitution of the F^- ion for OH^- are discussed.

TABLE V

Infrared Absorption Spectral Data (4000–300 Cm^{-1}) for Sulfate, Chromate, Molybdate, and Tungstate Minerals[a]

Mineral name, formula crystal system (and class)	Selected frequency ranges and absorption magnitudes (example mineral: shows all absorptions plus locality)	No. of samples
	Sulfates	
Barite,[b] $BaSO_4$ O(2/m 2/m 2/m)	1180–1170s, 1085–1079s, 614–607s (1170s, 1110sh, 1080s, 980w, 630sh, 607s, Versailles, Mo.)	5
Celestite,[b] $SrSO_4$ O(2/m 2/m 2/m)	1192–1172s, 1085–1076s, 613–605ms (1172s, 1110s, 1076s, 978m, 638m, 605ms, Dundas, Ont.)	3
Anglesite,[b] $PbSO_4$ O(2/m 2/m 2/m)	1108–1090s, 625m, 600–594m (1155xbrsh, 1100s, 1065xbrsh, 965xw, 625m, 595m, 460w, Minersville, Utah)	3
Anhydrite,[b] $CaSO_4$ O(2/m 2/m 2/m)	1159–1150s, 1130–1110s, 676–671m, 616–610m, 597–593m (1150s, 1110s, 671m, 655sh, 610m, 593m, Windsor, Nova Scotia, Canada)	4
Glauberite, $Na_2Ca(SO_4)_2$ M(?)	1142–1135s, 1108–1101s, 635–631m, 611–607m [1137s, 1127s, 1102s, 631m, 607m, 472w(?) locality]	3
Gypsum, $CaSO_4 \cdot 2H_2O$ M(2/m) var satin spar	3572–3555s, 1150–1135xs, 605–595m, 460–450m (3572s, 3400s, 1682m, 1616m, 1135xs, 1110xs, 662m, 595m, 450m, 365sh, 325sh, 300m, East Bridgeford, England)	3
Gypsum,[b] $CaSO_4 \cdot 2H_2O$ M(2/m) var. selenite	1150–1140s, 657–655ms, 598–595m (3610m, 3555m, 1640m, 1150s, 1112s, 1090s, 1005w, 657ms, 625w, 595ms, 485w, 410w, 305w, England)	2
Chalcanthite, $CuSO_4$, $5H_2O$ Tri($\bar{1}$)	1158–1153s, 1000–995s, 663–658s, 612–604m [3520m, 3420m, 1635m, 1200s, 1153s, 995s, 963m, 658s, 604m, 588m, 520w, 305sh, (?) locality]	2
Epsomite,[b] $MgSO_4 \cdot 7H_2O$ O(222)	1660–1655brm, 1100–1085s, 630–617s, 430–425m (3750sh, 3540sh, 3400s, 1655brm, 1110sh, 1085s, 1000w, 986w, 725brsh, 617s, 425m, 300w, near Ashcroft, B.C., Canada)	2
Brochantite, $Cu_4SO_4(OH)_6$	1119–1113s, 1088–1087, 873–868m, 602–597s, 421–415wm	2

TABLE V—(*Continued*)

MINERAL NAME, FORMULA CRYSTAL SYSTEM (AND CLASS)	SELECTED FREQUENCY RANGES AND ABSORPTION MAGNITUDES (EXAMPLE MINERAL: SHOWS ALL ABSORPTIONS PLUS LOCALITY)	NO. OF SAMPLES
M(2/m)	[3585m, 3565m, 3400sh, 3385m, 3265m, 1680–1610xw, 1160sh, 1140sh, 1113s, 1087s, 990w, 940w, 868m, 850sh, 793sh, 776wm, 695xxw, 635sh, 624xw, 605sh, 597s, 510w, 480w, 460w, 415wm, 395xw, 371w, 310xw, Arizona (?)]	
Antlerite, $Cu_3SO_4(OH)_4$	1157–1150ms, 1115–1105ms, 1075–1071ms, 990–987w, 517–514s, 462–460s	3
O(2/m 2/m 2/m)	[3574m, 3487m, 1150ms, 1105ms, 1072ms, 988w, 873m, 855sh, 790s, 775s, 760s, 693w, 635m, 614m, 514s, 460s, 395w, 370wm, 330w (?) locality]	
Alunite, $KAl_3(SO_4)_2(OH)_6$	1087–1086s, 1030–1027s	2
Trig($\bar{3}$ 2/m or 3m)	(3485?, 1170ms, 1086s, 1030s, 802m, 780m, 700m, 682m, 605m, 505sh, 475m, 450sh, Hungary)	
Jarosite, $KFe_3(SO_4)_2(OH)_6$	1100–1090s, 690–685mw, 638–623m, 512–510s	2
Trig($\bar{3}$ 2/m or 3m)	(3430sh, 3355m, 3260sh, 1190sh, 1100s, 1028s, 1018s, 790sh, 685mw, 638m, 512s, Luming, Nev.)	
	Chromate	
Crocoite,[b] $PbCrO_4$	858–850s, 840–835sh	2
M(2/m)	(850s, 840sh, 784sh, 388w, near Dumcas, Tasmania)	
	Molybdates and tungstates	
Wolframite,[b] $(Fe, Mn)WO_4$	625–605m, 530–525brsh	2
M(2/m)	(850brsh, 795m, 668brsh, 605m, 525brsh, Bagdad, Ariz.)	
Ferberite,[b] $FeWO_4$	818–800m, 625–615m, 535brsh	2
M(2/m)	[850brsh, 800m, 615m, 535brsh, 347m (?) locality]	
Scheelite,[b] $CaWO_4$	808–775s	2
T(4/m)	[850sh, 775s, 373w, 305wm (?) locality]	
Wulfenite,[b] $PbMoO_4$	785–772xs, 374–372m, 305–304m	3
T(4/m or 4)	[775xs, 374m, 360sh, 304m (?) locality]	

[a] See Table I, footnote [a] for abbreviations.

[b] Raman data in Chapter 12, Section IV,A,5.

TABLE VI

INFRARED ABSORPTION SPECTRAL DATA (4000–300 CM^{-1}) FOR SELECTED PHOSPHATE, ARSENATE, AND VANADATE MINERALS[a]

MINERAL NAME, FORMULA CRYSTAL SYSTEM (AND CLASS)	SELECTED FREQUENCY RANGES AND ABSORPTION MAGNITUDES (EXAMPLE MINERAL: SHOWS ALL ABSORPTIONS PLUS LOCALITY)	NO. OF SAMPLES
	Phosphates	
Monazite (Ce, La, Y, Th)PO_4	1050xbrs, 955–940s, 620w	2
M(2/m)	(1050xbrs, 992s, 940s, 620w, 530m, Minas Gerais, Brazil)	
Amblygonite (Li, Na)-Al(PO_4)(F, OH)	3375–3373s, 1085–1083s, 815–812m, 485–475ms, 324–318m	3
Tri($\bar{1}$)	(3375s, 1180s, 1083s, 1017s, 812m, 590w, 525sh, 475m, 410m, 388sh, 318m, near Keystone, S. Dakota)	
Lazulite (Mg, Fe)-$Al_2(PO_4)_2(OH)_2$	3405–3400ms, 1010–1004m, 812–804m, 652–646m, 430–426w	2
M(?)	(3405ms, 1196mw, 1111s, 1058sh, 1025sh, 1004m, 804m, 773m, 646m, 585sh, 548m, 519m, 460xw, 426w, 377wm, 313w, near Jacksonville, Ore.)	
Apatite,[b] $Ca_5(PO_4)_3$(F, Cl, OH)	1095–1090s, 1050–1038s, 605–600s, 575–560ms, 318–310ms	5
H(6/m)	(1090s, 1040s, 960m, 863sh, 600ms, 560ms, 465w, 358sh, 314ms, Durango, Mexico)	
Pyromorphite,[b] $Pb_5(PO_4)_3Cl$	1030–1020s, 970–960xs, 575–558s, 543–538s	4
H(6/m)	(1138sh, 1020s, 963s, 920sh, 868sh, 711sh, 670sh, 638sh, 572s, 542s, 475sh, 435sh, Germany)	
Wavellite, $Al_3(PO_4)_2(OH)_3 5H_2O$	3520s, 1025–1015s, 452–443ms	2
O(?)	(3520s, 3400sh, 1635w, 1600sh, 1095sh, 1050s, 1015s, 925sh, 852w, 640s, 587ms, 545ms, 482ms, 443ms, 375m, 348sh, 311m, Montgomery Co., Ark.)	
Autunite, $Ca(UO_2)_2(PO_4)_2 \cdot$ 10–12 H_2O	1125–1120s, 1023–1005s, 545–538m	3
T(4/m 2/m 2/m)	(3580, 3430, 3270, 1632m, 1120s, 1005s, 915m, 820w, 680–650sh, 538m, 428xw, 308m, Utah)	

TABLE VI (*Continued*)

MINERAL NAME, FORMULA CRYSTAL SYSTEM (AND CLASS)	SELECTED FREQUENCY RANGER AND ABSORPTION MAGNITUDES (EXAMPLE MINERAL: SHOWS ALL ABSORPTIONS PLUS LOCALITY)	NO. OF SAMPLES
	Arsenates	
Erythrite, $Co_3(AsO_4)_2 \cdot 8H_2O$	784–783s, 452–446m	2
M(2/m)	(3435s, 3390sh, 3150brxw, 3040s, 1622m, 1570m, 1050brm, 870sh, 815sh, 784s, 682w, 550sh, 475sh, 446m, 415m, 365xw, 310xw, Sonora, Mexico)	
Annabergite,[b] $Ni_3(AsO_4)_2 \cdot 8H_2O$	832–825w, 795m, 460–455w, 427–424w	2
M(2/m)	(3500–3000ms, 1630w, 1582w, 1095xw, 915w, 825w, 795m, 455w, 424w, 375w, 330w, Churchill Co., Nev.)	
Mimetite,[b] $Pb_5(AsO_4)_3Cl$	790–783s, 418–413ms	2
H(6/m)	(1008sh, 954m, 918sh, 808sh, 783s, 565wm, 538wm, 520sh, 413ms, 378m, Roughton Gill, Eng.)	
	Vanadates	
Vanadinite,[b] $Pb_5(VO_4)_3Cl$	810–801s, 745–739s	2
H(6/m)	(810s, 745s, Turkey)	

[a] See Table I, footnote [a] for abbreviations.
[b] Raman data in Chapter 12, Section IV,A,6.

H. Nesosilicates (Table VII)

The olivine group has been the subject of a good correlative study by Duke and Stephens (1964), which has led to more recent work by other investigators. Burns and Huggins (1972) discuss the vibrational spectra of Mg–Fe–Mn olivines as related to cation determinative curves. Huggins (1973) employs evidence obtained from the vibrational spectra and relates this to cation ordering in the comparatively simple olivine structure. Details on the effect of cation ordering on the far infrared spectra of olivines are given in Chapter 9, Section IV, B.

Goncharova *et al.* (1970) and Dowty (1971) review the crystal chemistry of certain garnets and relate them to spectral studies. An example of a detailed analysis of an infrared absorption spectrum of an almandine–pyrope garnet is given by Omori (1971a).

TABLE VII

INFRARED ABSORPTION SPECTRAL DATA (4000–300 CM^{-1}) FOR SELECTED NESOSILICATES[a]

MINERAL NAME, FORMULA CRYSTAL SYSTEM (AND CLASS)	SELECTED FREQUENCY RANGES AND ABSORPTION MAGNITUDES (EXAMPLE MINERAL: SHOWS ALL ABSORPTIONS PLUS LOCALITY)	NO. OF SAMPLES
Olivine[b] (Fe, Mg)$_2$SiO$_4$ O(2/m 2/m 2/m)	985–980s, 898–880s, 612–605ms, 505–504s, 416–414s, 360–355s–m (981s, 957sh, 884s, 838ms, 605ms, 522sh, 504s, 479sh, 416s, 400sh, 360s, 325sh, Jackson Co., N. Car.)	9
Willemite,[b] Zn$_2$SiO$_4$ Trig($\bar{3}$)	980–975s, 931–929s, 872–868s, 575–570s, 460wm (975s, 929s, 896s, 872s, 785m, 610ms, 570s, 460wm, 450sh, 415sh, 390sh, 378m, 350sh, Franklin, N.J.)	2
Andalusite (chiastolite), Al$_2$SiO$_5$ O(2/m 2/m 2/m)	1015–1010s, 950–930ms, 535–520s, 480–475s, 355–345w (1075sh, 1015s, 930ms, 830w, 750w, 700w, 535s, 475s, 412m, 355w, Mono Co., Calif., or Lancaster, Mass.)	3
Sillimanite, Al$_2$SiO$_5$ O(2/m 2/m 2/m)	1185–1175s, 915–900s, 545–544s, 510–509s, 490–488s, 440–434s (1175s, 956s, 900s, 885s, 815s, 740m, 690m, 670sh, 628m, 565m, 544s, 529sh, 509s, 490s, 434s, 371wm, 360sh, 332sh, 318w, Storrs, Conn.)	2
Kyanite, Al$_2$SiO$_5$ Tri($\bar{1}$)	1040–1030s, 1010–990s, 727–720s, 630–628s, 598–597s, 470–460s, 442–430s (1030s, 990s, 933s, 894s, 727s, 665s, 630s, 615s, 597s, 560s, 538s, 524sh, 498s, 460s, 430s, 390m, 377ms, 343m, 333m, 320m, 310m, Minas Gerais, Brazil)	2
Staurolite, Fe$_2$Al$_9$O$_6$-(SiO$_4$)$_4$(O, OH)$_2$ M(2/m) or pseudo-O	3420–3410wm, 1090–1080s, 695–685s, 595–586s, 435–430s (3670xw, 3470sh, 3410wm, 1162m, 1080s, 1020s, 850sh, 796m, 784sh, 775m, 685s, 625xbrs, 586s, 511sh, 478s, 460s, 430s, 392ms, 368ms, 310sh, Madagascar)	2
Topaz,[b] Al$_2$SiO$_4$(OH, F)$_2$ O(mm2 or 222)	1171–1158m, 1010–990s, 620–609s, 483–474ms, 445–443s (3635m, 1158m, 990m, 815s, 765w, 609s, 549m, 505ms, 474ms, 443s, 392m, 380sh, 330m, 305m, Villa Rica, Brazil)	2

TABLE VII—(*Continued*)

MINERAL NAME, FORMULA CRYSTAL SYSTEM (AND CLASS)	SELECTED FREQUENCY RANGES AND ABSORPTION MAGNITUDES (EXAMPLE MINERAL: SHOWS ALL ABSORPTIONS PLUS LOCALITY)	NO. OF SAMPLES
Almandine,[b] $Fe_3Al_2(SiO_4)_3$ I(4/m $\bar{3}$ 2/m)	1090–1085sh, 970–963s, 872–870s, 570–569ms, 480s, 455s (1085sh, 995sh, 963s, 896s, 870s, 650sh, 569ms, 530sh, 480s, 455s, 420s, 383s, 343m, 325m, 300sh, Gore Mtn, Warren Co., N.Y.)	3
Pyrope,[b] $Mg_3Al_2(SiO_4)_3$ I(4/m $\bar{3}$ 2/m)	1097–1080sh, 970–962s, 872–870s, 575–565ms, 482–472s, 384–378ms (1097sh, 1000sh, 965s, 897s, 867s, 572ms, 540sh, 480s, 456s, 384ms, 330m, Mexican Hat, Utah)	3
Grossularite,[b] $Ca_3Al_2(SiO_4)_3$ I(4/m $\bar{3}$ 2/m)	1080–1070sh, 960sh, 915–905s, 840s, 540–538ms, 450–444s (1070sh, 960sh, 905s, 840s, 760sh, 615wm, 538ms, 510sh, 470sh, 444s, 390s, 350sh, 323sh, Chihuahua, Mexico)	3
Andradite,[b] $Ca_3Fe_2(SiO_4)_3$ I(4/m $\bar{3}$ 2/m)	1080sh, 895–885s, 840–828s, 512–510s, 438–435ms (930sh, 885s, 828s, 815sh, 588w, 510s, 480ms, 435ms, 385s, 348m, 308s, Graham Co., Ariz.)	2
Uvarovite,[b] $Ca_3Cr_2(SiO_4)_3$ I(4/m $\bar{3}$ 2/m)	1080–1072sh, 960–945s, 910–900s, 855–840s, 470–455xs, 445–425xs [1072sh, 960s, 910s, 855s, 670w, 630sh, 618wm, 540s, 500sh, 470xs, 445xs, 392s, 335m, (?) locality]	2
Zircon,[b] $ZrSiO_4$ T(4/m 2/m 2/m)	1040–1000brs, 618–600brm, 438–430s (1000brs, 950s, 860sh, 600brm, 430s, 315m, Bancroft, Ont., Canada)	3
Sphene,[b] $CaTiSiO_5$ M(2/m)	960–953sh, 871–865s, 570–560m, 470–440ms, 433–424s (955sh, 870s, 690m, 566m, 480sh, 462ms, 424s, 360sh, 338s, 312s, Minas Gerais, Brazil)	3
Datolite, $Ca(OH)BSiO_4$ M(2/m)	3492–3490s, 1050–1034s, 1015–998s, 853–847s, 478–473s, 455–451m (3492s, 1154s, 1094s, 1034s, 998s, 914s, 879s, 847s, 784s, 686m, 572m, 522s, 496s, 485sh, 473s, 451m, 445sh, 440sh, 420m, 354m, 336w, 306sh, West Granby, Conn.)	2

TABLE VII—(*Continued*)

MINERAL NAME, FORMULA CRYSTAL SYSTEM (AND CLASS)	SELECTED FREQUENCY RANGES AND ABSORPTION MAGNITUDES (EXAMPLE MINERAL: SHOWS ALL ABSORPTIONS PLUS LOCALITY)	NO. OF SAMPLES
Dumortierite, $Al_7O_3(BO_3)(SiO_4)_3$	1380–1370m, 1078–1050s, 490–461s, 435–430w	2
O(2/m 2/m 2/m)	(1370m, 1170–1150sh, 1078s, 795xw, 777wm, 720xxw, 690wm, 520–495sh, 461s, 430w, 395m, 370m)	
Phenacite,[b] Be_2SiO_4	1040–1035sh, 985–972s, 945–935s, 780–774s, 645–635m, 427–423s	2
H(?)	(1035sh, 972s, 935s, 898s, 774s, 714s, 686s, 635m, 592s, 423s, 347w, Brazil)	
Danburite,[c] $CaB_2(SiO_4)_2$	1142–1130s, 1000–971s, 950–948s	2
O(?)	(1158sh, 1130s, 1095s, 1000s, 948s, 865s, 810s, 743m, 694m, 650m, 610s, 519m, 472m, 447w, 425m, 414m, 382m, Charcas, Mexico)	

[a] See Table I, footnote [a] for abbreviations.
[b] Raman data in Chapter 12, Section IV,B,1.
[c] Raman data in Chapter 12, Section IV,A,4.

I. *Sorosilicates (Table VIII)*

Aside from the spectra alone that occur in individual articles, not very much infrared absorption work has been accomplished regarding the sorosilicates.

J. *Cyclosilicates (Table IX)*

Plyusnina (1963) worked out Be–O assignments for beryl and other beryllium minerals. More recent work by Wood and Nassau (1968) and Povarennykha and Nefedov (1971) further characterized beryl by their absorption spectra. Vlasova and Topunova (1971) obtained spectra of beryls of different genetic types, and were able to distinguish these on the basis of the varying position of the H_2O molecule in the crystal structure, as this variation affected infrared frequencies.

K. *Inosilicates (Table X)*

A chemically analyzed series of 21 orthopyroxenes has been studied (Lyon, 1963) and their spectra clearly show the changing detail of an infrared spectrum with substitution in the Mg–Fe series.

In the diopside–jadeite series, an almost linear dependence in the frequency range from 748 to 674 cm^{-1} on the content of jadeite components provides an evaluation of the degree of isomorphic substitution (Kuznetsova and Moskaleva, 1968). Omori (1971b) analyzed the absorption spectrum of diopside in much detail.

Wilkins (1970) deals with Fe:Mg ratios and ordering scheme in the actinolite–tremolite series and attempts to relate them to environment of formation. Burns and Greaves (1971) attempt correlation of infrared and Mössbauer site population measurements of actinolite.

L. *Phyllosilicates* (*Table XI*)

More work has been done on the clay minerals than on any other group, and numerous references are listed in Lyon (1967, 1970), Grim (1968), and Stubican and Roy (1961 a, b, c); reference thereto will not be repeated herein.

TABLE VIII

INFRARED ABSORPTION SPECTRAL DATA (4000–300 CM^{-1}) FOR SELECTED SOROSILICATE MINERALS[a]

MINERAL NAME, FORMULA CRYSTAL SYSTEM (AND CLASS)	SELECTED FREQUENCY RANGES AND ABSORPTION MAGNITUDES (EXAMPLE MINERAL: SHOWS ALL ABSORPTIONS PLUS LOCALITY)	NO. OF SAMPLES
Idocrase (vesuvianite) $Ca_{10}Mg_2Al_4(Si_2O_7)_2(SiO_4)_5(OH)_4$	3570–3560w, 870–860m, 580–570ms, 450s	2
Tet (4/m 2/m 2/m)	(3560w, 1075w, 955s, 860m, 800m, 755wm, 615wm, 570m, 450s, 365sh, Lake Jaco, Mexico)	
Clinozoisite, $Ca_2Al_3Si_3O_{12}(OH)$	1108ms, 948s, 640s, 455s, 397s	1
M(2/m)	(3350m, 1108ms, 1035m, 948s, 882s, 830sh, 725sh, 640s, 590sh, 565sh, 515s, 455s, 397s, 352ms, 305m, Sonora, Mexico)	
Epidote, $Ca_2(Al, Fe)_3$-$Si_3O_{12}(OH)$	3360m, 1120–1100m, 950–944s, 650–640m, 455–450m, 398–389ms	3
M(2/m)	(3460wm, 3360m, 1100m, 1071m, 1031m, 944s, 875ms, 830sh, 715sh, 640m, 560sh, 510ms, 450m, 389ms, 340sh, Sulzer, Alaska)	

[a] Refer to Table I, footnote[a] for abbreviations.

TABLE IX

INFRARED ABSORPTION SPECTRAL DATA (4000–300 CM^{-1}) FOR SELECTED CYCLOSILICATE MINERALS[a]

MINERAL NAME, FORMULA CRYSTAL SYSTEM (AND CLASS)	SELECTED FREQUENCY RANGES AND ABSORPTION MAGNITUDES (EXAMPLE MINERAL: SHOWS ALL ABSORPTIONS PLUS LOCALITY)	NO. OF SAMPLES
Beryl,[b] $Be_3Al_2Si_6O_{18}$	3700–3698w, 1025–1010sh, 975–953s, 686–678ms, 497–490s, 440–432s	4
H(6/m 2/m 2/m)	[3698w, 3597w, 1198s, 1140sh, 1020sh, 953s, 805m, 740m, 678ms, 650m, 592m, 522m, 494s, 436s, 365m, 350sh (?) locality]	
Cordierite $(MgFe)_2Al_4Si_5O_{18}$	1170s, 965–960s, 768–765ms, 487s, 458–448s, 372s	2
O(2/m 2/m 2/m)	[1170s, 1080sh, 960s, 903s, 794sh, 765ms, 750sh, 669m, 614m, 588m, 570sh, 487s, 458s, 419sh, 372s, 360sh (?) locality]	
Tourmaline,[b] $Na(MgFe)_3Al_6(BO_3)_3(Si_6O_{18})(OH, F)_4$	3585–3555w, 1350–1330sh, 1285–1240m, 982–972s, 520–495s, 425–415sh,	3
Trig(3m) (var. schorl)	(3620sh, 3590w, 3555w, 3496w, 1335sh, 1285m, 1095s, 1050sh, 1020s, 972s, 840w, 775w, 758sh, 709w, 650sh, 628sh, 495s, 440s, 425sh, 360m, Minas Gerais, Brazil)	
Tourmaline, Li-tourmaline	1365–1355m, 1310–1290m, 1040–1030s, 1008–985s, 620–400xxbs	2
Trig(3m) (var. rubellite)	[3650, 3580, 3470, 1355m, 1290m, 1103sh, 1030s, 985s, 782m, 719m, 620–400xxbrs, 370m, (?) locality]	

[a] Refer to Table I, footnote[a] for abbreviations.
[b] Raman data in Chapter 12, Section IV,B,3.

The clay minerals, as a group, have been subjected to several additional studies involving their identification via structural and compositional differences (Ahrichs *et al.*, 1966; Boldyrev, 1966; Takeoka, 1970; Calvet and Prost, 1971; Larson *et al.*, 1972).

Parker (1969) classified kaolinites by infrared spectroscopy, and Ledon and White (1966) correlated the OH frequencies with the OH lattice sites. O–H stretching vibrations in kaolinite and minerals of the kaolinite group have been investigated by Van der Marel and Kröhmer (1969).

TABLE X

INFRARED ABSORPTION SPECTRAL DATA (4000–300 CM^{-1}) FOR SELECTED INOSILICATE MINERALS[a]

MINERAL NAME, FORMULA CRYSTAL SYSTEM (AND CLASS)	SELECTED FREQUENCY RANGES AND ABSORPTION MAGNITUDES (EXAMPLE MINERAL: SHOWS ALL ABSORPTIONS PLUS LOCALITY)	NO. OF SAMPLES
	Pyroxenes	
Enstatite, $MgSiO_3$	1060–1025s, 691–662wm, 504–482s, 350–340m	25
O(2/m 2/m 2/m)	(1090sh, 1045sh, 1010s, 971s, 870sh, 775w, 688wm, 660wm, 530m, 490s, 462sh, 448s, 390ms, 350m, Mitchell Co., N. Car.)	
Hypersthene (Fe, Mg)SiO_3	1070–1035s, 945–900s, 638–632m, 508–490s, 335–325m	3
O(2/m 2/m 2/m)	(1057s, 1040sh, 970sh, 940s, 865ms, 758w, 720w, 685s, 638wm, 560sh, 538sh, 502s, 450ms, 395m, 335m, 310sh, Lake St. John, Canada)	
Diopside, $CaMgSi_2O_6$	1080–1070s, 970–965s, 868–860s, 513–510s, 405–400s	3
M(2/m)	(1070s, 980sh, 965sh, 920sh, 860s, 670w, 630wm, 510s, 465s, 400s, 335m, Ontario, Canada)	
Hedenbergite, $CaFeSi_2O_6$	1078–1062s, 970–958s, 870–860s, 515–509s, 405–398ms	3
M(2/m)	(1066s, 1020sh, 963s, 918sh, 862s, 665w, 630wm, 510s, 468s, 398ms, 365sh, 325m, Island of Elba, Italy)	
Augite (Ca, Na, Mg, Fe, Mn) (Fe, Al, Ti)(Si, Al)$_2O_6$	1070–1062s, 970–963s, 862–860s, 515–505s, 410–401s	4
M(2/m)	(1062s, 962s, 862s, 671sh, 625wm, 505s, 470s, 401s, 325wm, 300m, Indian Wells, Ariz.)	
Pigeonite (Mg, Fe, Ca) (Mg, Fe)Si_2O_6	1066s, 960s, 503s, 465s, 325m	1
M(2/m?)	(1096sh, 1066s, 960s, 913s, 860sh, 780sh, 750w, 665xw, 627wm, 503s, 465s, 400m, 325m, Loudon Co., Va.)	
Acmite,[b] $NaFeSi_2O_6$	1060–1048s, 960–939s, 643–638wm, 505–502ms, 465–451ms	6
M(2/m)	(1048s, 1012s, 954s, 730w, 642s, 568m, 502m, 451s, Magnet Cove, Ark.)	
Jadeite,[c] $NaAlSi_2O_6$	1068–1053s, 995–980s, 668–655m, 590–580s, 475–462s, 437–429s	8
M(2/m)	(1065s, 995s, 940sh, 860m, 748w, 718sh, 668m, 615sh, 590s, 470s, 435s, 408m, China)	

TABLE X—(*Continued*)

MINERAL NAME, FORMULA CRYSTAL SYSTEM (AND CLASS)	SELECTED FREQUENCY RANGES AND ABSORPTION MAGNITUDES (EXAMPLE MINERAL:SHOWS ALL ABSORPTIONS PLUS LOCALITY)	NO. OF SAMPLES
Spodumene, $LiAlSi_2O_6$ M(2/m)	1095–1080s, 1035–1020s, 642–635m, 595–588s, 475–470s, 405–398wm (1125sh, 1084s, 1020s, 913s, 859m, 636ms, 590s, 540sh, 475s, 400m, 375m, 320sh, 305w, Dixon, Taos Co., New Mex.)	4
	Pyroxenoids	
Wollastonite, $CaSiO_3$ Tri($\bar{1}$)	1088–1085s, 968–964m, 570–567m, 475s, 455s (1085s, 1060w, 1020m, 964m, 927xw, 903w, 679m, 643m, 567m, 510w, 475s, 455s, 410w, 394w, 338m, 300sh, Devonshire, England)	2
Pectolite, $Ca_2NaHSi_3O_9$ Tri($\bar{1}$)	1060–1050s, 1010–995s, 677–669m, 450s, 416–412m [1050s, 995s, 970sh, 920xbrs, 900xbrs, 669m, 649m, 525m, 450s, 412m, 340w, 305xbrwm (?) locality]	2
Rhodonite, $MnSiO_3$ Tri($\bar{1}$)	1064–1060s, 946–943s, 578–576m, 455–451s, 420–415sh (1100sh, 1060s, 1030s, 1000sh, 943s, 895s, 870sh, 723wm, 690w, 665wm, 576m, 560sh, 535sh, 520sh, 490ms, 451s, 420sh, 388w, 360w, 306xw, Franklin, N.J.)	2
	Amphiboles	
Anthophyllite $(Mg, Fe)_7Si_8O_{22}(OH)_2$ O(2/m 2/m 2/m)	3680wm, 1010xs, 667s, 462xs, 448xs, 392s (3680wm, 1010xs, 880sh, 780w, 667s, 605w, 530ms, 500sh, 462xs, 448xs, 423s, 392s, 340sh, Bakersville, N. Car.)	1
Cummingtonite[b] $(Fe, Mg)_7Si_8O_{22}(OH)_2$ M(2/m)	978s, 690w, 492ms, 433ms (1126m, 1085ms, 1026sh, 1000s, 978s, 900m, 778w, 740xxw, 690w, 648w, 519m, 492ms, 455m, 433ms, New Zealand)	1
Tremolite, $Ca_2Mg_5Si_8O_{22}(OH)_2$ M(2/m)	3668w, 994s, 956s, 504s, 465s, 445s, 388m (3668w, 1093ms, 1043ms, 994s, 956s, 918ms, 760wm, 682wm, 545m, 530m, 504s, 465s, 445s, 422sh, 400m, 388m, 357m, 310wm, St. Lawrence Co., N.Y.)	1

TABLE X—(*Continued*)

MINERAL NAME, FORMULA CRYSTAL SYSTEM (AND CLASS)	SELECTED FREQUENCY RANGES AND ABSORPTION MAGNITUDES (EXAMPLE MINERAL: SHOWS ALL ABSORPTIONS PLUS LOCALITY)	NO. OF SAMPLES
Actinolite, $Ca_2Fe_5Si_8O_{22}(OH)_2$ M(2/m)	3670–3660w, 1105–1098s, 1000–993s, 952–950s, 508–505s, 465–461s, 356–355m (3660w, 1098s, 1050s, 1015sh, 993s, 950s, 920s, 758m, 720sh, 683m, 660sh, 640sh, 540sh, 508s, 461s, 420sh, 400m, 389m, 356m, 312sh, 300sh, Chester, Vt.)	4
Hornblende, $NaCa_2(Mg, Fe, Al)_5(Al, Si)_8O_{22}(OH)_2$ M(2/m)	965–955s, 510–505s, 465–450s (3680sh, 1060s, 965s, 745m, 661m, 510s, 450s, 315m, Faraday, Ont., Canada)	3
Glaucophane, $Na_2(Mg, Fe)_3Al_2$-$Si_8O_{22}(OH)_2$ M(2/m)	1105–1100s, 980–950s, 480–465s (3631w, 1105s, 1050sh, 980s, 900sh, 789w, 730sh, 555m, 475s, 450sh, 368m, 330sh, Sonoma Co., Calif.)	3
Riebeckite, $Na_2(Mg, Fe, Al)_5$-$Si_8O_{22}(OH)_2$ M(2/m)	1125–1124s, 1078–1070s, 980–964s, 538–536m, 453–450s (3612w, 3430w, 1125s, 1077s, 969s, 900sh, 750wm, 700w, 644wm, 538m, 490sh, 453s, 398sh, 314w, El Paso Co., Colo.)	3

[a] Refer to Table I, footnote [a] for abbreviations.
[b] Data for example mineral obtained from Lyon (1963).
[c] Data for example mineral obtained from Moenke (1962).

Reh (1970a) reviewed the infrared absorption studies of the degree of ordering of kaolinite, listing 15 references.

Chester and Elderfield (1971) concentrated on the identification of montmorillonite-type clays in marine sediments, whereas Shimoda and Brydon (1971) studied the intrastratification of mica and montmorillonite. More detailed analysis regarding the montmorillonite structure, especially the interaction between H ions and OH groups (Russell and Fraser, 1971) and between OH and Al (Gupta and Malik, 1969) has been accomplished also.

TABLE XI

Infrared Absorption Spectral Data (4000–300 Cm^{-1}) for Selected Phyllosilicate Minerals[a]

Mineral name, formula crystal system (and class)	Selected frequency ranges and absorption magnitudes (example mineral: shows all absorptions plus locality)	No. of samples
Kaolinite (no. 5),[b] $Al_4Si_4O_{10}(OH)_8$	3705–3691ms, 3672–3660m, 1038–1030s, 545–535s, 475–465ms	4
Tri($\bar{1}$)	(3691ms, 3665m, 3650m, 3619m, 1111m, 1095m, 1030s, 1003s, 932m, 909m, 789w, 749w, 690w, 535s, 465ms, 425m, 360wm, 340m, Lamar Pt., S. Car.)	
Halloysite (no. 12),[b] $Al_4Si_4O_{10}(OH)_8\cdot$ [xH_2O hydrated]	3710–3702wm, 3629–3620m, 1650–1640w, 1034–1020s, 550–530s, 474–465ms	3
M(2/m?)	(3704m, 3627m, 3530m, 1650w, 1034s, 910m, 800sh, 750w, 530s, 465ms, 430m, 360wm, 345wm, Bedford, Ind.)	
Montmorillonite (no. 20)[b] $(Na, Ca)_{0.33}$-$(Al, Mg)_2Si_4O_{10}$-$(OH)_2\cdot nH_2O$	3640–3620mw, 920–910morsh, 525–515s, 470–465s	3
M(2/m?)	(3635m, 1645w, 1060s, 990s, 910m, 836w, 675w, 520s, 470s, 425s, 335m, Polkville, Miss.)	
Serpentine (var. antigorite) $Mg_6Si_4O_{10}(OH)_8$	3690–3675s, 1080–1075s, 990–985s, 410–398s	3
M(2/m?)	(3675s, 1075ms, 985s, 965s, 620m, 560m, 497s, 398ms, 302w, Cardiff, Md.)	
Pyrophyllite, $Al_2Si_4O_{10}(OH)_2$	1120–1115s, 1070–1065s, 951–946ms, 542–535s, 485–480s	3
M(2/m)	[3691m, 3674m, 3660sh, 3620wm, 1120s, 1068s, 1045s, 1040sh, 1020sh, 946s, 918sh, 835w, 815w, 775w, 738wm, 680m, 580sh, 538s, 525sh, 483s, 465sh, 428sh, 395sh, 356m, 353sh, 340sh, (?) locality]	
Talc, $Mg_3Si_4O_{10}(OH)_2$	1020–1010xs, 674–663m, 466–457s, 428–420s, 388–385m	4
M(2/m)	(3680m, 1011xs, 872w, 785xw, 663m, 532m, 500sh, 460s, 446s, 420s, 388ms, 380sh, 340m, 320sh, Fowler, N.Y.)	
Muscovite, $KAl_2(AlSi_3)O_{10}(OH)_2$	1030–1016s, 540–525s, 480–467s, 348–345m	5
M(2/m)	(3625wm, 1067sh, 1016s, 945sh, 838w, 748w, 600sh, 540s, 472s, 405m, 345m, 310w, Keystone, S. Dak.)	

TABLE XI—(*Continued*)

MINERAL NAME, FORMULA CRYSTAL SYSTEM (AND CLASS)	SELECTED FREQUENCY RANGES AND ABSORPTION MAGNITUDES (EXAMPLE MINERAL: SHOWS ALL ABSORPTIONS PLUS LOCALITY)	NO. OF SAMPLES
Phlogopite $KMg_3(AlSi_3O_{10})(OH)_2$	1005–995s, 695–690wm, 466–460s	3
M(2/m)	(3700w, 3460brm, 1090sh, 1005s, 810wm, 730sh, 692wm, 505sh, 463s, 278w, 340wm, Hull, Quebec, Canada)	
Biotite $K(Mg, Fe)_3$-$(AlSi_3O_{10})(OH)_2$	1002–986s, 692–660shw, 465–440s, 352–335w-m	50
M(2/m)	3660w, 3550–3400w, 998s, 720m, 660sh, 446s, 405m, 335m, Redstone, N.H.)	
Lepidolite KLi_2Al-$(Si_4O_{10})(OH)_2$	1022–1001s, 535–525ms, 482–472s, 445–432m	4
M(2/m)	[3400sh, 3350w, 1109s, 1001s, 780w, 744w, 527ms, 480s, 432m, 380sh, 310m, (?) locality]	
Margarite Ca_2Al_4-$(Si_4Al_4)O_{20}(OH)_4$	3635–3634m, 940–915s, 697–690m, 490–485s, 390m	3
	(3635m, 990sh, 925s, 690m, 606m, 538ms, 488ms 435w, 400sh, 390m, Chester, Mass.)	
Chlorite $(Mg, Fe, Al)_6$-$(Al, Si)_4O_{10}(OH)_8$	980s, 760m, 438s, 368ms	1
M(2/m)	(3560wm, 3420wm, 980s, 940sh, 760m, 705m, 460sh, 438s, 368ms, 327sh, 300w, Chester, Vt.)	
Apophyllite,[c] $KFCa_4(Si_8O_{20})\cdot 8H_2O$	3560–3555m, 1695–1687wm, 1016–1010s, 502–498ms, 415–414wm	3
T(4/m 2/m 2/m)	(3555m, 3300–3200sh, 3050w, 1687wm, 1128wm, 1095wm, 1914s, 780sh, 762w, 597w, 530w, 502ms, 475m, 414wm, 316m, Patterson, N.J.)	

[a] Refer to Table I, footnote [a] for abbreviations.
[b] Code identification number refers to a specific STANDARD clay mineral, the spectral data for which are given as the example mineral.
[c] Raman data in Chapter 12, Section IV,B,4–6.

In two heating studies (Shimoda and Brydon, 1971; Hayashi *et al.*, 1969), changes in frequencies have been observed in a variety of sheet structures. Reh (1970b) showed that specific clay mineral associations exist in specific alteration deposits in connection with hydrothermal mineralization.

In a detailed study involving serpentine, Ptak (1970) interpreted the infrared spectrum of chrysotile.

Proton delocalization in micas was the subject of an investigation by Fripiat *et al.* (1965), which dealt with the effect of increasing temperature on OH stretching vibrations. Jorgensen (1966) studied the effect of varying amounts of Fe–Mg–Al–Li on frequencies in the O–H bond region. Epel'-baum *et al.* (1967), Plastinina *et al.* (1969), and Chaussidon (1970) used changes in the spectra of phlogopite to evaluate changes in the structure during compositional changes. Syritso *et al.* (1971) studied micas from metasomatically altered granites.

M. Tektosilicates (Table XII)

Wachter and Lyon (1970) provided a recent summary of research activity including their own extended discussion on two or more silica phases that are mixed; this was also done for the feldspars.

Dodd and Fraser (1967) show the variation of H-bonded OH in synthetic α-quartz. Bukanov *et al.* (1969) worked with natural quartz crystals with differing lithium content and attempted correlation with spectra. A possible use of infrared absorption analyses to determine the temperature range of formation of massive vein quartz was examined by Dolomanova *et al.* (1970). Opal was classified into three groups based on its structure, with a further subdivision being based on its texture, in the study by Jones and Segnit (1971). Gasoyan (1971) developed a method for determining alpha and beta modifications in quartz in vein melt aggregates. Plyusnina *et al.* (1970) studied cryptocrystalline silica, and Manghinani (1970) examined infrared characterisrics of fused silica and silicate under high pressures and temperatures.

Panov and Fedorova (1966) supplemented the studies made by Laves and Hafner (1956) and Hafner and Laves (1957) by continuing to study the degree of ordering in potassic and solic feldspars from granitic rocks. Bahat (1970) restricted his work to high-temperature alkali feldspars. Martin (1970) obtained unit cell parameters and infrared absorption data for high albite–low albite in order to prove his earlier claim of the direct hydrothermal synthesis of ordered albite. Kuznetsova (1971) gives a step-by-step procedure for determining the degree of ordering in K–Na feldspars. Structural transformations in K-feldspars are reflected by infrared data by a change from orthoclase–perthites (Precambrian) through intermediate orthoclase (Paleozoic) to maximum microcline (Mesozoic) selected from ultrametamorphic rocks from a specific region (Kuznetsova and Khoreva, 1972).

Thompson and Wadsworth (1957) and Thompson (1968) worked on the determination of the composition of the plagioclase feldspars by means

TABLE XII

INFRARED ABSORPTION SPECTRAL DATA (4000–300 CM^{-1}) FOR SELECTED TEKTOSILICATE MINERALS[a]

MINERAL NAME, FORMULA CRYSTAL SYSTEM (AND CLASS)	SELECTED FREQUENCY RANGES AND ABSORPTION MAGNITUDES (EXAMPLE MINERAL: SHOWS ALL ABSORPTIONS PLUS LOCALITY)	NO. OF SAMPLES
Quartz,[b] SiO_2	1170–1162ms, 1085–1078xs, 780–773m, 462–454s, 370–366m	57
Trig(32)	(1162ms, 1078xs, 792m, 773m, 689w, 506m, 454s, 390m, 366m, 330sh, Hot Springs, Ark.)	
Cristobalite, SiO_2	1100–1098s, 800–790wm, 485–475m, 435–428m	3
I(4/m $\bar{3}$ 2/m)	(1098s, 1030s, 790wm, 710sh, 640sh, 585m, 544wm, 475m, 428m, 333sh, Inyo Co., Calif.)	
Opal, $SiO_2 \cdot nH_2O$	1100–1085s, 790–785m, 476–470s	6
Amorphous (ess.)	(3700–3200w, 1650w, 1097xs, 785m, 470s, Mexico)	
Sanidine (K, Na)(Al, Si)$_4O_8$	1025–1015s, 639–632m, 546–537m, 430–423s	5
M(2/m)	(1110w, 1020s, 770w, 720m, 639m, 580m, 543m, 425s, 380w, 335w, Leadville, Colo.)	
Orthoclase (K, Na)$AlSi_3O_8$	1050–1030s, 647–638m, 542–534m, 430–420s	3
M(2/m)	(1133s, 1085w, 1050s, 1005s, 770m, 745m, 725m, 645m, 608w, 580m, 538m, 463w, 420s, 372m, 328m, Baveno, Piedmont, Italy)	
Microcline (K, Na)$AlSi_3O_8$	1050–1040s, 650–640m, 538–533m, 430–418s	45
Tri($\bar{1}$)	(1135s, 1050s, 1005s, 767m, 723m, 645m, 580m, 535m, 462m, 418s, 370s, 326m, Gunnison Co., Colo.)	
Albite, $NaAlSi_3O_8$	1010–997s, 650–641m, 541–529m, 428–412s	8
Tri($\bar{1}$)	(1150s, 1095s, 1033s, 1000s, 785m, 758m, 743m, 723m, 645m, 587s, 533m, 475mw, 462m, 425s, 398w, 385xw, 373w, 336m, Renfrew Co., Ont., Canada)	
Oligoclase (Na, Ca)-(Al, Si)$AlSi_2O_8$	1010–1000s, 643–625m, 538–532m, 430–420m	32
Tri($\bar{1}$)	(1145s, 1100s, 1000s, 750s, 625m, 580m, 533m, 460wm, 420m, 385m, 320sh, Tvedstrand, Norway)	
Andesine (Na, Ca)-(Al, Si)$AlSi_2O_8$	1011–995s, 639–632m, 543–528m, 430–420ms	3
Tri($\bar{1}$)	(1150s, 1092s, 1028s, 995s, 782wm, 758wm, 738wm, 718wm, 639m, 605sh, 582ms, 528m, 470m, 458m, 420ms, 410ms, 400ms, 386ms, 370ms, 328m, Risor, Norway)	

TABLE XII—(*Continued*)

MINERAL NAME, FORMULA CRYSTAL SYSTEM (AND CLASS)	SELECTED FREQUENCY RANGES AND ABSORPTION MAGNITUDES (EXAMPLE MINERAL: SHOWS ALL ABSORPTIONS PLUS LOCALITY)	NO. OF SAMPLES
Labradorite (Ca, Na)-(Al, Si)$AlSi_2O_8$	995–990s, 625–620m, 538–535m, 428–422m	3
Tri($\bar{1}$)	(1138sh, 1095s, 990s, 950s, 748w, 625m, 575m, 535m, 465wm, 380m, 310sh, Nain, Labrador)	
Bytownite (Ca, Na) (Al, Si)$AlSi_2O_8$	1000–987s, 628–620m, 542–538m, 425–422ms	8
Tri($\bar{1}$)	(1160sh, 1075s, 1000s, 798w, 775w, 752w, 624m, 540m, 520s, 467s, 422ms, 395m, 375m, 343m, Eyscott Hill, England)	
Anorthite, $CaAl_2Si_2O_8$	988–985s, 624–619m, 540–536m, 433–425m	3
Tri($\bar{1}$)	(1135ms, 1075ms, 985s, 925s, 768w, 752w, 740sh, 725w, 690sh, 660sh, 620m, 565sh, 536m, 482sh, 468m, 425sh, 378m, 345sh, 300sh, Placer Co., Calif.)	
Scapolite $(Na, Ca)_4$-$Al_3(Al, Si)_3Si_6O_{24}$-(Cl, CO_3)	1175–1160s, 1015–1000s, 545–541ms, 465–462ms	2
T(4/m)	(3620w, 3510w, 3470w, 1516m, 1419m, 1160ms, 1110sh, 1050sh, 1000s, 870sh, 850sh, 740sh, 690m, 605m, 541ms, 465ms, 415m, 360m, 300sh, Oaxaca, Mexico)	
Leucite, $KAlSi_2O_6$	1038–1028s, 722–720wm, 436–435ms	2
I(4/m $\bar{3}$ 2/m)	(1028s, 720wm, 635w, 540w, 435ms, 340w, Roccamonfina, Italy)	
Nepheline, $NaAlSiO_4$	1000–990s, 700–690ms, 513–509ms, 472–469s	2
H(6)	(1085sh, 990s, 690ms, 579xw, 509ms, 469s, 415ms, Bancroft, Ont.)	
Heulandite, $CaAl_2Si_7O_{18}\cdot 6H_2O$	1635–1632m, 1040–1025xbrs, 602–595m, 465–450s	3
M(2/m)	(3610m, 3400xbrsh, 1632m, 1190sh, 1025xbrs, 775sh, 715wm, 660wm, 595m, 520sh, 452s, 405m, Two Islands, Nova Scotia)	
Stilbite, $NaCa_2$-$(Al_5Si_{13})O_{36}\cdot 14H_2O$	1645–1640w, 1030–1020s, 449–435ms, 415–395sh	3
M(2/m)	(3620–3480xbrm, 1640w, 1140sh, 1020s, 608sh, 540m, 449ms, 395m, 370m, West Paterson, N.J.)	
Chabazite, $CaAl_2Si_4O_{12}\cdot 6H_2O$	1650–1645m, 1030–1010s, 630–625m, 465–460m, 415–408ms	2
Trig($\bar{3}$ 2/m)	(3740, 3640–3395ms, 1650wm, 1150sh, 1030–1010s, 765sh, 720sh, 625m, 512m, 460m, 430sh, 408ms, Paterson, N.J.)	

TABLE XII—(*Continued*)

MINERAL NAME, FORMULA CRYSTAL SYSTEM (AND CLASS)	SELECTED FREQUENCY RANGES AND ABSORPTION MAGNITUDES (EXAMPLE MINERAL: SHOWS ALL ABSORPTIONS PLUS LOCALITY)	NO. OF SAMPLES
Analcime, $NaAlSi_2O_6 \cdot H_2O$	1635–1630mw, 1040–1020s, 635–620wm, 465–445s	3
I(4/m $\bar{3}$ 2/m)	3612wm, 3580sh, 1630wm, 1108sh, 1020s, 890sh, 766m, 730m, 702sh, 620wm, 542sh, 445s, Parrsboro, Nova Scotia)	
Natrolite, $Na_2Al_2Si_3O_{10} \cdot 2H_2O$	1640–1630w, 980–975s, 633–620m, 417–416m	2
O(mm2)	(3545m, 3330wm, 1630w, 1083ms, 1058ms, 1035sh, 975s, 720sh, 682wm, 653sh, 620m, 575wm, 538wm, 506wm, 440wm, 416m, 400sh, 362wm, 355sh, Peckinsville, Pa.)	

[a] Refer to Table I, footnote [a] for abbreviations.
[b] Raman data in Chapter 12, Section IV,B,4–6.

of infrared spectroscopy. Angino (1969) investigated the far-infrared absorption region and discovered several well-resolved bands.

Schwarcz and Speelman (1965) tried to investigate the coordination of sulfur and carbon in scapolite via their spectra, but concluded that this was impossible owing to interference from Si–O bands. Wehrenberg (1971) discussed in detail the infrared absorption spectra of the same mineral.

Finally, Breger *et al.* (1970) made an infrared study of water in heulandite and other zeolites.

IV. Summary

In reviewing the literature it becomes immediately apparent that while much analytical work has been done relative to the correlation of infrared absorption spectra with mineral compositions and structures, not very much has been accomplished with respect to additional geological applications of these type data. In addition to the work that is referred to herein, Wachter and Lyon (pp. 53–54, 1970) summarize some of the applications that are more directly related to economic geology for both individual minerals and polymineralic rocks.

Finally, spectral data for 42 additional minerals have been obtained, but have not been included herein. These minerals, the spectral data for which can be obtained from the writer upon request, are as follows: adularia, anorthoclase, axinite, benitoite, bixbyite, borax, cancrinite, carnotite, celsian, cerargyrite, columbite, diaspore, dickite, dioptase, hausmannite, hectorite, hemimorphite, huebnerite, illite, laumontite, lawsonite, lechatelierite (glass), nacrite, neptunite, nontronite, paragonite, periclase, prehnite, pyrochlore, serpentine (variety chrysotile), sodalite, spessartite, stishovite, tantalite, thorite, torbernite, turquois, tyuyamunite, uraninite, vivianite, xenotime, and zoisite.

References

Adler, H. H. (1964). *Amer. Mineral.* **49**, 1002–1015.
Adler, H. H. (1965). *Amer. Mineral.* **50**, 1553–1562.
Adler, H. H., and Kerr, P. F. (1963a). *Amer. Mineral.* **48**, 124–137.
Adler, H. H., and Kerr, P. F. (1963b). *Amer. Mineral.* **48**, 839–853.
Ahrichs, J. L., Russell, J. R., Harter, R. D., and Weismiller, R. A. (1966). *Proc. Indiana Acad. Sci.* **75**, 247–255.
Anderson, A., and Wong, L. Y. (1969). *Can. J. Chem.* **47**, 2713–2715.
Angino, E. E. (1969). *Kans. State Geol. Surv., Bull.* **94**, 9–11.
Bahat, D. (1970). *J. Geol. Soc. Aust.* **17**, 93–102.
Bhatnagar, V. M. (1968). *Bull. Soc. Chim. Fr.* pp. 125–127.
Boldyrev, A. I. (1966). *Mineral. Sb.* **20**, 96–101.
Boldyrev, A. I., and Povarennykh, A. S. (1968a). *Mineral. Sb.* **22**, 11–21.
Boldyrev, A. I., and Povarennykh, A. S. (1968b). *Zap. Vses. Mineral. Obshchest.* **97**, 3–12.
Breger, I. A., Chandler, J. C., and Zubovic, P. (1970). *Amer. Mineral.* **55**, 825–840.
Bukanov, V. V., Kreyscop, V. N., and Markova, G. A. (1969). *Mineral. Sb.* **23**, 261–269.
Burns, R. G., and Greaves, C. (1971). *Amer. Mineral.* **56**, 2010–2033.
Burns, R. G., and Huggins, F. E. (1972). *Amer. Mineral.* **57**, 967–985.
Calvet, R., and Prost, R. (1971). *Clay Miner.* **19**, 175–186.
Chaussidon, J. (1970). *Bull. Groupe Fr. Argiles.* **22**, 105–106.
Chester, R., and Elderfield, H. (1966). *Chem. Geol.* **1**, 277–290.
Chester, R., and Elderfield, H. (1967). *Sedimentology* **9**, 5–21.
Chester, R., and Elderfield, H. (1971). *Chem. Geol.* **7**, 97–105.
DeCamargo, W. G. R., Nazario, G., and Leite, C. R. (1970). *Int. Mineral. Ass., Int. Ass. Genesis Ore Deposits, Gen. Meet.* **7**, 237.
Dodd, D. M., and Fraser, D. B. (1967). *Amer. Mineral.* **52**, 149–160.
Dolomanova, E. I., and Rudnitskaya, E. S. (1969). *Izv. Akad. Nauk SSSR, Ser. Geol.* **1**, 145–148.
Dolomanova, E. I., Gasoyan, M. S., Rudnitskaya, E. S., Tyutneva, G. K., and Uspenskaya, A. B. (1970). *Dokl. Akad. Nauk SSSR* **192**, 155–158.
Dowty, E. (1971. *Amer. Mineral.* **56**, 1983–2009.
Duke, D. A., and Stephens, J. D. (1964). *Amer. Mineral.* **49**, 1388–1406.

Elderfield, H. (1970). *Instrum. News* **21**, 14–15.

Elderfield, H., and Glasby, G. P. (1973). *Chem. Geol.* **11**, 117–122.

Engel, G., and Klee, W. E. (1970). *Z. Kristallogr., Kristallgeometric, Kristallphys., Kristallchem.* **132**, 425.

Epel'baum, M. B., Zhigun, I. T., and Bobrov, B. S. (1967). *Stroit. Mater. Betony* **2**, 167–173.

Fijal, J. (1970). *Przegl. Geol.* **18**, 22–26.

Fijal, J., and Klapyta, Z. (1971). *Przegl. Geol.* **19**, 76–80.

Forneris, R. (1969). *Amer. Mineral.* **54**, 1062–1074.

Fortsch, E. (1970). *Neues Jahrb. Mineral., Abh.* **113**, 219–250.

Fripiat, J. J., Rouxhet, P. P., and Jacobs, H. (1965). *Amer. Mineral.* **50**, 1937–1958.

Gasoyan, M. S. (1971). *Izv. Akad. Nauk SSSR, Ser. Geol.* **4**, 82–87.

Goncharova, E. I., Marchenko, E. Y., and Amirnov, G. I. (1970). *Geokhimiya* **7**, 864–866.

Gotman, Y. D., Pudovkina, I. A., and Solutseya, L. S. (1963). *Vop. Magmat., Metamorf. Rudoobrazov.* pp. 80–87.

Grim, R. E. (1968). *In* "Clay Mineralogy," 2nd ed., pp. 448–463. McGraw-Hill, New York.

Gupta, G. C., and Malik, W. W. (1969). *Amer. Mineral.* **54**, 1625–1634.

Hafner, S , and Laves, F. (1957). *Z. Kristallogr., Kristallgeometric, Kristallphys., Kristallchem.* **109**, 204–225.

Hayashi, H., Otsuka, R., and Imai, N. (1969). *Amer. Mineral.* **54**, 1613–1624.

Huggins, F. E. (1973). *Chem. Geol.* **11**, 99–109.

Jones, J. B., and Segnit, E. R. (1971). *J. Geol. Soc. Aust.* **18**, 57–67.

Jorgensen, P. (1966). *Proc. Nat. Conf. Clays, Clay Miner., 13th, 1964* pp. 263–273.

Karr, C., Jr., and Kovach, J. J. (1969). *Appl. Spectrosc.* **23**, 219–223.

Kislovskiy, L. D., and Knubovets, R. G. (1970). *Zap. Vses. Mineral. Obshchest.* **99**, 609–614.

Klee, W. E., and Engel, G. (1970). *J. Inorg. Nucl. Chem.* **32**, 1837–1843.

Kuznetsova, L. G. (1971). *Mineral. Sb.* **25**, 18–26.

Kuznetsova, L. G., and Khoreva, B. Y. (1972). *Zap. Vses. Mineral. Obshchest.* **101**, 486–489.

Kuznetsova, L. G., and Moskaleva, V. N. (1968). *Zap. Vses. Mineral. Obshchest.* **97**, 715–718.

Larson, S., Pardor, G. W. F., and Gebbie, H. A. (1972). *Amer. Mineral.* **57**, 998–1002.

Laves, F., and Hafner, S. (1956). *Z. Kristallogr., Kristallgeometric, Kristallphys., Kristallchem.* **113**, 265–296.

Ledon, R. L., and White, J. L. (1966). *Proc. Nat. Conf. Clays Clay Miner., 13th, 1964* pp. 389–315.

LeGeros, R. Z., Trantz, O. R., LeGeros, J. P., and Klein, E. (1968). *Appl. Spectrosc.* **22**, 357.

LeGeros, R. Z., LeGeros, J. P., Trantz, O. R., and Klein, E. (1970). *Mid-Amer. Spectrosc. Symp.* **7B**, 3–12.

Liese, H. C. (1974). *Appl. Spectrosc.* **28**, 135–139.

Lyon, R. J. P. (1962). "Minerals in the Infrared." Stanford Res. Inst., Stanford, California.

Lyon, R. J. P. (1963). "Evaluation of Infrared Spectrophotometry for Compositional Analysis of Lunar and Planetary Soils." Stanford Research Institute for NASA, Washington, D. C.

Lyon, R. J. P. (1967). *In* "Physical Methods in Determinative Mineralogy" (J. Zussman, ed.), pp. 371–404. Academic Press, New York.

Manghinani, M. H. (1970). *U. S. Off. Nav. Res. Contract, Final Rep., Hawaii Inst. Geophys.* **70–30,** 1–38.

Martin, R. F. (1970). *Contrib. Mineral. Petrol.* **26,** 72–74.

Moenke, H. (1962). "Mineral Specktren." *Deut. Akad. Wiss. Berlin,* Akademie-Verlag, Berlin.

Nyquist, R. A., and Kagel, R. O. (1971). "Infrared Spectra of Inorganic Compounds (3800–45 cm^{-1})." Academic Press, New York.

Omori, K. (1969). *J. Mineral. Soc. Jap.* **9,** 219–234.

Omori, K. (1971a). *Amer. Mineral.* **56,** 841–849.

Omori, K. (1971b). *Amer. Mineral.* **56,** 1607–1616.

Panov, E. N., and Fedorova, L. G. (1966). *Zap. Vses. Mineral. Obshchest.* **95,** 739–743.

Parker, T. W. (1969). *Clay Miner.* **8,** 135–136.

Plastinina, M. A., Lugovskaya, Y. S., and Kukovskiy, Y. G. (1969). *Mineral. Sb.* **23,** 314–316.

Plyusnina, I. I. (1963). *Geochemistry* (*USSR*) **2,** 174–180.

Plyusnina, I. I., and Zaytseva, L. A. (1969). *Vestn. Mosk. Univ., Ser. Geol.* **24,** 110–115.

Plyusnina, I. I., Maleyev, M. N., and Yefimova, G. A. (1970). *Izv. Akad. Nauk SSSR, Ser. Geol.* **9,** 78–83.

Povarrennykha, A. S., and Gevorkyan, S. V. (1970). *Dopov. Akad. Nauk. Ukr. RSR, Ser. B.* 118–121.

Povarennykha, A. S., and Gevorkyan, S. V. (1971). *Dopov. Akad. Nauk Ukr., RSR, Ser. B,* 306–310.

Povarennykha, A. S., and Nefedov, E. V. (1971). *Geol Zh.* (*Russ. Ed.*) **31,** 13–27.

Preud'homme, J., and Tarte, P. (1971). *Spectrochim. Acta, Part A* **27,** 1817–1835.

Ptak, W. (1970). *Pol. Akad. Nauk, Oddzial Krakowie, Kom. Ceram., Pr., Ceram.* **14,** 37–42.

Rao, C. N. R. (1964). "Chemical Applications of Infrared Spectroscopy." Academic Press, New York.

Reh, H. (1970a). *Int. Union Geol. Sci., Ser. A* **2,** 278–282.

Reh, H. (1970b). *Ber. Deut. Ges. Geol. Wiss., Reihe B* **15,** 345–361.

Russell, J. D., and Fraser, A. R. (1971). *Clays Clay Miner.* **19,** 55–59.

Rutstein, M. S., and White, W. S. (1971). *Amer. Mineral.* **56,** 877–887.

Schwarcz, H. P., and Speelman, E. L. (1965). *Amer. Mineral.* **50,** 656–666.

Seidl, V., Knop, O., and Falk, M. (1969). *Can. J. Chem.* **47,** 1361–1368.

Shcherban, S. A., and Nurmagambetov, K. N. (1970). *Zh. Prikl. Spektrosk.* **13,** 566–568.

Shimoda, S., and Brydon, J. E. (1971). *Clays Clay Miner.* **19,** 61–66.

Stubican, V., and Roy, R. (1961a). *Z. Kristallogr., Kristallogeometric, Kristallphys., Kristallchem.* 200–214.

Stubican, V., and Roy, R. (1961b). *J. Amer. Ceram. Soc.* **44,** 625–627.

Stubican, V., and Roy R. (1961c). *Amer. Mineral.* **46,** 32–52.

Syritso, L. F., Zalashkova, N. Y., and Zorina, L. N. (1971). *Int. Geol. Rev.* **13,** 717–729.

Takeoka, K. (1970). *Sci. Eng. Rev. Doshisha Univ.* **8,** 78–89.

Thompson, C. S. (1968). *Diss. Abstr. B* **28,** 2973.

Thompson, C. S., and Wadsworth, M. E. (1957). *Amer. Mineral.* **42,** 334–341.

Van der Marel, H. W., and Krohmer, P. (1969). *Beitr. Mineral. Petrol.* **22,** 73–82.

Vlasova, E. V., and Topunova, G. A. (1971). *Akad. Nauk SSSR* **20,** 40–46.

Wachter, B. G., and Lyon, R. J. P. (1970). "Infrared Mineralogical Techniques for Ore Search." Dept. Mineral Eng., Stanford University, Stanford, California.

Wehrenberg, J. P. (1971). *Amer. Mineral.* **56,** 1639–1654.

White, J. L. (1971). *Bauxite-Alumina-Aluminum* **2,** 115–122.

White, W. B. (1967). *Mater. Res. Bull.* **2,** 381–394.

White, W. B., and Keramidas, B. G. (1971). *In* "Solid State Chemistry: Novel High Temperature Materials," pp. 87–88. U. S. Dept. Comm., Nat. Bur. Stand. Inst. Mater. Res., Washington, D. C.

Wilkins, R. W. T. (1970). *Amer. Mineral.* **55,** 1993–1998.

Wood, D. L., and Nassau, K. (1968). *Amer. Mineral.* **53,** 777–800.

Yu, S. C., and Zoltai, T. (1972). *Amer. Mineral.* **57,** 1873.

Ziborova, T. A., and Khramtsov, V. N. (1969). *Geol. Polez. Iskop. Urala* **1,** 296–297.

Zorina, M. L., and Syritso, L. F. (1972). *Zh. Prikl. Spektrosk.* **16,** 1043–1045.

9

FAR-INFRARED SPECTROSCOPY OF MINERALS

John J. Kovach, Arthur L. Hiser, and Clarence Karr, Jr.

I. Introduction

A practical result that evolved out of the infrared analysis of lunar samples in our laboratory was a thorough examination of the spectral–structural correlations of a wide variety of terrestrial minerals in the low-frequency and far-infrared regions.

The far-infrared region of the spectrum is generally limited to the frequencies below 200 cm^{-1} (Triple Commission on Spectroscopy, 1962). However, many absorption bands in the 200–500 cm^{-1} region need to be considered since the very low frequency results are often difficult to correlate unless the higher frequency region is also considered. This is particularly true for many minerals, since the analytically useful absorption frequencies occur in the region below 500 cm^{-1}.

The availability of commercial far-infrared spectrophotometers capable of routine operation has provided the chemist with a powerful analytical tool. The accessibility of this region now makes it possible to characterize more completely many mineral classes since many vibrations that occur in this region are very sensitive to compositional as well as structural changes. The analytical application of the far-infrared (Karr, 1972) for minerals is limited, particularly for the silicates, since the data are sparse. This is due to the complexities involved in the silicates and the almost necessary use of normal coordinate treatment for a meaningful interpretation of the results.

This chapter will describe the application of the far-infrared region to the analysis of the olivine, pyroxene, and feldspar silicates and the carbonate minerals. Far-infrared spectral data on some representative lunar rocks and fines are given in Chapter 11, Section IV.

II. Experimental

Samples used were both natural and synthetic specimens. Phase purity was verified using both x-ray analysis and comparison with available infrared spectra. All samples were ground under alcohol to minimize sample alteration. Spectra were obtained on a Perkin-Elmer 621 grating infrared spectrophotometer equipped with a dry air purge in the region 500–200 cm^{-1} (equipment is named for identification only and does not necessarily imply endorsement by the U.S. Bureau of Mines). A Perkin-Elmer FIS-3 vacuum far-infrared spectrophotometer was used in the 400–30 cm^{-1} region. Cesium iodide was used as the matrix in the 500–200 cm^{-1} region and polyethylene was used from 400 to 30 cm^{-1}. Combination of the spectra obtained from both instruments is presented since the absorption bands to be discussed in some cases lie near the frequency limits of both instruments. As such, the spectra were combined to give the best spectral presentations for discussion. Frequencies within the mineral series to be covered were read consistently within each matrix to minimize any frequency effects that may result from the two different matrices.

III. Factors Influencing the Application of the Far-Infrared Region for Minerals

Normal coordinate treatment by a physical description of the vibrational modes involved can best compensate for the various factors that can affect the analysis of minerals. However, in many instances vibrational assignments have not been made or are incomplete. Thus in the application of the low-frequency infrared region to minerals, a number of factors need to be considered in the interpretation of the results.

The far-infrared frequencies are primarily dependent on the M—O bonds and will vary as the cation changes. Generally, for the silicates the lattice functions as a network orienting these various cations. As such, the chemical composition of a particular mineral is necessary, together with confirmation of a particular structure by x-ray analysis and comparison with existing infrared data. Polymorphism is common and lattice distortion is frequent. Lattice distortion can occur as a result of formation conditions or can result from the physical handling of the mineral. Lattice distortion can change the suggested symmetry, causing either the appearance or the disappearance of absorption bands. The most important factor to consider is the effect of isomorphous substitution, which is more closely related to ionic size rather than chemical similarity. Cation replacement often occurs, such as the substitution of aluminum and germanium for silicon in the tetrahedra. Aluminum can also replace some of the six-coordinated cations Mg^{2+}, Fe^{2+}, or Ca^{2+} in eightfold coordination. Since in many cases the type and extent of substitution is dependent on geological conditions, this too becomes an important consideration (see Chapter 8 for some examples of the effects of geological conditions). For the silicates, such as the pyroxenes, the cation distribution is both temperature and pressure dependent, and mineral origin becomes important in making spectral comparisons.

The commercially available synthetic minerals of known composition can now be obtained to supplement the use of natural minerals. While they are useful in making spectral comparisons, agreement with natural specimens is important because inhomogeneity of the sample and varying degrees of crystallinity can occur.

IV. Nesosilicates: Olivines

A. *Symmetry and Cation Coordination*

The olivine minerals have the composition M_2SiO_4, where M is the divalent cation Mg^{2+}, Fe^{2+}, Mn^{2+}, or Ca^{2+} (or combination of the cations).

The structure of the olivines consists of independent SiO_4 tetrahedra linked by the divalent cations in sixfold coordination. There are two distinct crystallographic six-coordinate cation positions in the structure—M1, M2—and each site is distorted from octahedral symmetry. The M1 site is elongated along the O_3–O_3 axis, and M2 is more irregularly distorted and is the larger site (Burns, 1970a). The structure is characterized by orthorhombic symmetry and belongs to the space group Pbnm. Olivine contains four stoichiometric units with a total of 28 atoms per unit cell. Substitution in natural olivines occurs in the octahedral sites, with only trace amounts in the tetrahedral positions (Deer *et al.*, 1967). The classification of crystal vibrations of olivine is presented in Chapter 13, Section II, B.

Below 500 cm^{-1} the olivine minerals exhibit a fairly uniform infrared spectrum which contains intense bands that are sensitive to varying cation content. The intense band in the forsterite spectra (shown in Fig. 1) at 415 cm^{-1} is attributed in part to the ν_2 vibration of the SiO_4 tetrahedra and an unspecified mode of the MgO_6 octahedra. The 415 cm^{-1} band is shifted to lower frequency with the substitution of Ni^{2+} and Ge in the octahedral and tetrahedral positions of the olivine structure, respectively.

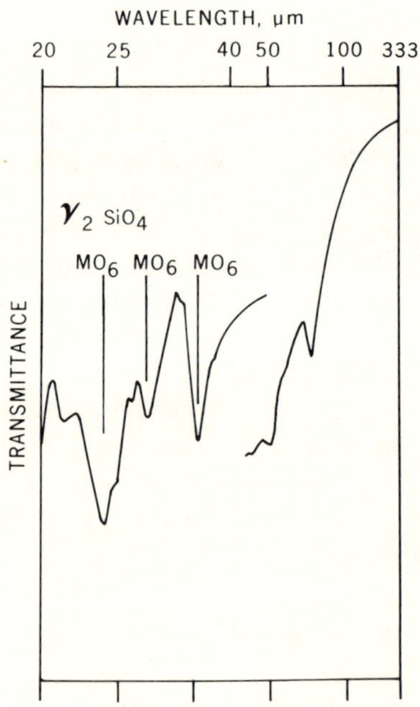

FIG. 1. Far-infrared spectrum for forsterite, Mg_2SiO_4.

The 360 and 292 cm^{-1} bands are due to vibrations associated with the MO_6 octahedra and are relatively unaffected by substitution in the tetrahedral positions (Tarte, 1963; Paques-Ledent and Tarte, 1973). The 224, 201, and 142 cm^{-1} bands are assigned to mixed vibrational modes involving both the SiO_4 and MO_6 coordination polyhedra (Oehler and Günthard, 1969). The effect of cation substitution in the octahedral position for the olivines can be seen in Table I. There is a progressive shift to lower frequencies for the 360 and 292 cm^{-1} band for olivine as the mass and size of the cation increases. However, the shift is more progressive with size since the masses of Ni^{2+}, Fe^{2+}, and Mn^{2+} are relatively the same. While the 201 cm^{-1} band shifts to lower frequency, it appears to be more influenced by the silicate lattice. The 142 cm^{-1} band shifts to lower frequency and appears to be mass dependent. However, intermediate peaks of Ni and Fe olivines could not be detected, probably because of weak intensity.

B. Effect of Cation Ordering on the Infrared Spectra of Olivines

1. *Forsterite,* Mg_2SiO_4; *Monticellite* $(CaMg)_2SiO_4$; *and* γ-Ca_2SiO_4

The vibrational frequencies for Mg_2SiO_4, $(CaMg)_2SiO_4$, and γ-Ca_2SiO_4 are shown in Table II. The frequencies at 360 and 292 cm^{-1} for forsterite for the MO_6 octahedra vibrations show only small shifts in frequency when compared with monticellite. While they shift slightly to higher frequencies, these values are in general agreement with those reported by Tarte. The relative invariance of these bands indicates that these vibrations are more closely associated with the MO_6 octahedra in the M1 position of the olivine structure. In monticellite the Mg^{2+} is ordered on the M1 sites, while the Ca^{2+} is ordered on the M2 site. The 142 cm^{-1} band in olivine is analogous to the 117 cm^{-1} band in monticellite and does appear to be influenced by the Ca^{2+} in the M2 site. The 117 cm^{-1} band does not shift in going to γ-Ca_2SiO_4 in which both sites are occupied by Ca^{2+}.

2. *Forsterite* (Mg_2SiO_4)–*Fayalite* (Fe_2SiO_4)

Linear frequency composition determinative curves for the forsterite–fayalite series have been obtained with the absorption band associated primarily with the fundamental vibrations (Duke and Stephens, 1964; Burns and Huggins, 1972). The linear dependency of the 415 cm^{-1} band of forsterite with increased iron substitution has been shown to be a useful analytical frequency for the compositional determination of lunar olivines (Estep *et al.*, 1971). If a vibration is not particularly cation-site sensitive, or the distribution of the different cations on the sites is random, the determinative curve would be linear and reflect bulk composition. However,

TABLE I

LOW-FREQUENCY ABSORPTION BANDS FOR OLIVINES

MINERAL	FREQUENCY (CM^{-1})												
	ν_2-SiO_4 MO6				MO6			MO6					
Forsterite													
Mg_2SiO_4	467(w)[a]	415(s)	403(sh)	380(w)	360(m)	319(w)		292(s)	277(sh)	224(w)	201(w)		142(w)
Ni_2SiO_4	480(s)	375(s)			330(s)		305(w)	270(w)		220(w)	198(w)		
Fayalite													
Fe_2SiO_4	475(s)	360(m)			310(m)			250(m)			196(w)	180(w)	
Tephroite													
Mn_2SiO_4	480(s)	340(m)			280(s)			230(m)			190(w)	178(w)	102(w)

[a] Abbreviations: w, weak; m, medium; s, strong; sh, shoulder.

TABLE II

LOW-FREQUENCY ABSORPTION BANDS FOR SYNTHETICS OF THE OLIVINE STRUCTURE

	FREQUENCY, (CM^{-1})												
Forsterite													
Mg_2SiO_4		380(w)[a]	360(m)	319(sh)		292(s)	271(sh)		224(w)	201(w)			142(w)
Monticellite													
$(Ca, Mg)_2SiO_4$			370(m)			295(s)		235(m)	214(sh)	202(m)		167(w)	117(w)
γ-Ca_2SiO_4[b]	394(sh)		332(s)		307(w)	265(m)		231(m)	215(m)		187(w)		117(w)

[a] Abbreviations: w, weak; m, medium; s, strong; sh, shoulder.
[b] Oehler and Günthard (1969).

if the vibration is site sensitive and appreciable ordering occurs, a significant deviation from linearity could result (Burns and Huggins, 1972). While essentially free replacement of the Mg^{2+} and Fe^{2+} in the forsterite–fayalite series occurs, ordering can occur on the two different crystallographic sites. The larger Fe^{2+} would occupy the larger M2 site, with the smaller Mg^{2+} on the M1 site. Figure 2 shows the determinative curves obtained for the 360, 292, and 142 cm^{-1} bands of forsterite attributed to the MO_6 octahedral vibration of the M1 site for the first two bands and M2 for the

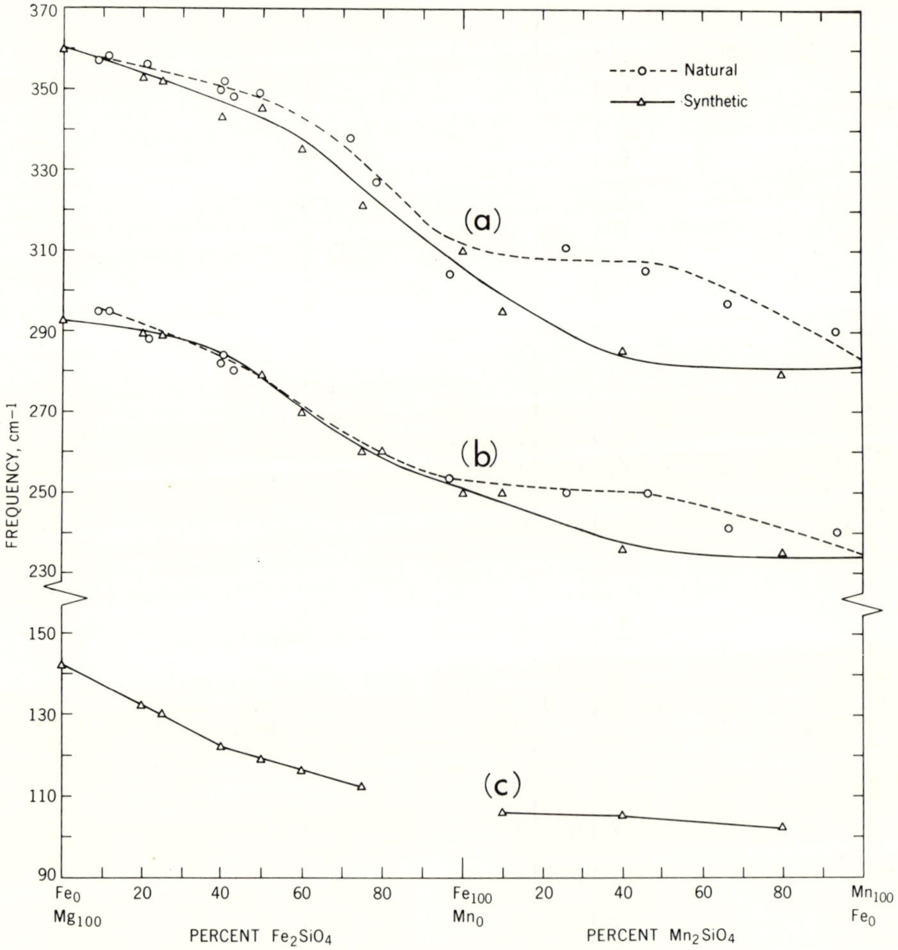

FIG. 2. Frequency variation with composition for the far-infrared absorption bands for the forsterite–fayalite–tephroite olivine series. Curves (a) and (b) show M2-site dependence and curve (c) shows M1-site dependence.

last band. If as the fayalite content increases the Fe^{2+} distribution is random, then the determinative curves would probably be linear. If the distribution is not completely random on both sites, deviation from linearity in a predictable manner would occur. Curves (a) and (b) of Fig. 2 suggest that for lower amounts of Fe^{2+} substitution on the M1 site occurs at a lower rate than that at higher iron concentrations. At higher concentrations, as a proportionately greater number of the M2 sites are occupied, additional substitution of iron results in more iron introduced into the M1 site but at a higher rate than at lower concentrations. The general trends observed for the synthetics for the 360 and 292 cm^{-1} bands agree well with data on the natural specimens (Burns and Huggins, 1972) also showing a more rapid decrease in frequency at higher iron contents. The inverse shape of curve (c) correlates with enrichment of the iron on the M2 site at lower iron contents. The shift to lower frequencies for the 142 cm^{-1} band with increased iron is accompanied by a decrease in intensity so that this band was not detectable beyond 75% Fe_2SiO_4. For the forsterite–fayalite series from crystal field spectra Burns (1970b) concluded that the occupancy of Fe^{2+} ions approaches a random distribution on the M1 and M2 sites, with only a very small tendency toward relative enrichment in the M2 position. Ordering in the natural forsterite–fayalite series can be affected by the presence of Ca^{2+} and Mn^{2+}. The larger Ca^{2+} and Mn^{2+} would occupy the larger M2 site, thus causing an enrichment of the Fe^{2+} in the M1 site. The nonlinearity of the frequency/composition curves of the low-frequency infrared absorption bands of olivines shown in Fig. 2 appears to be a particularly sensitive indicator of ordering on the M1 and M2 sites.

3. *Fayalite (Fe_2SiO_4)–Tephroite (Mn_2SiO_4)*

The observed frequencies for the synthetic Fe–Mn series are listed in Table III. The shift with increased Mn^{2+} content is to lower frequency, but

TABLE III

LOW-FREQUENCY BANDS FOR SYNTHETIC FE–MN COMPOUNDS OF THE OLIVINE STRUCTURE

COMPOSITION	FREQUENCY, CM^{-1}					
Fe_2SiO_4	362(m)[a]	310(s)	256(w)	196(m)	180(m)	—
$(Fe_{90}Mn_{10})_2SiO_4$	363(m)	295(s)	250(w)	194(m)	178(m)	106(w)
$(Fe_{60}Mn_{40})_2SiO_4$	360(m)	285(s)	236(w)	192(m)	175(m)	105(w)
$(Fe_{20}Mn_{80})_2SiO_4$	355(m)	280(s)	235(w)	188(m)	170(m)	102(w)

[a] Abbreviations: s, strong; m, medium; w, weak.

the overall frequency shift with composition was not as large as that observed for the Mg–Fe series. The 196 and 180 cm^{-1} bands in Fe_2SiO_4 show a linear shift in frequency with increased Mn^{2+} content together with a systematic change in the ratio of their intensities. The determinative curves for the 256 and 310 cm^{-1} bands of Fe_2SiO_4 show a trend for the natural Fe–Mn olivines similar to that for the natural Mg–Fe series—curves (a) and (b) in Fig. 2. The determinative curves for the natural specimen indicate that for low Mn^{2+} content up to 50% Mn_2SiO_4 there is little shift in frequency for these bands attributed to the MO_6 octahedra for the M1 site, while at higher concentrations of Mn^{2+} the shift in frequency is more pronounced. For the natural specimens in the Fe–Mn series, ordering occurs with the Fe^{2+} enriched on the M1 site (Burns and Huggins, 1972; Burns, 1970b). If at low concentration of Mn^{2+} (less ca. 50%) iron is preferentially in the M1 position, little frequency shift would be expected. At higher concentrations of Mn^{2+} (with more M2 sites already filled) additional Mn^{2+} occupies the M1 site, thus causing a more pronounced shift in frequency. The determinative curves (a) and (b) for the synthetic Fe–Mn olivines indicate that Mn^{2+} is enriched in the M1 site at lower concentrations; the difference in ordering compared with the natural specimens may be attributed to sample history. Since the Mn^{2+} and Fe^{2+} are quite similar in size and mass, the distribution of the cations on the sites would be sensitive to formation conditions. For the natural specimens the distribution is probably the result of equilibrium conditions, while for the synthetic, other parameters such as ionic mobility may have the greater influence. The close agreement between curves (a) and (b) for the natural and synthetic olivines of the Mg–Fe series may result from relatively large differences in Mg and Fe in both size and mass such that the distribution of the cations in the synthetic samples could approximate that observed for the natural specimens. Curve (c) for the synthetic Fe–Mn series has the inverse shape of curves (a) and (b), indicating that Mn^{2+} entered the M2 site at a slightly higher rate at the higher concentrations.

V. Chain Silicates: Pyroxenes

A. Symmetry and Cation Coordination

The pyroxenes are chain silicates of composition $(SiO_3)_n$ formed by the sharing of two of four corners of the SiO_4 tetrahedra. The chains are linked laterally by cations of Ca, Mg, Fe, Na, and Al. The pyroxenes in which calcium predominates in the M2 site (diopside–hedenbergite) have the symmetry of the space group C2/c. The calcium is eight-coordinated by oxygen and all silicate chains are crystallographically equivalent and are relatively straight. Pyroxenes containing little or no calcium have the M2 sites occupied by Mg or Fe, which are in sixfold coordination with oxygen. Reduced coordination of the M2 site occurs in pigeonite, clinoenstatite, and clinoferrosilite by contorting half the silicate chains. The silicate chains are no longer crystallographically equivalent and the space group is $P2_1/c$. The orthorhombic pyroxenes differ from $P2_1/c$ in the stacking sequence along the *a* axis; the Pbca orthorhombic stacking sequence appears stable when the Ca content is less than about 5 mole% $CaSiO_3$ (Smyth and Burnham, 1972).

B. Diopside–Hedenbergite Series

Diopside is the high calcium pyroxene and has the space group C2/c. The unit cell contains four formula units with four equivalent M1 positions and four equivalent M2 positions. Diopside forms a solid solution along the diopside–hedenbergite tie line. Ca^{2+} occupies the M2 site, with Fe^{2+} substituting for the Mg^{2+} in the M1 position. Iron orders on the M2 site when calcium is deficient.

The far-infrared spectra for diopside and hedenbergite are shown in Fig. 3. For a series of hedenbergite specimens of composition $(Ca_{0.50}Mg_{0.23}Fe_{0.27})SiO_3$ to $(Ca_{0.50}Mg_{0.10}Fe_{0.385})SiO_3$ and in which the iron is ordered on the M1 site (Bancroft and Williams, 1971) and the M2 site is completely occupied by calcium, no significant frequency shifts from the diopside spectrum were observed in the region 500–200 cm^{-1} with increased iron. However, the hedenbergite of composition $(Ca_{0.45}Mg_{0.10}Fe_{0.425})SiO_3$, in which 10% of the M2 sites are occupied by iron as a result of deficient calcium, showed frequency shifts from that of diopside. The 395 cm^{-1} band was no longer split and the 200 cm^{-1} band increased in intensity. The 395 cm^{-1} band in diopside was assigned (Omori, 1971) to a bending vibration of the chain associated with the calcium and the nonbridged oxygens and the 220 cm^{-1} band to a stretching vibration involving both the Mg^{2+} and Ca^{2+} and the nonbridged oxygen of the chain. The 395 cm^{-1} band occurs at the same frequency in hedenbergite. However, the 400 cm^{-1} band disappears in hedenbergite and is probably due to a bending vibration of the silicate chain associated with the magnesium. The 220 cm^{-1} band of diopside does not appear to be site dependent because no frequency shifts were observed with substitution in either the M1 or M2 sites.

The region below 220 cm^{-1} for both diopside and hedenbergite shows intense bands that appear to be calcium dependent. The intense band

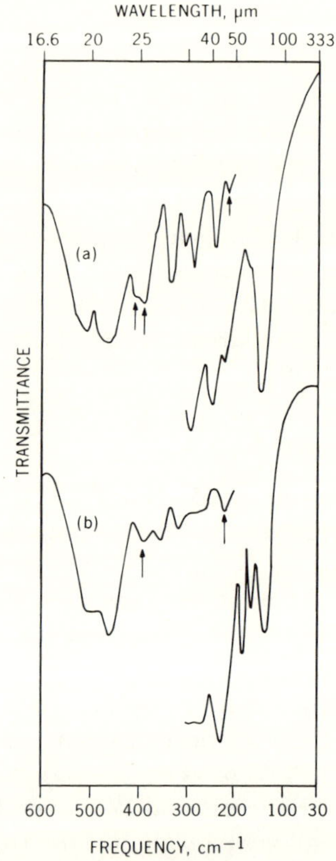

FIG. 3. Far-infrared spectra of (a) diopside $(Ca_{0.50}Mg_{0.50})SiO_3$ and (b) hedenbergite $(Ca_{0.45}Mg_{0.10}Fe_{0.425})SiO_3$.

near 150 cm^{-1} for diopside splits into three for hedenbergite. Synthetic specimens of composition $(Ca_{0.10}Fe_{0.22}Mg_{0.68})SiO_3$, $(Ca_{0.20}Fe_{0.22}Mg_{0.58})SiO_3$, $(Ca_{0.30}Fe_{0.22}Mg_{0.48})SiO_3$, and $(Ca_{0.40}Fe_{0.22}Mg_{0.38})SiO_3$ all show a band near 150 cm^{-1} that increases in intensity as the amount of calcium increases and in diopside at $Ca_{0.50}$ this band is most intense. Because essentially no frequency shift is observed and because in the high-calcium pyroxenes such as diopside the silicate chains are relatively straight and equivalent, this low-frequency band may be due to chain deformations that are dependent on the Ca^{2+} content. Synthetic samples of composition $(Ca_{0.10}Fe_{0.45}Mg_{0.45})SiO_3$, $(Ca_{0.20}Fe_{0.45}Mg_{0.35})SiO_3$, $(Ca_{0.30}Fe_{0.45}Mg_{0.25})SiO_3$, and $(Ca_{0.40}Fe_{0.45}Mg_{0.15})SiO_3$, approaching that of hedenbergite, show a low-frequency band near 140 cm^{-1} that increases in intensity with increasing calcium contents and splits into a triplet at $Ca_{0.40}$ similar to that seen for the hedenbergite in curve (b) of Fig. 3.

C. Monoclinic Pyroxenes: Pigeonite

The low-calcium monoclinic pyroxenes belong to the space group $P2_1/c$ and differ from the high-calcium pyroxenes in that a greater portion of the M2 sites are occupied by Mg^{2+} and Fe^{2+}. The $CaSiO_3$ content ranges from 5 to 15% in natural samples and the Ca^{2+} is ordered on the M2 site. The clinopyroxenes of the pigeonite series have relatively few far-infrared absorption bands. However, the bands near 400 and 350 cm^{-1} that appear in the spectra of the pigeonites can be used for compositional determination for the series. The effects of calcium on the low frequency analytical bands can be seen in Fig. 4. The absorption bands near 400 and 350 cm^{-1} in curve (b) for a clinohypersthene of composition $Fe_{0.3}Mg_{0.7}SiO_3$ shift to lower frequency with increased iron substitution, as shown for curve (a)

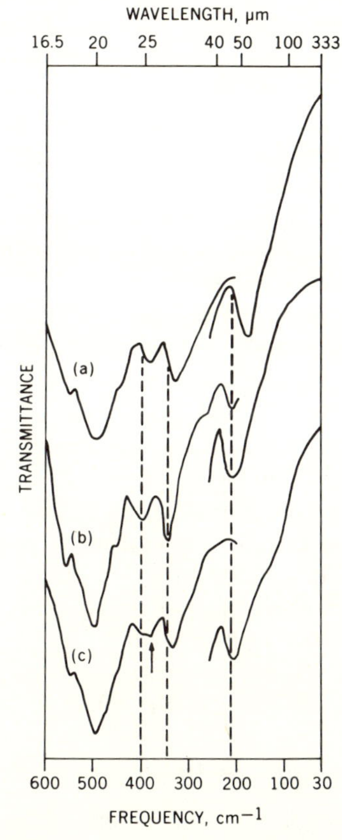

FIG. 4. Far-infrared spectra of synthetic clinopyroxenes of composition: (a) $Mg_{0.5}Fe_{0.5}SiO_3$, (b) $Mg_{0.7}Fe_{0.3}SiO_3$, and (c) $Ca_{0.1}Mg_{0.6}Fe_{0.3}SiO_3$.

for a clinohypersthene of composition $Fe_{0.5}Mg_{0.5}SiO_3$. In addition, the 220 cm^{-1} band shifts to lower frequency. The effect of the addition of 10% $CaSiO_3$ into the structure maintaining constant iron can be seen by the comparison of curves (b) and (c). All three low-frequency bands shift slightly to lower frequency. The new band at 395 cm^{-1} is probably due to the same chain vibration observed for diopside at 395 cm^{-1} associated with the calcium of the M2 site.

The lower frequency band of this doublet is the new component upon going from clinohypersthene with $(Fe_{0.30}Mg_{0.70})SiO_3$ to $(Ca_{0.10}Fe_{0.30}Mg_{0.60})SiO_3$, and this would seem to be due to calcium, which has a greater mass than magnesium. Similarly, in going from hedenbergite $(Ca_{0.45}Mg_{0.10}Fe_{0.425})SiO_3$ to diopside $(Ca_{0.50}Mg_{0.50})SiO_3$, the higher frequency band of the doublet appears as the new component and this would seem to be due to magnesium, which has a lesser mass than calcium. In clinopyroxenes, the calcium would essentially order on the M2 sites. At low ferrosilite content, ordering of the iron on the two nonequivalent sites would also occur. The determinative curves in Fig. 5 for the pigeonite series show variations of frequency with composition that are influenced by the relative distribution of the iron on the sites. The 400 cm^{-1} band for the pigeonite series shifts to lower frequency with increasing iron content. A more rapid decrease in frequency is observed with composition at low ferrosilite (Fs) content, with an inflection indicating much less rapid decrease near Fs_{40}. If this vibration is site M2 dependent and at low iron concentration there is a preferential ordering of the iron on the M2 site, a rapid shift in frequency would be expected. As proportionately greater amounts of M2 sites are occupied, the distribution of the iron over M1 and M2 becomes more random and the frequency change with composition would be different. The 400 cm^{-1} band, therefore, appears to be M2 site dependent. The inflection near Fs_{40} occurs near the point of high occupancy of the M2 site by iron and calcium. The 350 cm^{-1} band appears to be more site M1 dependent since the initial part of the curve shows relatively slight change in frequency and the inflection of curve (b) appears to be the reverse of that for curve (a), with a very rapid decrease in frequency with composition occurring at ferrosilite contents above 30%. The 220 cm^{-1} band—curve (c)—shifts linearly over most of the composition range and may be similar to the 220 cm^{-1} band in diopside.

D. Orthopyroxenes: Enstatite–Orthoferrosilite Series

The orthopyroxenes of the enstatite–ferrosilite series have absorption bands in the far-infrared that shift systematically to lower frequency with increased iron. The absorption bands near 450, 400, and 350 cm^{-1} in the

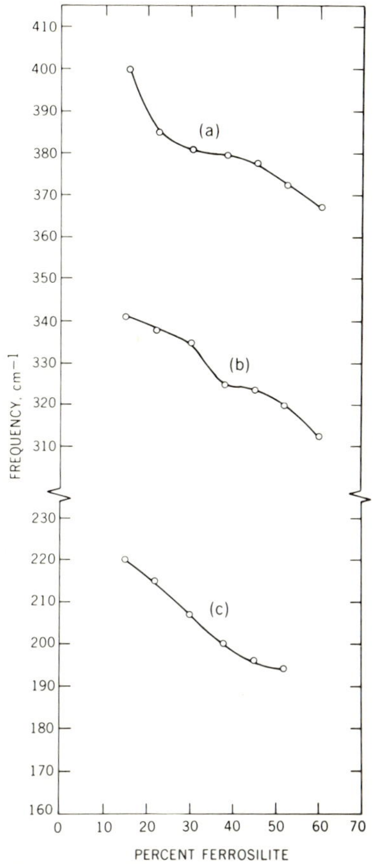

FIG. 5. Frequency variation for synthetic pigeonites ($Ca_{0.10}$) with change in composition for the absorption bands (a) 400 cm^{-1}, (b) 340 cm^{-1}, and (c) 220 cm^{-1}.

spectrum of enstatite shown in Fig. 6 occur in the spectra of all members of this series. The 450 cm^{-1} band is highly specific for the orthopyroxenes and can be used to distinguish from the monoclinic analogs. The systematic shift to lower frequency for the absorption bands near 400 and 350 cm^{-1} in the spectrum of enstatite with increased iron can be used to distinguish individual members of the series.

The orthopyroxenes contain little Ca, less than 5% $CaSiO_3$, and the two nonequivalent sites are occupied by Mg^{2+} and Fe^{2+}. The low-frequency spectra for the orthopyroxenes show systematic variations of frequency with composition that can be attributed to the ordering of the iron on the two nonequivalent sites. Mössbauer studies have shown that for metamorphic orthopyroxenes a more disordered Fe^{2+} distribution occurs over the nonequivalent octahedrally coordinated sites M1 and M2 as the amount of iron increases (Virgo and Hafner, 1970). The intensity of the 450 cm^{-1} band in the spectrum of enstatite decreases as the amount of iron increases. The reduced intensity of the 450 cm^{-1} band correlates with increased disorder of the iron over the two nonequivalent sites induced by temperature and shock as observed by Mössbauer spectroscopy (Estep *et al.*, 1972). For the orthopyroxene series $Fs_{16.4}$ to Fs_{80} (Lyon, 1963) in addition to a progressive decrease in intensity of the band near 450 cm^{-1}, a systematic shift to lower frequency is observed with increasing iron content. The frequency shift of the 450 cm^{-1} band with composition is similar to that observed for the absorption bands near 400 and 350 cm^{-1}, as shown in Fig. 7.

For Fs_6 these bands shift systematically to lower frequency with increased ferrosilite content. For orthopyroxenes with low ferrosilite content, the distribution of the iron is greater on the M2 site. As the Fe content increases and more of the M2 site becomes occupied, more iron is distributed onto the M1 site. Thus, if the vibration is M2 site dependent for low ferrosilite content, the frequency shift would be more rapid and not as

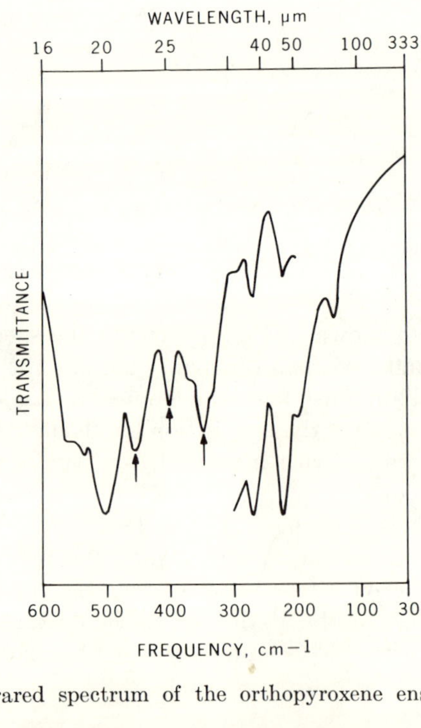

FIG. 6. Far-infrared spectrum of the orthopyroxene enstatite ($MgSiO_3$).

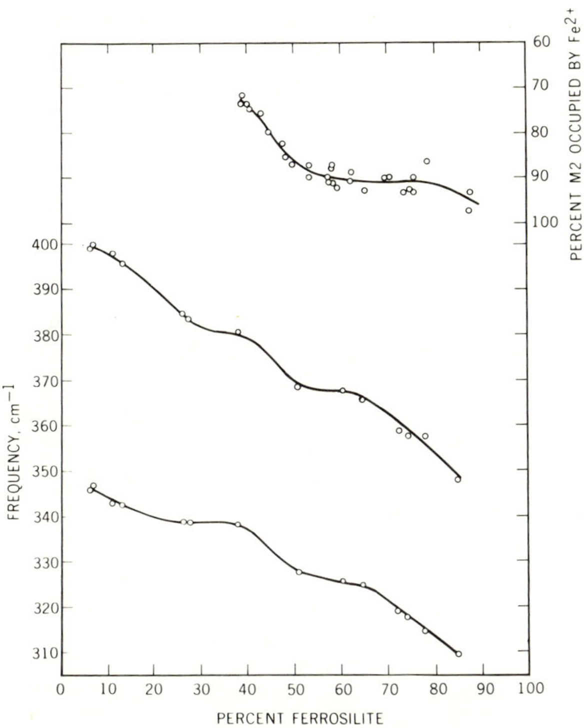

FIG. 7. Frequency variation of the far-infrared absorption bands of orthopyroxene with composition, compared with percent M2 occupied by Fe^{2+}, as determined by Mössbauer spectroscopy.

great as more of the M2 sites are occupied. The curves in Fig. 7 show that for both absorption bands the rate of shift with ferrosilite content is small at Fs_{30}–Fs_{40} and Fs_{50}–Fs_{65}, but the pattern of both curves is similar. Comparison of these determinative curves with that of percent M2 occupied by Fe^{2+} as determined by Mössbauer spectroscopy for the higher ferrosilite content (Virgo and Hafner, 1970) shows similar trends. This indicates that the two absorption bands are influenced by the M2 site occupancy of the Fe because no correlation was obtained with percent Fe on the M1 site.

VI. Framework Silicates: Feldspars

A. Structure of Lattice

The feldspar minerals have a structure which consists of a three-dimensional network linked by $(Al,Si)O_4$ tetrahedra. The interstices are occupied by the cations K, Na, and Ca. The potassium feldspars ($KALSi_3O_8$) sanidine and orthoclase are monoclinic and microcline is triclinic. The plagioclase feldspars albite ($NaAlSi_3O_8$) and anorthite ($CaAl_2Si_2O_8$) have a triclinic crystal structure and form a solid solution.

B. Potassium Feldspars

The feldspar group of minerals has a number of far-infrared absorption bands that are useful for characterization. The potassium feldspars show systematic frequency shifts and intensity changes that can be used to characterize the series sanidine, orthoclase, and microcline. The 329 cm^{-1} band shown in Fig. 8, curve (b), for microline shifts to higher frequency (12 cm^{-1} shift) as the Al and Si disorder increases in going from microline to sanidine. The splitting index (Laves and Hafner, 1965; Bahat, 1970), determined for a series from a higher frequency band, has been used as a measure of the Al/Si order in the feldspar group. The linear shift in the 329 cm^{-1} band of microcline with the splitting index is a means of characterizing the intermediate members of the series. The 329 cm^{-1} band is a Si—O—Si(Al) deformation and torsion and the shift is of the same order of magnitude (12 cm^{-1}) as that observed for the higher frequency band.

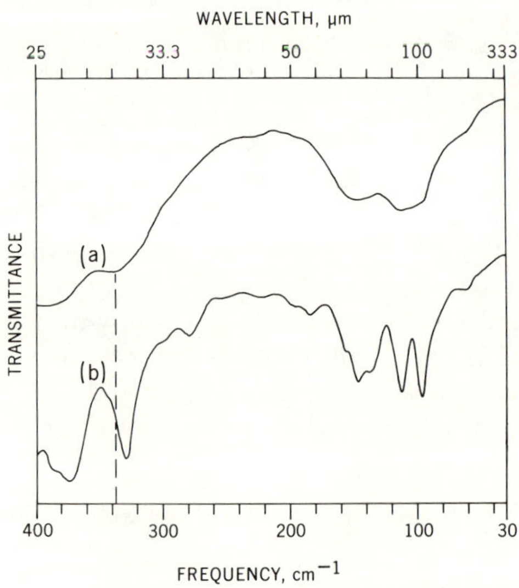

FIG. 8. Far-infrared spectra of $KAlSi_3O_8$: (a) sanidine, Germany, and (b) microcline, Canada.

Below 200 cm^{-1}, the potassium feldspars show significant splittings in absorption bands as the structure becomes more ordered. Two broad bands near 154 and 114 cm^{-1} for sanidine progressively split and increase in intensity as the structures become more ordered. The low frequency bands in the feldspars are assigned to K–O vibrations (Iiishi *et al.*, 1971) and little frequency shift occurs with increasing ordering. The increased resolution of the low-frequency region for microline results from increased ordering in the series. The increased resolution and the higher degree of splitting could be used to characterize further the intermediate feldspars.

C. Plagioclase Feldspars (Low Albite–Anorthite)

The plagioclase feldspar series, albite to anorthite, shown in Fig. 9, shows a strong band near 187 cm^{-1} for albite, which shifts linearly to higher frequency with increased anorthite content (Estep *et al.*, 1971). The shift to higher frequencies for bands assigned to a Na–O vibration in albite and to a Ca–O vibration in anorthite (Iiishi *et al.*, 1971) can be attributed to the distortion of the (Si, Al)O_4 tetrahedra in the lattice. While the calcium ion is slightly larger and the M–O distances are approximately the same, the mass of calcium is nearly twice that of sodium. However, there is a progressive distortion of the tetrahedra with the substitution of calcium for sodium.

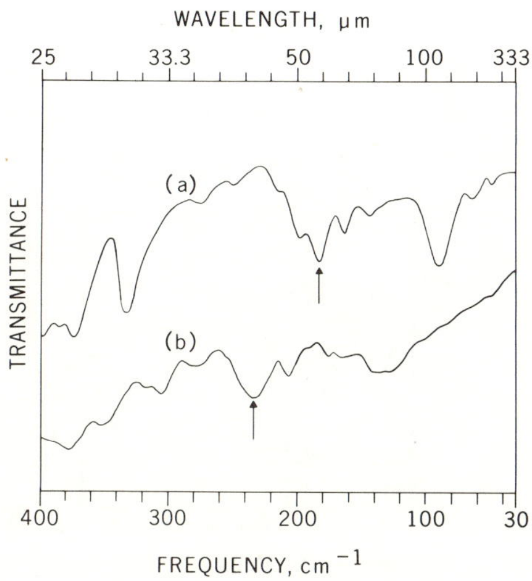

FIG. 9. Far-infrared spectra of (a) albite ($NaAlSi_3O_8$) and (b) anorthite ($CaAl_2Si_2O_8$), California.

TABLE IV

Far-Infrared Frequencies for the Ca-, Sr-, Ba-, and Pb-Feldspars

Compound	Cation mass (amu)	Cation radius (Å)	Frequencies[a] (cm^{-1})						
$CaAl_2Si_2O_8$	40.08	1.04	237(m)[b]	208(w)	191(w)	164(w)	136(w)	115(w)	100(w)
$SrAl_2Si_2O_8$	87.62	1.20	226, 220(vw)	202(vw)	176(w)	156(w)	124(vw)	101(vw)	93(w)
$BaAl_2Si_2O_8$	137.34	1.38	222(w)	195(w)	154(m)	130(m)	116(m)	84(w)	77(w)
$PbAl_2Si_2O_8$	207.19	1.26	213, 209(w)	170(w)	151(w)	116(w)	104(w)	—	76(w)

[a] Iishi *et al.* (1971).

[b] Abbreviations: m, medium; w, weak; vw, very weak.

Isomorphous substitution in anorthite for the calcium by Sr, Ba, and Pb shown in Table IV shows a decrease in frequency with increase in cationic size and mass for the M–O vibrations in anorthite (Iiishi *et al.*, 1971), the shift in the 237 cm^{-1} band for a Ca being 28 cm^{-1} with the substitution of Pb. This is less than would be expected since the mass of the lead is approximately five times that of calcium. However, the cationic size of the Pb is intermediate between Sr and Ba, and as the ionic radius of the cation becomes smaller and the distortion of the SiO_4 and AlO_4 tetrahedra increases, this could result in a smaller shift in frequency than would be expected if the vibrations were effected solely by mass of the cations.

VII. Carbonates: Calcite Group Minerals

The carbonates of the calcite group minerals belong to the space group D_{3d}^{6}: The rhombohedral cell contains two $CaCO_3$. The carbonate minerals with the calcite structure contain cations Mg^{2+}, Zn^{2+}, Mn^{2+}, Fe^{2+}, and Cd^{2+}.

The calcite group minerals show intense absorption bands in the far-infrared region useful for characterization. The systematic variation in the frequencies with cation substitution is shown in Table V. The intense bands at 380 cm^{-1} and 295 cm^{-1} shown for magnesite shift systematically with changes in the cation. These two lattice vibrations are due to translation of the metal cation, similar to those occurring in the isostructural nitrates and are motions perpendicular and parallel to the plane of carbonate anions (Hellwege *et al.*, 1970; Nakagawa and Walker, 1969), respectively.

The systematic shift to lower frequency with change in mass of the cation for both absorption bands is shown in Fig. 10, with the largest shift occurring for the 295 cm^{-1} band of magnesite. The largest deviation from the observed trend is for the metal translational mode of calcite perpendicular to the carbonate anion plane. While the mass of Ca^{2+} is intermediate between that of Mg^{2+} and Fe^{2+}, the ionic radius (1.04 Å) is larger and the frequency of this vibration is lowered. The lattice vibration in which the metal translation is parallel to the anion layer shows a more systematic shift with mass and is less affected by the ionic size of the calcium. In addition, the cationic within-plane vibrations occur over a much lower frequency range than for the between-plane vibrations, such that the low mass Mg^{2+} vibrates parallel to the plane at nearly as low a frequency as the high mass Cd^{2+} perpendicular to the plane.

TABLE V

Far-Infrared Frequencies for the Calcite Group Minerals

Mineral	Frequency (cm^{-1})							
Magnesite, $MgCO_3$	380(s)[a]	295(m)			247(m)		230(s)	
Siderite, $FeCO_3$	320(s)	216(m)	205(m)			182(m)		165(w)
Smithsonite, $ZnCO_3$	308(s)	200(br, s)						155(w)
Rhodochrosite, $MnCO_3$	305(s)	205(m)		180(m)				160(m)
Calcite, $CaCO_3$	308(s)	227(m)						105(s)
Otavite, $CdCO_3$	285(s)	150(m)						109(s)

[a] Abbreviations: s, strong; m, medium; w, weak; br, broad.

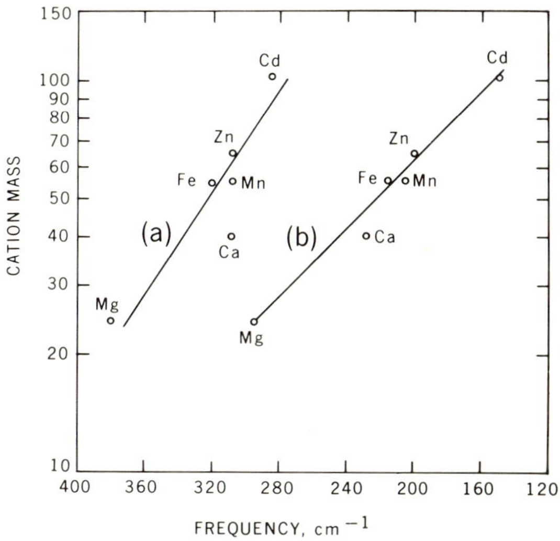

FIG. 10. Correlation between cation mass and frequency of the cation translationa vibrations perpendicular (a) and parallel (b) to the carbonate anion plane for six different calcite group carbonates.

References

Bahat, D. (1970). *J. Geol. Soc. Aust.* **17,** 93–102.

Bancroft, G. M., and Williams, P. G. L. (1971). *Amer. Mineral.* **56,** 1617–1625.

Burns, R. G. (1970a). *In* "Mineralogical Applications of Crystal Field Theory" (W. B. Harland, *et al.*, eds.), Earth Sci. Ser., Chapter 5, p. 82. Cambridge Univ. Press, London and New York.

Burns, R. G. (1970b). *Amer. Mineral.* **55,** 1608–1632.

Burns, R. G., and Huggins, F. E. (1972). *Amer. Mineral.* **57,** 967–985.

Deer, W. A., Howie, R. A., and Zussman, J. (1967). "Rock Forming Minerals," Vol. 1. Longmans, Green, New York.

Duke, D. A., and Stephens, J. D. (1964). *Amer. Mineral.* **49,** 1388.

Estep, P. A., Kovach, J. J., and Karr, C., Jr. (1971). *Proc. Lunar Sci. Conf., Geochim. Cosmochim. Acta, 2nd, 1971* Vol. 3, Suppl. 2, pp. 2137–2151.

Estep, P. A., Kovach, J. J., Waldstein, P., and Karr, C., Jr. (1972). *Proc. Lunar Sci. Conf., Geochim. Cosmochim. Acta, 3rd, 1972* Vol. 3, Suppl 3, pp. 3047–3067.

Hellwege, K. H., Lesch. W., Plihal, M., and Schaack, G. (1970). *Z. Phys.* **232,** 61–86.

Iiishi, K., Tomisaka, T., Kato, T., and Umegaki, Y. (1971). *Neues Jahrb. Mineral., Abh.* **115,** 89–119.

Karr, C., Jr. (1972). *Prog. Nucl. Energy, Ser. IX* **11,** 109–134.

Laves, F., and Hafner, S.(1965). *Z. Kristlalogr., Kristallgeometric, Kristallphys., Kristallchem.* **108,** 52–63.

Lyon, R. J. P. (1963). NASA *Tech. Note* **NASA TN D-1871.**

Nakagawa, I., and Walker, C. S. C. (1969). *J. Chem. Phys.* **51**, 1389–1397.
Oehler, O., and Günthard, Hs. H. (1969). *J. Chem. Phys.* **51**, 4719–4728.
Omori, K. (1971). *Amer. Mineral.* **56**, 1607–1616.
Paques-Ledent, M. T., and Tarte, P. (1973). *Spectrochim. Acta, Part A* **29**, 1007–1016.
Smyth, J. R., and Burnham, C. W. (1972). *Earth Planet. Sci. Lett.* **14**, 183–189.
Tarte, P. (1963). *Spectrochim. Acta.* **19**, 25–47.
Triple Commission on Spectroscopy (1962). *J. Opt. Soc. Am.* **52**, 476.
Virgo, D., and Hafner, S. S. (1970). *Amer. Mineral.* **55**, 201–223.

10

FAR-INFRARED OBSERVATION OF LUNAR ROCK

M. J. Pugh and J. A. Bastin

I. Introduction

There are two quite distinct experimental ways in which the far-infrared properties of lunar rock can be investigated. The moon, by virtue of its temperature and thermodynamic properties, is a thermal source; and the properties of the far-infrared radiation that it emits are in part determined by the optical characteristics of lunar surface rocks. These characteristics can thus to some extent be inferred by a study of the far-infrared radiation that we receive from the moon. The second, more recent, more direct, and in principle more complete, method is the direct study of lunar rocks in the laboratory. In this method we have at our disposal all the techniques not only for passive spectroscopy, but also for active selection of the infrared sources that are currently available for use in far-infrared research. An additional advantage of laboratory investigations of lunar rock is that the observations are not limited by the problems connected with the earth's atmosphere. At the same time it should be realized that we have at our disposal samples from a very restricted selection of sites and it seems likely that this situation will continue for many years to come.

We do not regard measurements in the far-infrared (here restricted to the region from 5 to 50 cm^{-1} approximately) as an end in themselves, and therefore we begin by enumerating the purposes or objectives in making such observations.

(1) *Thermal conductivity.* For nearly half a century (Epstein, 1929) it has been known that the thermal conductivity of the immediate surface layer of the lunar soil is very low, corresponding to that of a powdered rock in a vacuum. Observations over many years made in the mid-infrared range 8–14 μm determine the parameter

$$\gamma = (k\rho c)^{-1/2} \tag{1}$$

where k is the conductivity, and ρ and c, respectively, the density and specific heat of the powdered rock. The latter two quantities are scarcely dependent on the state of fragmentation of the rock, so that a determination of γ as a result of mid-infrared observations throughout a lunar diurnal cycle enables the conductivity of the immediate surface layer to be determined. One serious objection to such a procedure we can now see in retrospect; the lunar surface on a submetre scale is not homogeneous and as is now well known (Allen, 1971) a very small fractional coverage of the surface by rocks of high thermal conductivity can appreciably alter the value of the conductivity deduced from Eq. (1). This follows essentially since the deductions are made in a wavelength region where, for lunar temperatures, a strong Wien approximation to the Planck law is operative. In 1970 Bastin *et al.* suggested a method which obviates this difficulty since it depends on the far-infrared observations where the Rayleigh-Jeans linear-temperature dependence obtains for a blackbody source at lunar temperatures. From these far-infrared observations the following parameter can be determined:

$$\delta_\lambda = \frac{1}{\kappa_\lambda}\left[\frac{\omega c}{2k\rho}\right]^{1/2} \tag{2}$$

where ω is the angular frequency of the moon's synodic period, κ_λ the absorption coefficient of lunar rock, δ_λ can be determined from observations throughout the lunar phases, and κ_λ from laboratory measurements with lunar rock, so that Eq. (2) makes possible a determination of the mean thermal conductivity of the surface layer.

(2) *Scattering and absorption processes in the regolith.* Equation (2) results from a simple model of the lunar surface, in which the medium is

assumed continuous and only absorption processes take place. While this is probably a fair approximation more exact treatments are required if the measured brightness temperature $B_\lambda\ (T)$ is to be related to the thermodynamic temperatures in the top layers of the regolith. Such a relation clearly is theoretically complex, but does requires measurements of both the absorption and scattering coefficients of lunar rock at the wavelength in question (Linski, 1972). Laboratory measurements of both these coefficients are therefore required for lunar rock, and a detailed account of the way in which both may be experimentally inferred is considered in Section II.

(3) *Surface and subsurface texture.* Measurements of the fractional linear polarization of far-infrared and microwave thermal radiation from the lunar limb are ambiguous in that they are consistent with a family of models having different degrees of surface roughness and dielectric constant (Clegg and Carter, 1970). Direct laboratory measurements of the dielectric constant at any one wavelength, however, virtually remove the ambiguity, so that surface roughness parameters for the lunar surface may be determined. With the increase in techniques (especially telescope size) in far-infrared astronomy coupled with the decreasing likelihood of continued direct lunar exploration this technique may become a valuable tool in the investgation of individual areas of the lunar surface.

Exactly the same reasoning applies to the analysis of radar reflection observations. At present these have been carried out at longer wavelengths than those at which far-infrared passive measurements can be made. Indeed the most interesting recent measurements have been made at long radar wavelengths although even in this case it is valuable to have dielectric constant measurements at many wavelengths in order to infer the most accurate value to be adopted at the wavelength of observation.

(4) *Radio observations of vertical temperature gradient.* Measurements of the thermal gradient in the lunar subsurface layers are of importance in determining the magnitude of internal lunar heat sources. Direct measurements on the lunar surface using thermocouples have, as an alternative, the determination of the variation of brightness temperature of the moon with wavelength. The longer the wavelength of observation, the greater the depth to which the measurements relate. Most of the important measurements showing the temperature gradient are at wavelengths of several centimeters or greater. However, it is important to have a reliable estimate of the near-surface temperature. Unfortunately, this is complicated by the rectifying effect of the uppermost layer, which in turn is consequent upon the diurnal heating cycle combined with the strong temperature variation of the thermal conductivity of the regolith. The

separation of this effect, the Muncey effect (Muncey, 1958), from the effect of internal heat sources, the Fremlin effect (Fremlin, 1959), requires accurate measurements of far-infrared and microwave lunar thermal radiation at as many wavelengths as possible.

(5) *Mineralogical investigations.* As will be discussed later in this section we expect considerable differences in the far-infrared absorption of glasses and crystalline materials. At wavelengths in excess of 1 mm appreciable amounts of radiation can be transmitted through several millimeters of material. There is thus the possibility of distinguishing between crystalline and glassy material within an assembly (e.g., a breccia) without actually cutting sections of the sample.

(6) *Standard source for far-infrared astronomy.* Linski (1972) has pointed out the possibility of using the moon's thermal radiation as a calibrating source for far-infrared and microwave astronomy. Because of uncertainties in the terrestrial atmospheric absorption it is not possible to calibrate accurately the lunar source by reference to standard sources at the observing site. A theoretical estimate of the brightness temperature, in the wavelength range in question, is therefore required. Such an estimate requires in turn knowledge of the optical properties of lunar rock in this wavelength range; these can be obtained by laboratory observations of lunar rock.

We have outlined the goals of far infrared observations of lunar rock. So far these have been achieved to varying degrees, and the present status of these results will be discussed in the final section.

A. Attenuation, Scattering, and Absorption

A sample of lunar fines or a block of breccia or basalt may absorb far-infrared radiation by a process in which the energy of the photon is converted into lattice vibration in the form of a phonon. This process will be described later on in this section. In addition, energy may be removed from the beam by scattering. Such scattering processes may take place at the edge of a particle, or at the interface between two particles. If we consider an individual particle that is larger than the wavelength of the radiation, the process may be described as a reflection process, the distribution of directions of the final photons being highly directional. However, for a sample with a haphazard array of particle orientations the process will appear macroscopically as one of scattering since the emergent directions of photons from the whole sample will be randomized. The scattering and absorption coefficients σ_λ and κ_λ are related to the coefficient of attenuation α_λ

$$\alpha_\lambda = \sigma_\lambda + \kappa_\lambda \tag{3}$$

Using a narrow beam of radiation and a detector situated at a distance from the sample such that the accepted beam has a small solid angle, then essentially the coefficient of attenuation α_λ is measured, since none of the absorbed, and virtually none of the scattered, radiation is registered by the detector.

Although reflection measurements from blocks of basalt have already been carried out and will be described in the next section, no measurements of radiation scattered from a divided sample have yet been made.

Such scattering measurements are probably feasible although it will doubtless require good optical designing to ensure that the recorded scattered radiation is sufficiently intense to be recorded without undue noise in the detector. Fortunately the effects of absorption may, in some cases, be determined by measuring the attenuation coefficient as a function of temperature and density.

B. Absorption Mechanisms

The energy of a photon in the far-infrared is small compared with electronic transitions in atoms and ions, small even relative to the typical energies of vibrations in molecules; the energy is of the order of the fine structure of vibrational spectra, corresponding to rotational transitions (e.g., water vapor). In analyzing the effects in rock materials, however, one is not dealing with isolated atomic groups, but with a whole lattice structure, be it perfect crystal or glassy.

Absorption of radiation in a lattice structure is a cooperative phenomenon, whereby the momentum and energy of a photon are imparted to the whole lattice structure. This energy and momentum are stored as a wave motion in the lattice, which wave may be treated in a manner analogous to a photon (it possesses quantized energy and momentum); hence it is named a phonon.

The allowed phonons of a lattice may be displayed on a dispersion diagram (Fig. 1) which shows the energy in a phonon of specified wavelength. Such curves should be constructed for both longitudinal and transverse modes of vibration, and for all directions within the structure under consideration. For any structure possessing more than one atom per unit cell of the lattice, there exist several possible energies for a given wavelength of vibration, corresponding to the modes of interaction possible within the unit cell. The lowest energy mode (corresponding to no relative motion within the cell) is the acoustic branch of the dispersion curve; the higher energy modes are called optic branches. The energy of a phonon is proportional to the frequency of vibration.

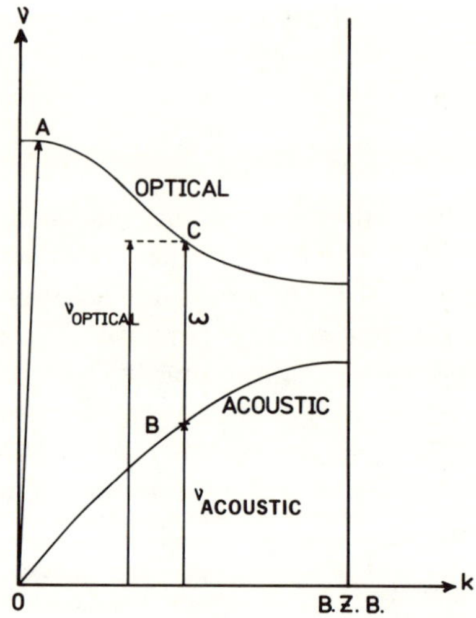

FIG. 1. The diagram displays two dispersion curves for the allowed relation of ν to k. Energy is proportional to ν, while k represents space-frequency or reciprocal wavelength. Direct single-phonon processes arise from absorption of photons such as OA, while the two-phonon process is represented by the photon BC, of energy $hw = h\nu_{\text{optical}} - h\nu_{\text{acoustic}}$ The line labeled B.Z.B. represents the Brillouin zone boundary.

Photons, compared with the lattice, have high energy and very low space-frequency; if represented on the dispersion diagram they lie close to the energy axis. The lattice is, then, capable of absorbing directly only those photons whose energy is such that they lie on an optic branch. This would give the material a series of sharp absorption lines. In practice, disorder effects, and the summation over all directions of propagation within the crystal (since the dispersion diagram changes with orientation relative to the lattice) will broaden these lines into curves. This absorption process converts one photon into one phonon, which is referred to as a single-phonon process.

The acoustic branch of the dispersion curve is populated with phonons according to a Boltzmann distribution in energy. It is possible for a photon to interact with a phonon, thereby producing a phonon of higher energy by absorbing the photon. This is referred to as a two-phonon process. For the diagram given it is apparent that this will broaden the low-frequency edge of absorption curves. The amount of increase in absorption will depend strongly on the temperature of the material. In particular, as the temperature approaches absolute zero, there will be no population of acoustic phonons, and this two-phonon process may not occur.

For any type of material, the absorption curve will be broadened as disorder occurs in the lattice, and further broadened if sufficient comminution of the sample allows size effects to influence the system. In the far-infrared region the wavelengths of radiation are comparable with particle sizes in the fines and breccias that make up the bulk of the lunar surface layer, so that one might expect considerable size effects. Further the frequencies and hence photon energy are so low that one is concerned solely with absorption in the wings of the lines.

For absorption to occur in samples one requires that single-phonon processes still be responsible. The class of material that best produces this condition is a glass (Hadni *et al.*, 1965). Evidence that absorption still occurs at low temperatures would indicate the presence of glassy materials. On the other hand, an observed reduction of absorption as temperature is decreased would indicate the presence of crystalline materials, for which only the two-phonon process would permit absorption in the far-infrared. In principle the relative proportions of these two effects would allow a determination of the relative amounts of crystalline and glassy materials without the necessity to sort samples by hand, or cut into breccias which contain both crystalline and glassy fragments.

II. Experimental Procedure and Results

A. Observations of Lunar Thermal Radiation

Observations have been made using telescopes with large aperture (> 1 m) of which the primary is either of optical tolerance in the visual wavelength range or has been carefully machined from a metal blank. Detectors which have been used are the germanium bolometer or the indium antimonide photoconductor; both only operate efficiently at liquid helium temperatures. Wavelength selection has been achieved using interference filters made from metal mesh. The detection of linear polarization has been carried out using unidirectional wire grids as analyzers.

Most observations consist in scanning across the lunar disk and including portions of blank sky at either limb in the scan so that the effect of atmospheric radiation can be eliminated. Successive scans made at various lunar latitudes enable a contour map of lunar brightness to be constructed. The magnitude of the response of the detector is usually compared to that from the sun. This has the advantage of allowing for the effects of atmospheric absorption; in addition, because the sun and moon have approximately the same angular diameter, the effects of finite resolution of the telescope can to some extent be eliminated. However, there are some disadvantages in the procedure. Even if the measurements are taken at times when the sun and moon have the same elevation in the sky, the assumption that the atmosphere does not vary azimuthally in water vapor content is only approximate and may lead to appreciable error when applied to observation from high-level peaked mountain sites. Second, the mean solar brightness temperature is only known approximately throughout the far-infrared region, so that there is an inevitable error if the sun is used as an absolute calibration source. Finally, there are variations in the solar far-infrared output; these are associated with plage and sunspot regions and may be particularly serious at times of sunspot maximum.

Measurements of the solar–lunar brightness temperature ratio made in the far-infrared (see Bastin *et al.*, 1964) indicate a value of δ_λ in the range around 1 mm given by the expression

$$\delta_\lambda = 0.5 \times 10^{-3}\, \lambda \tag{4}$$

Mapping of the lunar disk has shown that when the lunar surface is sunlit the marial regions have a higher brightness temperature than the adjacent upland regions (Clegg and Carter, 1970). At least part of the effect results from the lower albedo of the marial areas, although there are some indications (see Bastin and Gough, 1969) that there is more small-scale (centimeter size) roughness in the marial areas.

Clegg and Carter (1970) investigated the linear polarization of radiation from the lunar limb using far-infrared techniques. They analyzed their results in terms of a model surface composed entirely of furrows of slope $\pi/4$. As a consequence they estimate values for the dielectric constant of $\geqslant 6$ for the crystalline material.

Numerous solar–lunar comparison measurements have been made. Table I shows some data collected by Gezari *et al.* (1973), found in the

TABLE I

SOLAR–LUNAR COMPARISON MEASUREMENTS[a]

λ (μm)	$T_{\lambda S}/T_{\lambda M}$
1000	15.8 ± 1.1
450	12.6 ± 1.0
350	11.9 ± 1.0

[a] From Table I of Gezari *et al.* (1973).

various wavelength ranges that are accessible from high altitude sites such as Mauna Kea.

B. Measurements of Attenuation Coefficient

The attenuation of electromagnetic radiation by lunar rock in the far-infrared range has been determined using the cyanogen laser, the sun, or a mercury arc lamp as sources; broad-band filters (Ade and Bastin, 1972), and Fourier transform techniques are employed for wavelength selection; bolometers and photoconductive detectors as sensors. In the case of the cyanogen laser, which is a monochromatic source at 338 μm, no ancillary wavelength selection is required. Such is the intensity of this source (10^{-4}W) that even for quite strongly absorbing samples the power transmitted is sufficient to enable determination of the fractional transmission using the relatively insensitive Golay cell as a detector.

Figure 2 shows the variation of fractional transmission with areal density fitted to the expected exponential relation for fines sample number 10084,111. Deviations from this relation at low values of ρx were attributed to reflections in the sample cell, but may also be due to density effects in the sample. A measured absorption of 2.5 ± 0.2 $(\text{kg-m}^{-2})^{-1}$ is observed.

Solar source measurements have been made using relatively broad-bandpass filters centered at 1200 and 1700 μm yielding absorptions of $0.18 \pm .09$ and 0.3 ± 0.1 $(\text{kg-m}^{-2})^{-1}$, respectively (Clegg *et al.*, 1972). For these measurements a helium-cooled detector was employed, with the sample at the focus of a 1.6-m Cassegrain flux collector.

The above measurements have an element of simplicity of method about them. However, the Fourier transform technique is also of great value, and can be employed throughout the whole far-infrared range (see Chapter 11, Section II, A). This method produces spectra of reasonable quality with presently available sources and detectors. Such spectra would not be available so easily with a conventional monochromator spectrometer. Most of the spectra displayed in this section have been obtained with the polarizing interferometer of Martin and Puplett (1969). The radiation source used is a mercury arc lamp, in conjunction with a helium-cooled bolometer as detector.

Early measurements were taken to resolutions as great as 0.1 cm^{-1} in search of any fine structure in the attenuation spectra. In the absence of any such structure, later workers have reduced resolution to 0.5 or 1.0 cm^{-1} in order to reduce time taken per intererogram, and this allows correspondingly more interferograms to be taken in the course of any experimental run. Averaging the resulting spectra then produces a more accurate spectrum.

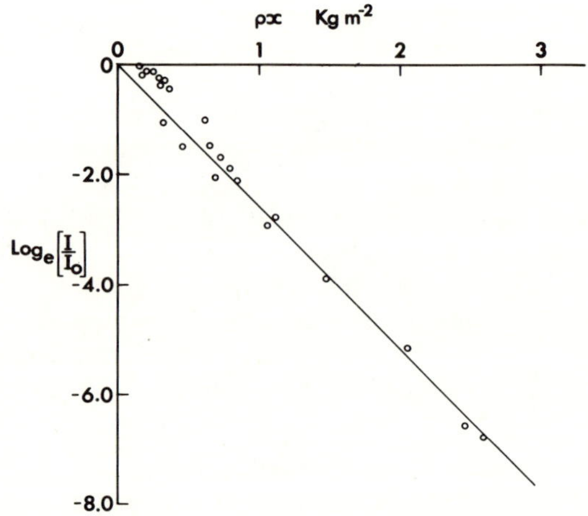

FIG. 2. Transmission through various thicknesses of lunar fines of sample 10084,111 at 338 μm wavelength. The intensities I, and I_0 refer to those after and before, respectively, transmission through the sample. ρ is the density, and x the thickness of the sample.

Figure 3 shows a series of room-temperature spectra taken on a number of different thicknesses of sample. The variation of absorption coefficient in the spectral range investigated implies that the mean curve is obtained by weighting the samples differently in the different spectral regions: The lower frequency absorption is determined by using thick samples which provide a measurable attentuation, and the higher frequency absorption comes from the thinner samples which are sufficiently transmitting. Above 40 cm^{-1} the absorption is so high that sample thicknesses of the order of 0.2 mm must be used to obtain sufficient transmitted signal. Since the fines contain a fair fraction of particles of about these dimensions experimentation in this range requires careful thought and interpretation.

C. Temperature and Density-Dependent Measurements

Following the discussion in Section I, B, and our knowledge of the nature of lunar fines we would expect the attentuation by such material in the far-infrared wavelength to be temperature dependent. Sample 10084,111 of lunar fines has been investigated for such dependence. The sample-holder module of the interferometer described above was adapted to provide a holder capable of holding the sample to various temperatures determined by the boiling points of refrigerant fluids (e.g., liquid N_2). The range of accessible temperatures was further increased by mounting the sample in the helium-cooled detector to take measurements near absolute zero.

Results from this series of measurements are shown in Fig. 4. There is an apparent marked variation of attenuation with temperature. The scatter of experimental points gives some idea of the problem with noise at the low-frequency end of the curves. Reexamination of the original spectra indicates that they are not trustworthy above 35 cm^{-1}. However, within the reliable spectral range the measurements show an attenuation coefficient which increases rapidly with frequency. If the coefficient of attenuation is written in the form

$$\alpha_\lambda = A\lambda^n \tag{5}$$

then at room temperature the value of n varies from -3.2 at 200 μm to -1.3 at 1500 μm. Both the constants A and n are temperature dependent, and Table II gives values for these constants at various temperatures.

To investigate the variation of the extinction coefficient with density the cyanogen laser has been used together with the simple Golay detector

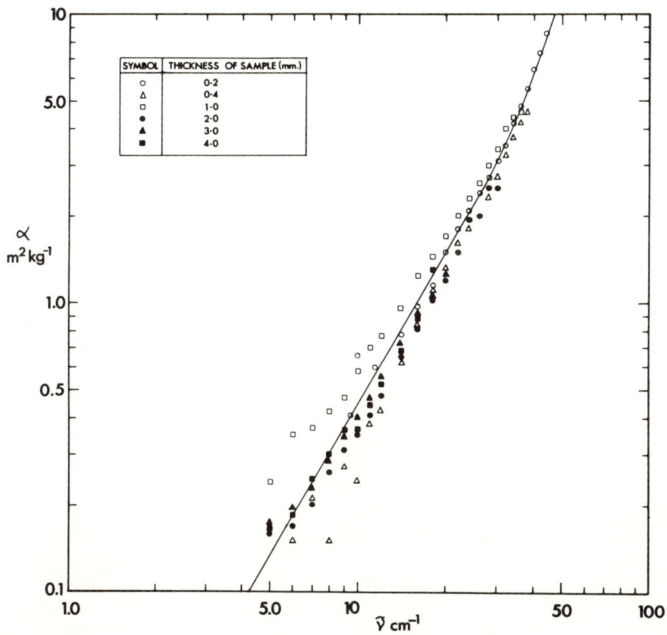

FIG. 3. Spectroscopic determination of the absorption coefficient for sample 10084,111. Absorption coefficient α, frequency in units of cm^{-1}, $\bar{\nu}$.

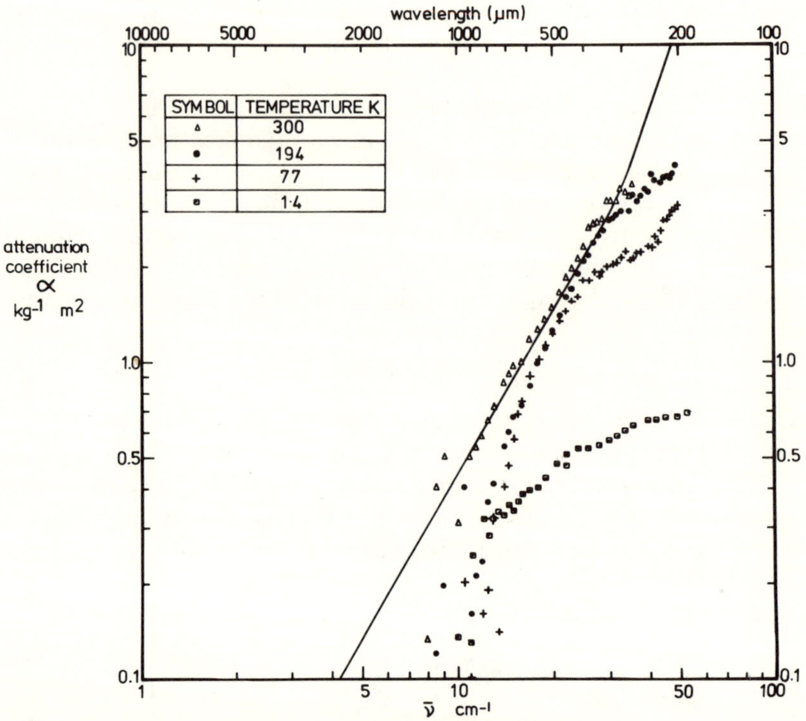

FIG. 4. Spectroscopic determination of the absorption coefficient at various temperatures. The continuous curve is the same as that in Fig. 3.

(Clegg *et al.*, 1972). In this work the samples were confined by Rigidex polyethylene disks against spacers of known thickness. The results are displayed as mass attenuation coefficients in Fig. 5, where it is apparent that the attenuation tends to a limit at density 3.2×10^3 kg-m^{-3}.

Attempts have also been made to explore this effect throughout the far-infrared range (Clegg *et al.*, 1972). The results are shown in Fig. 6. From Fig. 5 it is apparent that the attenuation at specific gravity 1.9 is not much greater than the limiting attenuation. Subtraction of the curve for relative density 1.90 from those for lower densities allows, in principle, a determination of the behavior of the scattering term itself as a function of frequency. Standing waves on the spectrum prevent this being very accurate, but show that the scattered contribution to the extinction for a sample at density 1.5 kg-m^{-3} is almost equal to the observed extinction for a sample at 1.4°K temperature, indicating a negligible true absorption at this temperature.

TABLE II

VARIATION OF ATTENUATION WITH TEMPERATURE

T(K)	30 cm^{-1} A	n
1.4	1.37×10^{-1}	−0.42
77	1.41×10^{-1}	−0.78
194	5.94×10^{-2}	−1.12
300	1.64×10^{-2}	−1.54

The scattering contribution σ_λ may be represented as a function of relative density ρ and wavelength λ by the expression.

$$\sigma_\lambda = \eta(\rho)\,\lambda^{-1} \tag{6}$$

The results shown in Fig. 6, while giving the wavelength dependency of the scattering coefficient, do not explicitly determine its variation with

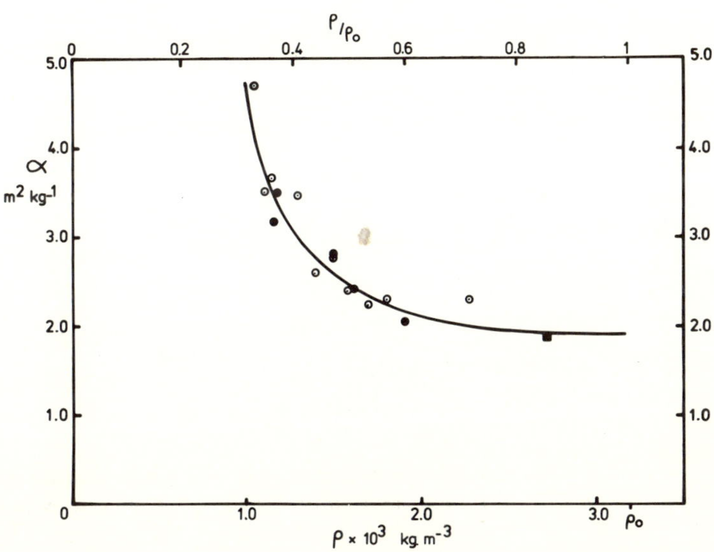

FIG. 5. Mass attenuation coefficient α for sample 10084,111, and microbreccia 10065,15 (square symbol). The extrapolated limit for voidless material is 1.88 kg^{-1} m^2, at density 3.2×10^3 kg m^{-3}.

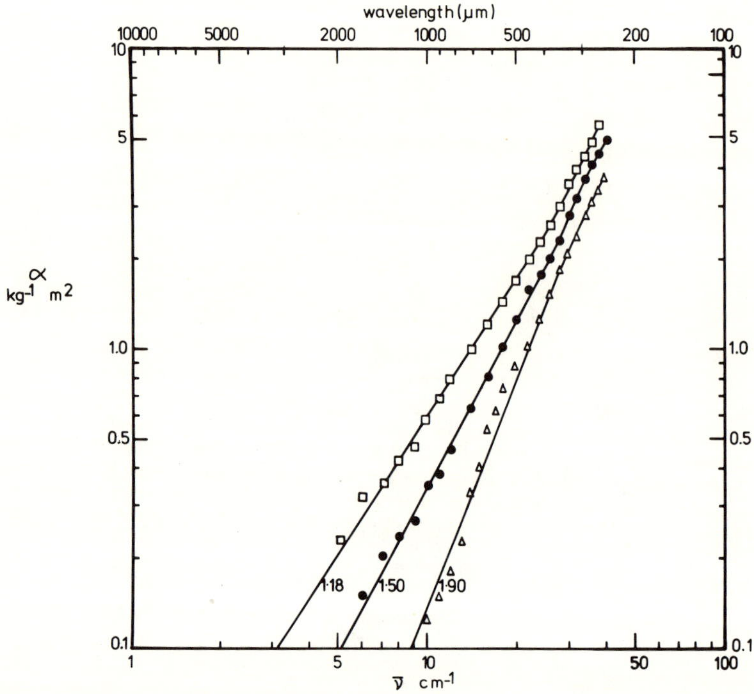

FIG. 6. Spectroscopic determination of the variation of absorption coefficient α with relative density. The curves are identified by the relative density of the samples in units of gm cm^{-3}.

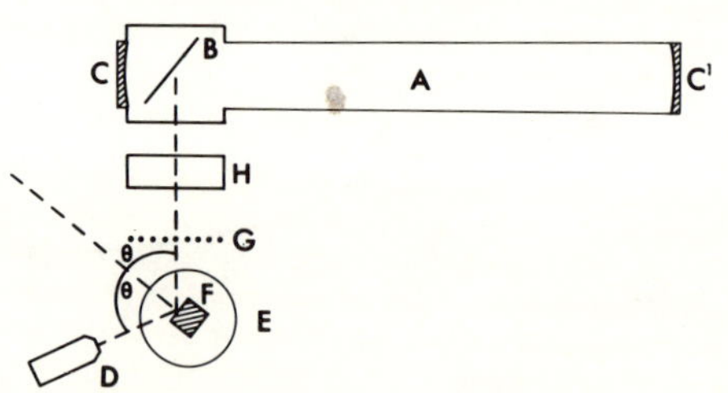

FIG. 7. Determination of the reflection coefficient of a block sample. A is the laser tube with end mirrors C and C′; the beam is brought out from the beam-splitter B; D is the Golay cell detector; E is the rotating table holding the sample F; G is a wire grid to select the required polarization. Since the laser output is partially plane-polarized it is necessary to rotate the plane of polarization for S-type measurements by means of the block H.

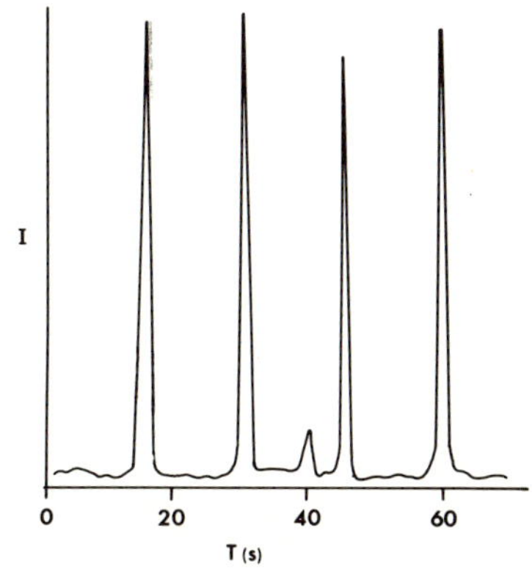

FIG. 8. The output curve obtained from block 12063,96 for some angle of incidence. The small peak is introduced by a facet on one corner which could not be removed without undue reduction of the sample size.

void ratio. However, Fig. 5 shows that, at 338 μm

$$\eta(\rho) = A\left[\frac{\rho - \rho_0}{\rho_0}\right]^m \tag{7}$$

where $A = 5.75.10^{-3}\text{m}^3\ \text{kg}^{-1}$, $m = 5.0$, and ρ_0 is density of the material of the fines particles.

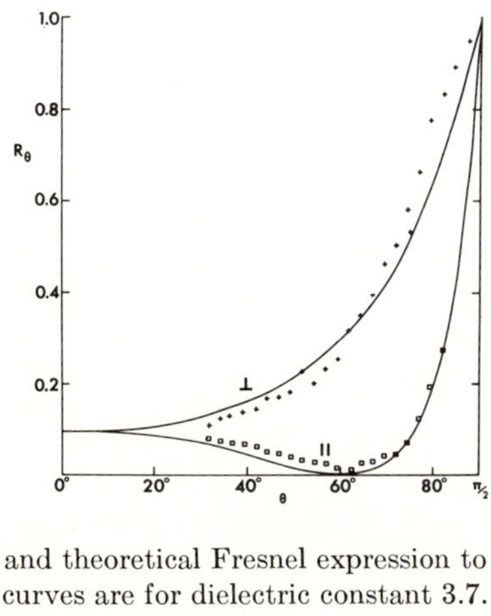

FIG. 9. Measurements and theoretical Fresnel expression to fit the data from breccia 10065,30. The computed curves are for dielectric constant 3.7.

TABLE III

DENSITY CORRECTIONS FOR LUNAR ROCK SAMPLES

NO. OF SAMPLE	ROCK TYPE	FREQUENCY OR WAVELENGTH	ϵ_{obs}	$\epsilon_{corrected}$
10017,64	Type A basalt	338 μm laser	7.05 ± 0.3	—
12063,96	Type A basalt	338 μm laser	6.25 ± 0.2	—
10065,30	Breccia	338 μm laser	3.7 ± 0.2	—
14259,38	Type D fines[a]	10^{11} Hz	4.0	2.7
14161,36	Fines, 2–4 mm[a]	10^{11} Hz	5.9	4.7

[a] Values for these samples, and for other Apollo 14 and 15 lunar fines, are listed in Chapter 11, Table II.

D. Dielectric Constant

The dielectric constant in the far-infrared region may be determined by a number of methods. For samples whose absorption coefficient is small the variation of reflected power with angle of incidence may be used to fit the Fresnel expressions. This method has been employed using the cyanogen laser as source, selecting the required orientation of polarization with a unidirectional wire grid. The results from this are listed in Table II, reproduced from Ade *et al.* (1971) (see also Figs. 7–9).

Another approach is employed by Perry *et al.* (1972), who have applied Kramers–Kronig analysis to broad-band reflectance spectra over the region 20–2000 cm^{-1} (see Chapter 11, Section III, A). The necessity for a smooth surface on the sample has required that fines samples be compressed to $2.4–2.6 \times 10^3$ kg-m^{-3} density. The dielectric constants thus obtained must be corrected to the density of material lying on the lunar surface. This correction may be applied by either some suitable theoretical expression (e.g., those listed in Clegg and Carter, 1970), or experimentally derived as discussed by Basset and Shackelford (1972).

The latter authors, working at 9.3, 24, 35, and 60 GHz, find a direct linear variation with density for the relative dielectric constant, for densities in the range $1.5–2.4 \times 10^3$kg-m^{-3}, and this would allow density corrections as shown in Table III.

III. Discussion

In conclusion we return to the objectives outlined at the beginning of Section I and assess what worthwhile information has been found in each case as a result of the measurements described in Section II.

The thermal conductivity of the immediate top few centimeters of the lunar regolith may be determined by using the values summarized by Eq. (4) using the simple absorption model from which Eq. (2) is deduced. This procedure using measured values at attenuation coefficient for lunar rock was first used by Bastin *et al.* (1970; see, also, Ade *et al.*, 1971). This comparison gives a conductivity for the lunar surface layer which is considerably lower than that deduced by either mid-infrared results or direct measurements of the thermal conductivity of lunar fines in the laboratory. Critics of the utility of far-infrared results will immediately find fault with the model used in deducing Eq. (2). However, it seems to us that a more exact model is unlikely to remove the discrepancy. Indeed for the reasons stated in Section I, there are grounds for believing that the infrared measurements give a mean of the conductivity of the lunar surface heavily weighed in the direction of the few highly conducting surface rocks. Similarly the effects of packing and adsorbed layers in the case of laboratory measurements may give higher values for the conductivity than would be expected for the material *in situ* on the lunar surface. Work on density effects has been reported by Cremers (1972) while effects of atmospheric contamination may be estimated from the work of Wechsler and Glaser (1965). To remove these discrepancies, a more exact form of the model leading to Eq. (2) is required, together with an updated attempt to measure δ_λ using the techniques that have been developed in the last ten years. In this context, the determination of individual scattering and absorption coefficients is of considerable importance in improving our models for radiation transfer in the regolith. The experimental techniques and data for carrying out this investigation have been described in detail in the previous section, and show that they are already producing significant quantitative results.

The dielectric constant measurements that have been made with lunar rock throughout the microwave and infrared range (see also Chapter 11, Section IV) begin now to form a reliable set of data which are of value in connection with models for surface and subsurface lunar structure. These are of particular importance in analyzing the nature, if any, of the regolith–bedrock interface. It also seems that a reexamination of the polarization of the lunar far-infrared radiation should now be made, especially in view of the more exact data that are available about void ratios and dielectric constants of lunar rock.

The problem of the measurement of absolute lunar brightness temperatures in the infrared, as a means of making more accurate thermal gradient measurements at various points on the lunar surface, is one of great significance. The various complications in the necessary theoretical model, as outlined and treated by Linski (1972), do not represent a barrier to this approach. The advent of absolute far-infrared bolometry and work from airborne platforms suggests that progress within this field will shortly be made.

The rapidly increasing interest in far-infrared astronomy makes it likely that the use of the moon as a calibration source will continue. It would be valuable to be able to decrease the errors in our knowledge of far-infrared lunar brightness temperatures, if possible by as much as an order of magnitude. Developments in the understanding of the properties of the lunar surface and the rapid increase in far-infrared observing techniques suggest that this hope is not unreasonable.

ACKNOWLEDGMENT

We are indebted to Dr. P. E. Clegg and Dr. P. A. R. Ade for helpful advice, information, and discussion during the preparation of this chapter.

References

Ade, P. A. R., and Bastin, J. A. (1972). *In* "Infrared Detection Techniques for Space Research" (V. Manno and J. Ring, eds.), pp. 257–264. Reidel Publ., Dordrecht, Netherlands.

Ade, P. A. R., Bastin, J. A., Marston, A. C., Pandya, S. J., and Puplett, E. (1971). *Proc. Lunar Sci. Conf., Geochim. Cosmochim. Acta, 2nd, 1971* Vol. 3, pp. 2203–2211.

Allen, D. A. (1971). *Moon* **2,** 435–462.

Basset, H. L., and Shackelford, R. G. (1972). *Proc. Lunar Sci. Conf., Geochim. Cosmochim. Acta, 3rd, 1972* Vol. 3, Suppl. pp. 3157–3160.

Bastin, J. A., Gear, A. E., Jones, G. O., Smith, H. J. T., and Wright, P. J. (1964). *Proc. Roy. Soc. Ser. A* **278,** 543–573.

Bastin, J. A., and Gough, D. O. (1969). *Icarus* **11,** 289–319.

Bastin, J. A., Clegg, P. E., and Fielder, G. (1970). *Proc. Lunar Sci. Conf., Geochim. Cosmochim. Acta, 1st, 1970* Vol. 3, 1987–1991.

Clegg, P. E., and Carter, B. S. (1970). *Mon. Not. Roy. Astron. Soc.* **148,** 261–274.

Clegg, P. E., Pandya, S. J., Foster, S. A., and Bastin, J. A. (1972). *Proc. Lunar Sci. Conf., Geochim. Cosmochim. Acta, 3rd, 1972* Vol. 3, pp. 3035–3045.

Cremers, C. J. (1972). *Proc. Lunar Sci. Conf., Geochim. Cosmochim. Acta, 3rd, 1972* Vol. 3, pp. 2611–2617.

Epstein, P. S. (1929). *Phys. Rev.* **33,** 269.

Fremlin, J. H. (1959). *Nature (London)* **183,** 239.

Gezari, D. Y., Joyce, R. R., and Simon, M. (1973). In press.

Hadni, A., Claudel, J., Gerbaux, X., Morlot, G., and Munier, J. M. (1965). *Appl. Opt.* **4,** 487.

Linski, J. L. (1972). *Astrophys. J., Suppl. Ser.* **25,** No. 216.

Martin, D. H., and Puplett, E. (1969). *Infrared Phys.* **10,** 105–109.

Muncey, R. W. (1958). *Nature (London)* **181,** 1458–1459.

Perry, C. H., Agrawal, D. K., Anastassakis, E., Lowndes, R. P., and Tornberg, N. E. (1972). *Proc. Lunar Sci. Conf., Geochim. Cosmochim. Acta, 3rd, 1972* Vol. 3, pp. 3077–3095.

Wechsler, A. E., and Glaser, P. E. (1965). *Icarus* **4,** 335–352.

11

ELASTIC AND INELASTIC LIGHT SCATTERING SPECTRA OF LUNAR ROCKS AND SOILS

Clive H. Perry

and

Robert P. Lowndes

I. Introduction

In this chapter we shall discuss the results of mid- and far-infrared and laser Raman investigations in the spectral frequency range 20–4000 cm^{-1} (5000–2.5 μm wavelength) on a number of lunar rocks and soils obtained from the different landing sites of the Apollo 11, 12, 14, and 15 missions. In general, studies of the interaction of electromagnetic radiation with lunar materials using elastic and inelastic light-scattering* techniques have several important consequences. They can be used to determine the charac-

*Elastic and inelastic light scattering are described in Chapter 12, Section II, C.

teristic lattice vibration frequencies of the lunar materials and then, by similar control measurements on terrestrial materials, the results of these experiments can be used to infer the chemical constituents of the lunar material together with their associated chemical or mineralogical abundances. In addition, the chronicling and identification of the lattice vibration frequencies may yield crystallographic structural information about the lunar materials. Finally, data from spectroscopic investigations of lunar material can be applied to composition mapping of the lunar surface; the uniqueness of the method suggests the possibility of performing remote infrared petrology using spectral matching techniques (Lyon, 1963; Aronson *et al.*, 1967; Perry and Lowndes, 1970; Perry *et al.*, 1971, 1972). All of these important consequences provide direct or indirect information concerning the history, evolution, and petrology of the moon, not only as a single entity but for different geographical regions of the lunar surface.

In achieving the above-mentioned aims, the infrared and Raman studies can be considered to be in many ways complementary. The strength of the Raman investigations stems primarily from the comparatively small wavelength and focused beam size of the exciting source which allows (nondestructive) probes of the smaller grains and inclusions of the lunar material and provides a relatively uncomplicated identification and interpretation of the modes of vibration of such lunar material. Similar experiments on grains or inclusions using infrared spectroscopy are restricted by the comparatively large wavelength and beam size of the incident radiation; in some cases the resulting spectra may often be complicated by the presence of radiative surface modes which are difficult to distinguish from bulk lattice modes. The strength of the infrared measurements lies more in the fact that they generate information on some important macroscopic properties that are characteristic to somewhat larger surface areas (>0.25 cm^2) of the lunar materials than those of grains or inclusions. One of the more important of these macroscopic properties determined from the infrared investigations is the dispersion of the complex dielectric response (and hence of the refractive index and absorption coefficient) for the lunar material. The dielectric constants are, of course, independent of the method of measurement and hence they provide fundamental input parameters for synthesizing, for instance, emission spectra of lunar materials to provide a comparison with any future remote sensed spectra or radiometric measurements of the lunar surface (Mendell and Low, 1972). Furthermore, the far-infrared dielectric constants determined from lunar materials are far less sensitive to sample dryness than values determined via low-frequency capacitative techniques, and a knowledge of the temperature and pressure (density) dependence of these dielectric constants, when extrapolated to lower frequencies, has important applications for radar studies of the depth and composition of the various subsurface layers of the moon.

II. Experimental

A. Far-Infrared and Infrared Measurements

In making the infrared, or elastic light scattering, measurements we have used a variety of diffraction grating and interferometric spectrophotometers to cover the frequency range 20–4000 cm^{-1}. A block layout of the Michelson Fourier transform interferometer, used in conjunction with a liquid helium-cooled, gallium-doped germanium bolometer, is shown in Fig. 1 (Perry *et al.*, 1966). This instrument provided spectra in the far-infrared range of 20–250 cm^{-1}, while Perkin-Elmer diffraction grating spectrometers (models 180, 301, and 520) were used to record the spectra in the frequency range 200–4000 cm^{-1} (see also Chapter 9, Section II). Both reflection and transmission scattering geometries were used with the incident radiation striking the samples at near normal incidence to facilitate and simplify the spectral analysis. The measurements could be recorded in the temperature range 4°–300°K using conventional liquid helium dewars and miniature helium or hydrogen Cryotip refrigerators.

The scattering geometry used in the infrared measurements was pri-

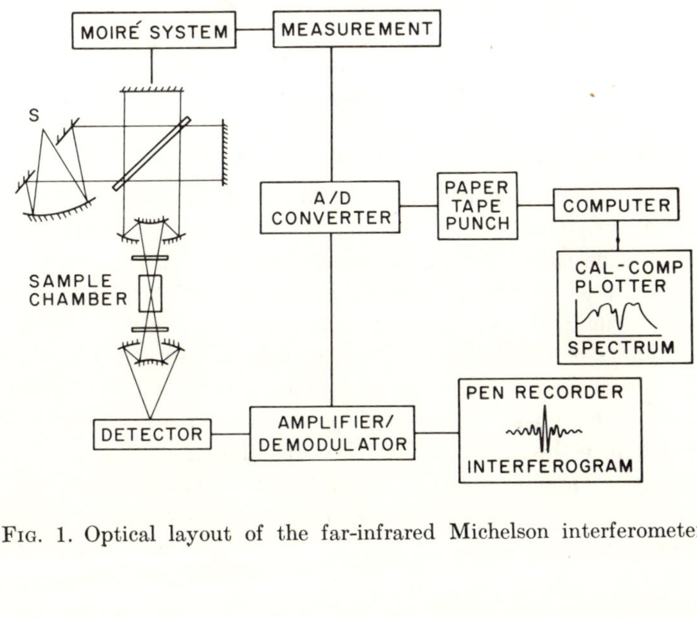

FIG. 1. Optical layout of the far-infrared Michelson interferometer.

marily determined by the nature of the lunar sample concerned. Our investigations were involved with three basic types of lunar sample: bulk samples in the form of slices or butt ends, pressed disks prepared by compressing fine samples under about 5 kbar pressure in an evacuable die, and suspensions of lunar fines or crushed lunar rocks in KBr, CsI, or polythene matrices. The reflection measurements were performed on the polished faces of the bulk samples and pressed disks, while transmission measurements were made on the suspension matrices.

B. Raman Measurements

The Raman scattering measurements were performed using excitation radiation at 488.0 and 514.5 nm from an Ar ion laser and at 632.8 nm using a He–Ne laser. The Raman scattered light was analyzed using a Spex 1401 double spectrometer incorporating a cooled ITT FW-130 photomultiplier and pulse counting electronics. The laser–Raman system (Perry *et al.*, 1972) is similar to that shown in Fig. 1, Chapter 12.

In investigating inclusions or grains embedded in lunar rocks, the focused laser beam spot, with a diameter of approximately 50 μm, could be unambiguously observed on grains about 100 μm across using a microscope system and observing the sample in the focal plane of the slit. A comparison of spectra recorded with the incident laser spot on and just off the sample grain or inclusion was always made as a precautionary measure to detect any spurious additional spectral information due to the surrounding materials. For further details on the principles and practice of Raman spectroscopy, see Chapter 12, Section II.

III. Data Analysis

A. Eigenfrequencies of Macroscopic Systems

The object of optical experiments such as reflectance, transmittance, or emission measurements is to determine the optical or dielectric parameters that characterize the system under investigation. In this section we shall discuss the case where the dimensions of the system are significantly larger than the wavelength of the incident radiation used in the experiment. In our experiments, the reflectance spectra measured for the lunar bulk samples in the far-infrared and infrared regions can be used to determine the dispersion of the optical or dielectric constants that characterize the lunar materials because the system obeys the causality principle.

The normal incidence reflectivity can be written as $R(\omega) = r(\omega)^2$, where the complex amplitude of the wave reflected is given by $r(\omega)e^{-i\theta(\omega)}$ and $\theta(\omega)$ is the phase difference caused by the reflection. Providing $R(\omega)$ is known, and because the system obeys the causality principle, the phase angle $\theta(\omega)$ can be obtained from the Kramers–Kronig relation (Robinson, 1952):

$$\theta(\omega) = \frac{2\omega}{\pi} \int_0^\infty \frac{\ln[r(\omega')]\, d\omega'}{\omega^2 - \omega'^2} \tag{1}$$

Once $r(\omega)$ and $\theta(\omega)$ are known, then the optical constants characterizing the medium, that is, the refractive index, n, and the extinction coefficient, k, are determined via an appeal to the Fresnel equations as

$$n = \frac{1 - r^2}{1 + r^2 - 2r \cos \theta} \tag{2}$$

$$k = \frac{2r \sin \theta}{1 + r^2 - 2r \cos \theta} \tag{3}$$

If n and k are known, then the absorption coefficient $\alpha = 4\pi k\omega$ and the real and imaginary parts of the dielectric constant, $\epsilon' = n^2 - k^2$ and $\epsilon'' = 2nk$, are readily determined. In practice we determine the infinite integral in Eq. (1) by representing $\ln[r(\omega')]$ by straight line segments and then evaluate the different parameters on a high-speed computer (CDC Cyber 72).

We use the dispersion of ϵ' and ϵ'' to locate the $q \approx 0$ transverse and longitudinal optic modes for the system (Lowndes, 1970). In dealing with long wavelength transverse optic modes it is the electric field that drives the system and hence the use of the dielectric response function $\epsilon(\omega)$ to describe these modes is the more appropriate. In dealing with long wavelength longitudinal optic modes, however, it is the dielectric displacement that drives the system and the inverse dielectric response function $\eta(\omega) = 1/\epsilon(\omega)$ is now the more appropriate response function. Both $\epsilon(\omega)$ and $\eta(\omega)$ are, of course, complex quantities following causal behavior. Providing the dielectric response for the system can be described by a first- or second-rank tensor, causality requires that the poles and zeros of $\epsilon(\omega)$ and $\eta(\omega)$ be symmetrically distributed about the imaginary axis of the ω-plane. A pole of $\epsilon(\omega)$ and a zero of $\eta(\omega)$ exists wherever a frequency occurs such that the driving electric field is zero and the dielectric displacement is nonzero: this frequency is a $q \approx 0$ transverse optic frequency. Analogously, a zero of $\epsilon(\omega)$ and a pole for $\eta(\omega)$ exists wherever a frequency occurs where the dielectric displacement is zero and the electric field is nonzero: this frequency is a $q \approx 0$ longitudinal optic frequency. Specifically it can be shown that, providing the mode damping is not too large, a maximum in $\omega\epsilon''$ or ϵ'' will locate a $q \approx 0$ transverse optic mode and that a minimum in $-\omega\eta''$ or $-\eta''$ will locate a $q \approx 0$ longitudinal optic mode.

B. Surface Modes for Small Particles

In some experiments size analysis is needed in order to deduce the intrinsic, shape-independent parameters of the solid. Even in reflectivity measurements of large samples certain care must be taken in the interpretation of the spectrum. As mentioned before, a Kramers–Kronig analysis is necessary to obtain the dielectric dispersion from reflectance data. However, Berreman (1967) has shown that extremely small bumps and pits on the surface of the crystal can give small, anomalous *reststrahlen* peaks due to localized resonances of the vibrations in the immediate neighborhood of the defects. The inverse is, of course, true in emission measurements as bulk samples may contain many small crystallite inclusions or the material may be particulate in nature. Again, anomalous bands could be caused by the excitation of the surface vibrational modes of these tiny crystallites. In transmission experiments, shape effects are always involved since the sample must have at least one very small dimension. The method of transmission through thin films has been employed for many years as a tool for the direct determination of the transverse frequency, ω_{TO} (at normal incidence) (Jones *et al.*, 1960) and of the longitudinal frequency, ω_{LO} (at nonnormal incidence) (Lowndes, 1970). Powdered crystals suspended in a convenient matrix (e.g., Nujol, KBr, CsI, polyethylene, etc.) are frequently used for transmission measurements, but the spectra can be misinterpreted if suitable care is not taken to discriminate among the various shape-dependent characteristic frequencies of long-wavelength modes. Such effects were first pointed out by Fröhlich (1949) and in the more recent work of Ruppin and Englman (1970) and Genzel and Martin (1972).

Qualitatively, the explanation of surface modes lies in the fact that the dipole moment accompanying these vibrations gives rise to a macroscopic electric field, which provides additional restoring forces. These forces depend on the direction of the wave vector and increase the frequencies of the longitudinal modes but decrease the frequencies of the transverse modes relative to those calculated from the short-range forces alone. The vibrational modes of particles of comparable or smaller dimensions than the natural excitation wavelengths of the bulk material can no longer be strictly classified as transverse (TO) and longitudinal (LO) optic modes. These fundamental modes (or surface modes) depend strongly on the size, shape, and density of the particles, but are constrained to approximately lie within a region $\omega_{\mathrm{TO}} \leq \omega \leq \omega_{\mathrm{LO}}$. For weak transitions, e.g., combination bands, impurity modes, electronic transitions, and even many fundamental internal molecular vibrations, the difference between ω_{TO} and ω_{LO} is often only slight and the bulk and surface modes are essentially degenerate. However, for materials where the bulk lattice modes do make significant contributions to the low-frequency static dielectric constant, ϵ_0, investigation of such samples in particulate form can give small-particle absorption spectra which will consist of contributions of modes distributed over a wide range of frequencies and are consequently not directly interpretable in terms of bulk mode frequencies representative of the macroscopic materials. This can be simply illustrated for crystalline particles in the form of spheres having a dielectric constant $\epsilon(\omega)$, radius R, and surrounded by a medium with a dielectric constant ϵ_M. The scalar Helmholtz equation $(\Delta + k^2) = 0$ will have transverse electric and magnetic spherical solutions of the form

$$\mathbf{M} = \text{curl } (a_r r\psi) \tag{4}$$

$$\mathbf{N} = \frac{1}{k} \text{curl curl } (a_r r\psi) \tag{5}$$

In terms of spherical coordinates, r, θ, φ,

$$\psi = y_{\text{lm}}(\theta, \varphi) \begin{cases} j_1(k_i r) & r < R \quad (6) \\ h_1(k_0 r) & r > R \quad (7) \end{cases}$$

where $k_0 = \epsilon_M{}^{1/2}\omega$, $k_i = \epsilon^{1/2}\omega$. The frequency in cm^{-1} is ω, y_{lm} are spherical harmonics, and j_1 and h_1 are spherical Bessel and Hankel functions, respectively (Morse and Feshbach, 1953, Chapter 5). For a simple ionic crystal having one infrared active transverse optic mode, ω_{TO},

$$\epsilon(\omega) = \epsilon_\infty + \frac{S^2}{\omega_{\text{TO}}^2 - \omega^2 - i\omega\gamma} \tag{8}$$

where γ is a frequency-independent damping function and S^2 is the oscillator strength such that $S^2 = (\epsilon_0 - \epsilon_\infty))\omega_{\text{TO}}^2 = \epsilon_\infty)(\omega_{\text{LO}}^2 - \omega_{\text{TO}}^2)$. ϵ_0 and ϵ_∞ are the low- and high-frequency dielectric constants, respectively. The frequencies of the electric modes with $\mathbf{E} \propto \mathbf{N}$ and $\mathbf{H} \propto \mathbf{M}$ are given by

$$\epsilon_{j1}(k_i R)\{k_0 R h_1(k_0 R)\}' = \epsilon_M h_1(k_0 R)\{k_i R_{j1}(k_i R)\}' \tag{9}$$

The optical properties can be expressed in terms of cross sections (i.e., in units of the geometrical cross section πR^2). The total (or extinction) cross section as given by Mie (1908) is

$$\sigma_t = -\frac{2}{(k_0 R)^2} \sum_{1=7} (21 + 7) Re(b_1) \tag{10}$$

where the Mie coefficient b_1 is given by

$$b_1 = -\epsilon_M j_1(k_0 R)\{k_i R_{j1}(k_i R)\}'$$

$$= \frac{-\epsilon_{j1}(k_i R)\{k_0 R_{j1}(k_0 R)\}'}{\epsilon_M h_1(k_0 R)\{k_i R_{j1}(k_i R)\}' - \epsilon_{j1}(k_i R)\{k_0 R h_1(k_0 R)\}'} \tag{11}$$

Note that for the electric modes [Eq. (9)] the denominator of Eq. (11) vanishes at the frequency of the electric eigen modes, and peaks in the optical spectra will occur at the frequencies of the radiative polariton modes. The transmission, reflection, emission, and absorption coefficients are given by

$$T = |t|^2, \qquad R = |r|^2, \qquad \epsilon = 1 - |r|^2, \qquad A = 1 - T - R \quad (12)$$

where

$$t = \frac{2\epsilon\epsilon_M k_i k_0 \exp(-2ik_0R)}{2k_0k_i\epsilon\epsilon_M \cos 2k_iR - i(k_i^2\epsilon_M^2 + k_0^2\epsilon^2) \sin 2k_iR}, \quad (13)$$

and

$$r = \frac{i(k_i^2\epsilon_M\epsilon_M' - k_0^2\epsilon^2)}{(2k_0k_i\epsilon\epsilon_M) \cot (2k_iR) - i(k^2\epsilon_M^2 + k_0^2\epsilon^2)}. \quad (14)$$

Poles in t and r occur at the frequencies where

$$\tan (k_iR) = \frac{-i\epsilon k_0}{\epsilon_M k_i}, \quad \text{and} \quad \cot (k_iR) = \frac{i\epsilon k_0}{\epsilon_M k_i} \quad (15)$$

have their solutions.

Using a simple model of a cubic ionic crystal with one infrared active mode having a dielectric function $\epsilon(\omega)$ given by Eq. (8), Fröhlich has shown that, for uniformly polarized spherical samples which are much smaller than the exciting radiation, the lowest energy surface mode has triple degeneracy and is infrared active. It has a frequency related to the transverse optic frequency of the macroscopic sample, ω_{TO}, by the relation

$$\omega_S^2 = \left(\frac{\epsilon_0 + 2\epsilon_M}{\epsilon_\infty + 2\epsilon_M}\right)\omega_T^2, \quad (16)$$

where ϵ_M is the dielectric constant of the homogeneous medium in which the sphere is embedded (e.g., $\epsilon_M = 1$ for air or vacuum). Ruppin and Englman (1970) have extended this treatment to include uniaxial and multimode crystals. The surface mode frequencies of materials having more than one infrared active transverse optic mode still have the form given by Eq. (16) but now the dielectric function given by Eq. (8) must be replaced by

$$\epsilon(\omega) = \epsilon_\infty + \sum_{j=1}^{n} \frac{S_j^2}{\omega_{TOj}^2 - \omega^2 - i\omega\gamma_j}, \quad (17)$$

where n is the number of infrared active modes and S_j^2 is now the oscillator strength of the jth mode. (This equation is strictly only valid along one of the three crystallographic directions for a nonisotropic material.)

The frequencies ω_{Sj} of a given type of surface mode can be written in a generalized form such that

$$\prod_j \left(\frac{\omega_{S_j}}{\omega_{\mathrm{TO}j}}\right)^2 = \frac{\epsilon_0 + 2\epsilon_{\mathrm{M}}}{\epsilon_\infty + 2\epsilon_{\mathrm{M}}} \tag{18}$$

As an example, Hunt *et al.* (1964) have measured the transmittance of powdered $KMgF_3$ dispersed in polyethylene films. $KMgF_3$ has the perovskite structure and possesses three infrared active modes of F_{1u} symmetry corresponding approximately to F–Mg–F stretching, Mg–F out of plane bending, and a K–MgF_3 lattice vibrations. In transmittance, three strong minima are indeed observed but their frequencies are intermediate between the resonance frequencies ω_{TO_j} and ω_{LO_j} derived from Hunt *et al.*'s analysis of the reflection spectra. A weaker minimum also appeared below 300 cm^{-1}

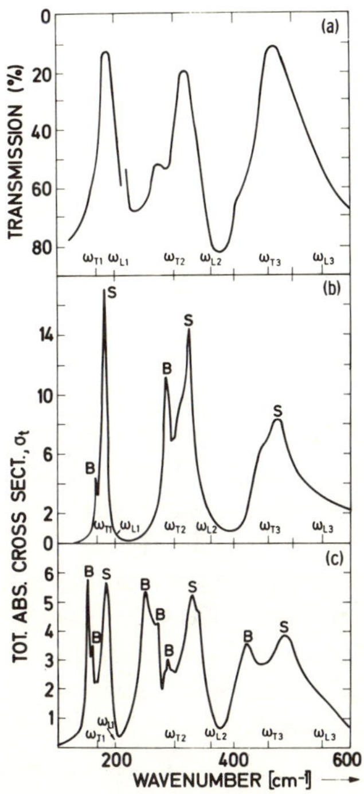

FIG. 2. (a) $KMgF_3$, small particle in poleythylene, experimental; (b) total absorption coefficient calculated for 4 μm diameter spherical particle; (c) total absorption coefficient calculated for 8μ diameter spherical particle; both bulk and surface modes (labeled B and S, respectively) are observed. (Ruppin and Englman, 1970.)

which could not be explained in terms of a bulk phonon mode. Ruppin and Englman (1970) calculated the extinction cross section for spherical $KMgF_3$ crystals embedded in polyethylene using the dielectric function [Eq. (17)] and parameters for ϵ_∞, S_j^2, ω_{TO_j}, and γ_j for $KMgF_3$ given by Perry and Young (1967). The theoretical curve for a spherical crystal of radius 4 μm shows qualitative features similar to those of the experimental curve and especially reproduces the minimum around 300 cm^{-1}. These curves are compared in Fig. 2 together with similar calculations for a particle having a spherical shape of radius 8 μm. In this case the complexity of the spectrum increases as the particle size becomes more comparable to the excitation wavelength, as both bulk and surface modes (labeled B and S, respectively) are observed. These examples are used to illustrate the caution needed in interpreting infrared spectroscopic data of dispersed powder samples, as the observed frequencies may not necessarily correspond to the mode frequencies characteristic of the same bulk material.

IV. Discussion and Results

Infrared and Raman measurements are concerned with the excitation of specific types of vibration of atoms relative to each other. For any group of atoms the maximum number of vibrational motions that may occur depends on the number of atoms present in each repeating group. The kind of motions depend on the geometry of these atoms relative to each other and on the values of the interatomic force constants between them. The appearance of a spectral feature corresponding to a particular type of motion depends on the specific selection rules, which are determined by the space group of the system. Vibrations involving a change in dipole moment are found in the infrared and contribute to the dielectric dispersion, whereas a motion which involves a change in the polarizability is observed in the Raman effect. Consequently, in the study of materials (whether lunar or terrestrial in origin), the complete determination of the symmetry, structure, and chemical nature of the bonds between atoms usually involves both spectroscopic techniques. As the structures become more complicated and the structural symmetry is lowered, bands can appear simultaneously in both measurements, thus often confirming particular assignments.

In the case of rock samples which are composed of a variety of materials and compositions, the spectra are frequently complex but these spectral features offer identification of each species once they have been characterized. Studies of the infrared spectra of mineral surfaces for application in remote compositional sensing have been reported by Lyon (1963), Aronson *et al.* (1967), Salisbury *et al.* (1970), and by Perry and Lowndes (1970). Hovis and Callahan (1966) have made measurements of the infrared reflectance spectra of some igneous rocks, etc., in the mid-infrared and Perry and Wrigley (1967) indicated that the observation of compositional differences in tektites could be used to approximately determine their geographical location of origin on the earth's surface. Conel (1969) has studied the emissivities of silicates in the 8–12 μm region and Lyon (1965) has shown that the stretching vibration of the Si—O bonds in the silicate SiO_4 tetrahedra which occur in the 800–1200 cm^{-1} frequency range can be correlated with SiO_2 content. Generally the substitution of the heavy cations like those of titanium and iron lowers the stretching vibration compared with the lighter cations like those of calcium, magnesium, and aluminum. Consequently, plagioclase-type materials which have moderate amounts of iron and a higher aluminum content have frequencies higher than those of the pyroxenes, which are more iron rich and contain smaller amounts of aluminum. This effect can be seen in both the spectra and the dielectric response of such materials and will be discussed more specifically in relation to particular samples.

The geographical locations of the landing sites on the lunar surface are shown in Fig. 3(a); (b), (c), and (d) of Fig. 3 show the locations of the individual samples discussed here for the Apollo 12, 14, and 15 missions, respectively.

The infrared reflectance spectra from 10–1700 cm^{-1} of some representative lunar rock samples from Apollo 11, 12, and 14 are shown in Fig. 4. From these spectra the real and imaginary parts of the dielectric constant were calculated and the corresponding plots are given in Fig. 5. Figure 6 comprises the spectra of five Apollo 14 fines [14163,31; 14161,36; 14259,38; 14230,95 (core sample); and 14141,37 (Cone Crater)] and one powdered rock sample 14321,263 (Cone Crater).

The reflectance spectra of six Apollo 15 fines are shown in Fig. 7. The samples investigated were 15091,65 and 15221,71 from Station 2 at the foot of the Apennine Front; 15071,61 from Station 1 (Elbow Crater ejecta); 15021,159 from the LM area; 15471,68 from Dune Crater (Station 4); and 15531,63 from Station 9a at the rim of Hadley Rille. The geographical location of the Apollo 14 and 15 samples on the lunar surface can be seen in (c) and (d) of Fig. 3. The spectra shown in Figs. 4, 6, and 7 represent the bulk mineralogical and chemical properties of the materials averaged over the area of the sample examined (0.025–1 cm^2). They form a basis for the comparison of these samples with other lunar samples and with known terrestrial materials.

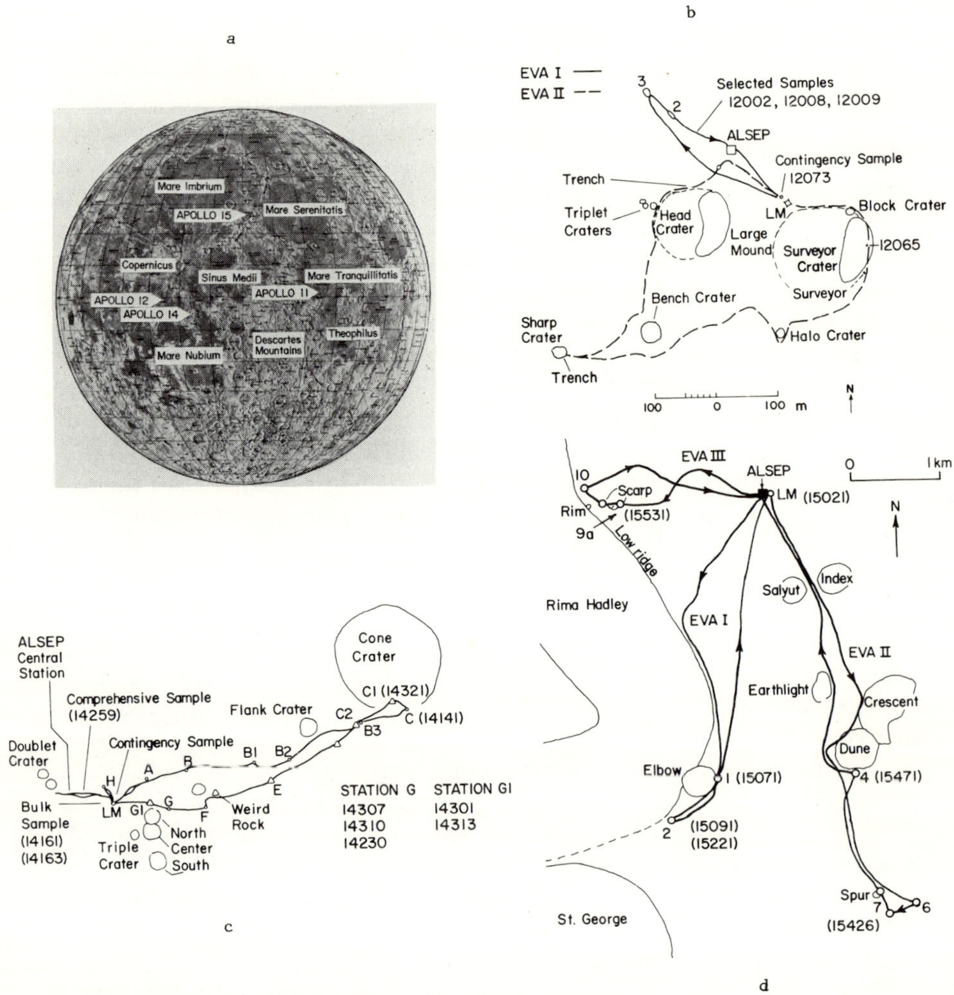

FIG. 3. (a) Map of the Apollo 11, 12, 14, 15 landing sites; (b) ,(c) and (d) show the geographical locations of the samples described in the text for Apollo 12, 14, 15, respectively.

In Fig. 8 the Raman spectra show some of the variations of lunar pyroxenes, found in samples 14310,76; 10058,56; 14310,184; 12002,18; and 12065,115. The systematic frequency shift of the stretching vibration at $\sim$1015 cm^{-1} (e.g., Estep *et al.*, 1971a,b, 1972), the bending vibration at $\sim$650 cm^{-1} and another diagnostic band around 400 cm^{-1} (Chapter 9, Section V,C) can be correlated with the ferrosilite content. In sample

14310,76, the 650 cm^{-1} band is almost a resolved doublet indicating the presence of both ortho- and clinopyroxene. Sample 10058,56 contains mostly clinopyroxene ranging from pigeonite to augite (Fs_{20}–Fs_{40}). Samples 12002,186 and 12065,115 show broader weaker spectra and the evaluation consequently is less well-defined but again both probably contain clino- and orthopyroxene.

Tables I and II summarize the spectroscopic and dielectric characteristics of most of the lunar rock and soil samples that we have investigated from the Apollo 14 and 15 missions.

Several researchers (Conel, 1969; Salisbury *et al.*, 1970) have suggested that the frequency of the peak in the emissivity can be used as a diagnostic feature for determining the silica abundance in fine-powder emissivity spectra. This peak is directly related through the dielectric function

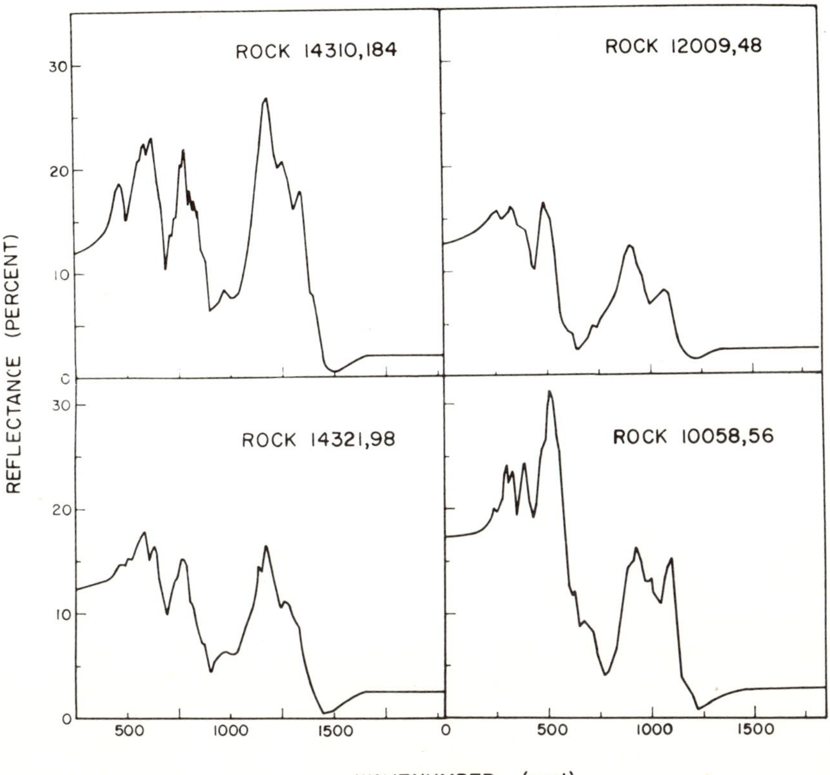

FIG. 4. Infrared reflectance spectra of rock samples 14310,184; 12009,48; 14321,98; and 10058,56 from 0 to 1600 cm^{-1}.

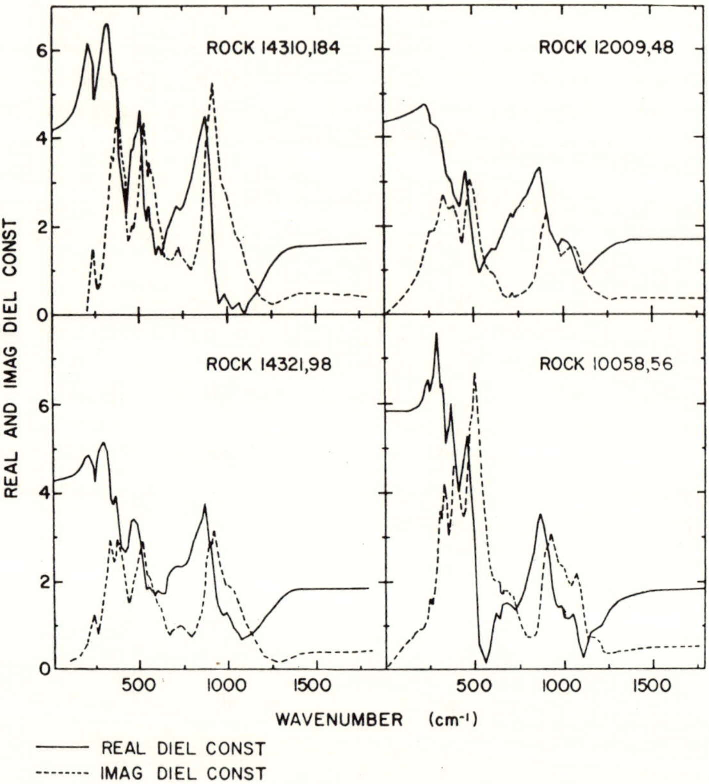

FIG. 5. The real (ϵ', dashed curve) and imaginary (ϵ'', solid curve) parts of the dielectric constant for rock samples 14310,184; 12009,48; 14321,98 and 10058,56 from 0 to 1600 cm^{-1}.

($\epsilon' = n^2 - k^2$ and $\epsilon'' = 2nk$, where n is the real part of the refractive index and k is the extinction coefficient) to the high-frequency minimum in the reflectivity, and in fact our reflectance spectra are simply inverted emission spectra referenced to blackbody emissivity. For powder spectra both curves are complicated by the sample particle size and packing fraction because of additional diffuse scattering mechanisms (Aronson and Emslie, 1972). This frequency (known as the Christiansen frequency and referred to as ω_c in the tables)* occurs in the region where $n \approx 1$ and $k \approx 0$. Perry *et al.* (1971) have shown that for lunar materials ω_c varies over the narrow range from

* Additional information on the Christiansen frequency is given in Chapter 5, Section III.

about 1195 cm^{-1} ($\sim$40% SiO_2) to about 1260 cm^{-1} ($\sim$50% SiO_2). Both Conel (1969) and Salisbury *et al.* (1970) have shown that particulate minerals can be characterized by their emission maxima (corresponding to ω_c). Comparisons of the ω_c frequencies for the lunar samples with those of various terrestrial minerals falling in the intermediate to ultrabasic range are shown in Table III. Differences in composition and particle size contribute to the errors in the measurement and the quoted values are estimated to be accurate to $\pm$5 cm^{-1}. The emissivity peak determined for 14259 soil by Logan *et al.* (1972) at 1216 cm^{-1} (Chapter 5, Fig. 10) is in good agreement with the ω_c values (1225 cm^{-1}) obtained from our reflectance measurements as given in Table II. Birkebak and Dawson (1972) obtained a value of 1210 cm^{-1} for 14163 soil having a density of 1.6 gm/cm^3 which is within experimental error of our values of 1205 cm^{-1} for 14163,31 (LM sample) with a density of $\sim$2.5 gm/cm^3. It is interesting to note that the infrared emissivity measurement of surface features in various lunar regions from balloon observations by Salisbury *et al.* (1970) and Murcray *et al.* (1970) vary from 1195 to 1235 cm^{-1}, again within this narrow region. Again, comparison with

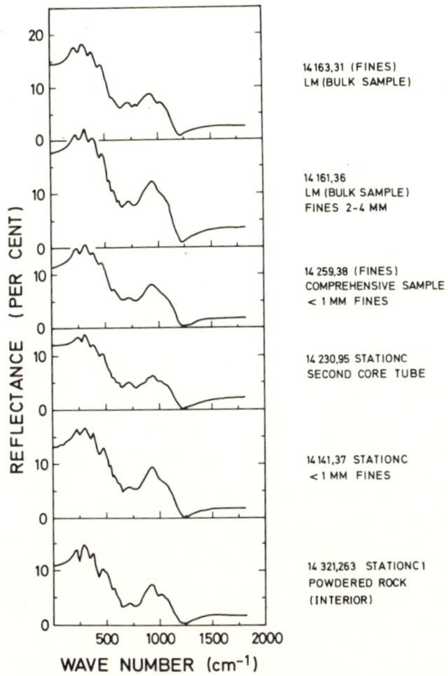

FIG. 6. The infrared reflectance spectra of five Apollo 14 fines [14163,31; 14161,38; 14259,38; 14230,95 (core sample); and 14141,37 (Cone Crater)] and one powdered rock sample 14321,263 (Cone Crater) from 0 to 1700 cm^{-1}.

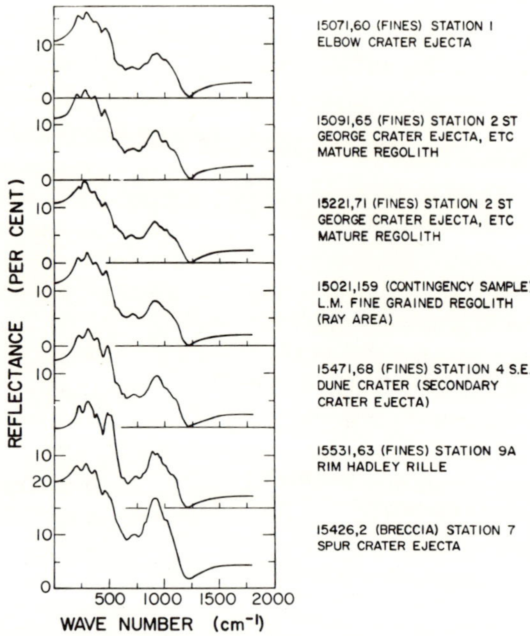

FIG. 7. The infrared reflectance spectra of six Apollo 15 fines (15091,65; 15221,71; 15071,60; 15021,159; 15471,68; and 15531,63) and one Apollo 15 rock sample 15426,2 from 0 to 1700 cm^{-1}.

Table III can be made, but any interpretations should be undertaken with caution due to possible systematic errors in the data. However, the emission peak frequencies are encompassed by the Christiansen frequencies of the materials investigated in this work and there are no obvious inconsistancies. Also the 14310,184 sample, which we estimate to have a high anorthite content, compares with noritic gabbros and anorthitic basalts in terms of the Christiansen frequency ($\omega_c \approx 1250$ cm^{-1}). However, portions of the specimens used to obtain the elemental analyses are not the same as the ones used to measure the ω_c values, and variations can be expected within the same parent sample. Consequently, correlation of the ω_c values with SiO_2 wt.% may be better than the data presented in earlier work by Perry *et al.* (1971), although particle size and packing of the material also effect the scattering components and can cause variations in the measured Christiansen frequency.

In Table II we have also shown the values of the far-infrared dielectric constant at $\sim 10^{11}$ Hz which were calculated from the Kramers–Kronig analysis (see, also, Perry *et al.*, 1971, 1972). Our values for the lunar rocks appear to be in general agreement with Ade *et al.*'s (1971) far-infrared measurements at 337 μm where a cyanogen laser was used. For sample 12063,96 (type A basalt) and for 10065,30 (breccia) they obtained values of 6.25 and 3.7, respectively. Based on the results of Perry *et al.* (1971) one may estimate that both these samples contain 25–30 mole % of titaniferous oxide. Sample 10017,64, another Apollo 11 type A basalt, has a high dielectric constant value of 7.05 (see Chapter 10, Table III) which compares with our measured value of 6.7 for 12008,23, and consequently it probably has a comparable or slightly higher ilmenite content (i.e., 30–35%). Comparison can be made with the dielectric response determined by capacitance measurements at low frequencies (10^2–10^6 Hz) by Chung *et al.* (1971, 1972), Collett and Katsuba (1971), and Lowndes et al. (1974), and at 430 MHz by Gold *et al.* (1972). The earlier values quoted by Chung *et al.* (1971) appear to be in error due to contamination by terrestrial water (Chung *et al.*, 1972). Our value for a 14310 sample of 5.2 (10^{11} Hz) and 5.99 (10^2–10^6 Hz) compares favorably with Chung and Westphal's (1972) value of 6–9 and the Gold *et al.*'s (1972) value of 6.6. However, there are still striking differences in sample 12002 where we obtained a dielectric constant (at 10^{11} Hz) of 4 which is to be compared with Chung *et al.*'s (1971) result of 8–10 and Collett and Katsuba's (1971) value of 9 at low frequencies. Similar differences were observed in our own meas-

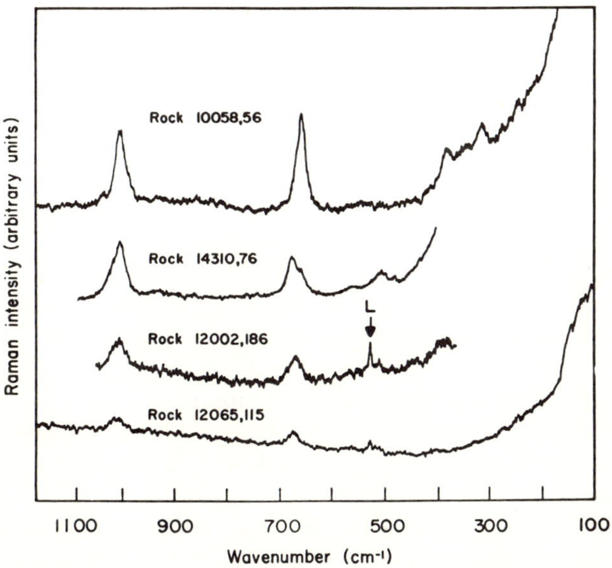

FIG. 8. Raman spectra of some lunar pyroxenes, found in samples 10058,56; 14310,76; 12002,186; and 12065,115.

TABLE I

COMPARISON OF INFRARED TRANSVERSE OPTICAL PHONON FREQUENCIES FOR LUNAR ROCK SAMPLES FROM APOLLO 14 AND 15

SUGGESTED ASSIGNMENT	14310,184	14321,198	14301,20	14307,18	14313,51	15426,2
Combinations	1065(w)					1040(w)
M–O stretch					1015(m)	
ν'''_3	990(m)	1010(w)	1010(w)	1005(m)		
		980(w)				
ν''_3	920(s)	920(m)	940(m)	930(s)	925(s)	
ν'_3		890(m)				850(m)
ν'_1						810(w)
Cation–oxygen	725(w)		745(w)	745(w)	730(w)	710(w)
vibrations		620(w)				640(w)
	580(w)		570(w)			610(w)
	560(w)					570(w)
M–O–M bend	530(m)		530(m)			
ν'_4		520(m)		525(m)	520(m)	
ν''_4	470(m)		475(m)	485(m)	470(s)	450(w)
ν', ν''''_4	390(m)	390(m)	390(m)	380(m)	380(m)	370(w)
ν''_2	345(m)	335(m)	345(w)		345(m)	340(w)
Lattice modes	240(w)	240(w)	240(w)		225(w)	280(s)
ϵ_0 at 10^{11} Hz	5.2	4.3	4.0	5.5	4.8	6.5
ω_c (cm^{-1})	1250	1205	1230	1230	1215	1220

urements in 12008,23 where the dielectric constant varies between 6.7 at 10^{11} Hz and 33 at 10^6–10^2 Hz. As both these samples contain free iron and other high-conductivity materials, they may give rise to additional relaxation mechanisms or free carrier absorption which could account for this strong dispersion of the dielectric constant. Density corrections for the dielectric constant are given in Chapter 10, Section II,D.

In our earlier report, Perry *et al.* (1971) found that sample 12008,23 has a strong vibration at $\sim$890 cm^{-1}. The high value of the measured static dielectric constant of this sample indicates that it has a very high ($>$30%) FeO content; it has also been reported that the major phase of the parent rock may be ilmenite (Warner, 1970). Sample 12009,48 has a lower FeO content ($\sim$20%) and the stretching vibration is now situated at $\sim$920 cm^{-1}. With the increase in plagioclase content, as, for example, in 12002,186 ($\sim$30% plagioclase), there is a continued shift of this band to higher frequencies (~930 cm^{-1}), and in 12065,115 (breccia), with about 50–50 plagioclase–pyroxene content, the trend is further indicated (~960 cm^{-1}).

In Fig. 4 we show some of the greatest variations that we have observed in the reflection spectra of lunar samples. The 14310,184 spectrum has features that clearly demonstrate that it is a high calcium oxide, alumina, and silica basalt; it has the highest ω_c frequency compared with the other lunar samples examined. It has high stretching vibrations at ~990 cm^{-1} and ~920 cm^{-1} and bending vibrations at ~470 cm^{-1} and 530 cm^{-1}. The lattice band at 240 cm^{-1} is characteristic of a high plagioclase material and is indicative of its high anorthite (An_{90}–An_{100}) content. For comparison the Cone Crater sample (14321,98) from Station C1 (which has comparable silica, alumina, and calcium oxide content) shows a lowering of the fre-

TABLE II

COMPARISON OF THE INFRARED OPTICAL PHONON FREQUENCIES, $\omega_c{}^a$ AND $\epsilon_0{}^b$ FOR APOLLO 14 AND 15 LUNAR FINES

SUGGESTED ASSIGNMENT	FINES (LM) 14163,31	FINES (2–4 MM) (LM) 14161,36	COMP. <1 MM FINES 14259,38	CORE SAMPLE 14230,95	STATION C <1 MM FINES 14141,37	STATION C1 POWDERED ROCK 14321,263
Combinations				1060	1050	
M–O stretch	1020	1020	1010	1010		1020
		930	930	900	900	920
	860		870			
		810				
Cation–oxygen	780			780		800
lattice vibra-	690	690	700			690
tions		610		660	660	
				610	600	610
	560	560	570	560	550	560
		540	540		530	
M–O–M bend					510	490
	450	450	440	460	430	450
	370	360	370	370	360	360
						330
Lattice modes	290	290	300	280	260	270
		230	200	230	220	220
ω_c (cm^{-1})	1195	1210	1225	1215	1230	1225
ϵ_0	4.9	5.9	4.0	4.2	4.5	3.8

TABLE II—(*Continued*)

	STATION 2 ST. GEORGE CRATER EJECTA		STATION 1 ELBOW CRATER EJECTA	RAY AREA (LM)	STATION 4 DUNE CRATER	STATION 9A HADLEY RILLE
SUGGESTED ASSIGNMENT	15091,65	15221,71	15071,60	15021,159	15471,68	15531,63
Combinations	1090	1100		1040	1100	1070
M–O stretch	1000	1000	1000		1000	
					950	960
						930
			920	890		910
	880	890	870		860	870
			820			
Cation–oxygen lattice vibrations	790	790	780		770	790
	670	670	670	660	680	700
	600	610	610		620	630
	550	550	580		560	
M–O–M bend						510
	440	440	460	470	450	450
	360	370	370	370	370	380
Lattice modes	280	280	290	300	280	280
	220	230	230	210	230	220
ω_c (cm^{-1})	1220	1220	1225	1220	1210	1205
ϵ_0	4.0	4.2	3.8	4.0	4.2	4.9

[a] Christiansen frequency.
[b] Dielectric constant for sample densities 2.4–2.6 gm/cm^3.

quency of the band at $\sim$500 cm^{-1}. This band appears to be related to the slightly higher orthopyroxene content. The spectrum of 12009,48 with its reduced silica and alumina content shows a further lowering of the bending vibration to $\sim$490 cm^{-1} which appears associated with the increase in FeO content. Generally, as the dominance of the framework silicates change gradually to the more open chain silicates of the pyroxenes, a lowering of the internal frequencies may be expected and is apparently observed. Finally in Fig. 4, the new features in the spectrum of sample 10058,56, which has FeO, SiO_2, Al_2O_3, and CaO contents comparable with those of 12009,48, indicate that they can be directly associated with the increasing TiO_2 present and the subsequent reduction in MgO.

The three Apollo 14 breccias (or fragmental rocks) examined, namely, 14307,18; 14313,51 and 14301,20, show unmistakable spectral similarities (Perry *et al.*, 1972). All are typical of the Fra Mauro feldspar-rich breccias and their spectra, although slightly enhanced in intensity, possess features that correspond closely with those observed in the Apollo 14 fines sample of which the bulk sample 14161,36 from the LM vicinity is a typical example.

The Fra Mauro breccias generally have lower silica and alumina contents than sample 14310, which is more likely of highland origin since it is rich in plagioclase (anorthosites). However, they are intermediate between the highland material and the mare material of Apollo 11 and 12 and these findings have been confirmed in the X-ray fluorescence experiment by Adler *et al.* (1972). These breccias are presumably representative of the ejecta blanket deposited by the impact that formed the Imbrium Basin and contain materials from the upper levels of the crust. The Apollo 14 soils also show considerable homogeneity in the spectra (see Fig. 6); this is confirmed in the analyses (Compston *et al.*, 1972), where the average soil content is 48% SiO_2, 17% Al_2O_3, 9% MgO, 10% FeO, 11% CaO, and 2% TiO_2. These major chemical abundances are very similar to those found in the Apollo 14 breccias and the spectral similarities between these and the Fra Mauro soils indicate that the latter were mainly produced by local impacts and consist of particulate debris from the rocks with a small admixture of other material (see, for example, Adams and McCord, 1972). From the infrared and Raman frequencies it would appear that these samples contain plagioclase An_{70}–An_{90}, olivine Fa_{20}–Fa_{35}, and pyroxene Fs_{25}–Fs_{35}, and are in general agreement with the report on 14313 (Floran *et al.*, 1972).

The Cone Crater breccia sample from Station C, 14321,98, has spectral similarities that lie between 14310 and the other three Apollo 14 rock samples. It has the highest magnesium content and lowest calcium content and although the silica abundance is similar to 14310 its alumina and FeO contents are more like the Fra Mauro breccias. It may in fact be part of an earlier regolith that suffered from the Imbrian ejecta and was later ejected during the Cone Crater event. The increase in iron content in 14321 over the 14310 samples is exhibited in a shifting to lower frequencies of the band $\sim$500 cm^{-1}. Its lower Christiansen frequency probably results from its high MgO content. This reverse behavior of ω_c is apparent in 12009,48, which again has a comparable abundance of MgO even though the SiO_2 content is less and the FeO content is increased.

In Figs. 6 and 7 we can see a direct comparison between the Apollo 14 and 15 soil samples. Again reference should be made to Fig. 3 for the relative geographical location of the samples. Both Sample 14259,38, which is

TABLE III

COMPARISON OF THE CHRISTIANSEN (ω_c) FREQUENCIES CHARACTERISTIC OF LUNAR AND TERRESTRIAL SAMPLES

SAMPLE	MINERALOGICAL DESCRIPTION	$\omega_c{}^a$ (CM^{-1})	TERRESTRIAL MATERIAL[b]	PERCENT OF WEIGHT OF SIO_2	ω_c	TERRESTRIAL MATERIAL[c]	ω_c
			Basalt, USBM (#57)	50.62	1285	Anorthite	1280
			Basalt (Vesicular) (#5)	57.94	1280	Basalt, Little Lake	1273
			Basalt (Hornblende) (#7)	51.82	1275	Quartz, basalt porphyr.	1272
			Basalt, USBM (#58)	54.66	1265	Basalt, pisgah	1270
			Gabbro, Ontario (#75)	48.20	1260	Basalt	1266
						Anorthite	1265
			Basalt, Col. #2	46.98	1260	Basalt, Red cinders	1260
			Basalt (Olivine porphyr.) (#3)	44.28	1260	Gabbro	1260
						Gabbro, San Marcus	1258
14310,184	Basaltic cryst. rock	1250	Basalt, N.J. (#4)	49.29	1250	Basalt, Pisgah cinders	1243
12073,42	Microbreccia	1240	Gabbro (Norite) (#74)	46.36	1250	Chondrite	1225
12009,48	Porphyritic basalt						
14301,20	Coherent breccia						
14307,18	Coherent breccia	1230	Basalt, Hawaii (#166)	46.70	1250		1213
14141,37	Station C < 1 mm fines						
14259,38	Comp. < 1 mm fines						
14321,63	Station C1 (powdered rock)	1225	Basalt, Germany (#6)	42.81	1237	Achondrite, Norton Co.	1209
15071,60	Station 1 fines Elbow Crater ejecta						
12002,186	Basalt						

10058,56	Ophitic basalt						
15426,2	Breccia						
15021,159	Ray area (LM)	1220					
15091,65	Station 2 fines						
15221,71	St. George Crater ejecta						
14313,51	Coherent breccia	1215	Chondrite, Bruderheim (#165)	39.83	1190	Eclogite	1205
14230,95	Fines (core sample)						
14161,36	Fines (2–4 mm) (LM)	1210					
15471,68	Station 4 fines Dune Crater						
14321,98	Breccia (Cone Crater)	1205	Peridotite, Ark. (#71)	38.55	1175	Peridotite	1192
15531,63	Station 9a fines Hadley Rille						
12008,23	Cumulate	1200	Serpentinite, USBM (#60)	38.78	1153	Olivine	1145
12065,115	Variolitic basalt						
14163,31	Fines (LM)	1195	Peridotite (Olivine) (#63)	37.78	1135		

[a] Value of ω_c is estimated from the reflectivity minimum where $n \approx 1$ and $k \approx 0$.
[b] Data taken from Salisbury *et al.* (1970).
[c] Data taken from Conel (1969).

representative of the regolith, and sample 14163,31, which is from a depth of a few centimeters, gave very similar spectra. Sample 14161,36, which consists of 2–4 mm fines, showed some slight sharpening of the reflectivity peaks. The sample 14230,95 from the second core tube again showed general overall spectral similarities and it was concluded that this sample was also representative of the soil in the area and confirmed its homogeneity. Some small spectral differences were observed with the <1 mm fines sample from Station C (14141,37) in the vicinity of Cone Crater. The most noticeable feature was the slight strengthening of the band about 450 cm^{-1} which showed that the Fra Mauro soil in this area was contaminated with the more pyroxene-rich Cone Crater soil ejecta. Some rock fragments from the interior of 14321 (associated with piece 14321,263) again showed some spectral enhancement of the features but they were not as pronounced as 14321,98, which was a polished butt end (presumably from the exterior). Again it would appear that the interior pieces of this rock more closely resembled the surrounding soil. The Apollo 15 soils show a larger variety of spectral differences that can be associated with their geographical locations and the corresponding mineralogical compositions. The two samples from the Apenine Front at Station 2 (15091,65 and 15221,71) comprise St. George Crater ejecta and other ejecta materials. The two showed some very slight spectral variations but in fact more closely resembled the Apollo 14 soils. Consequently, it would appear that the soils in the vicinity of the St. George Crater rim are substantially intermixed. To some extent this was also true of 15071,60 from Station 1 which might be expected to be more representative of Elbow Crater ejecta. Evidently, this is not the case and here again the soil is well intermixed. In the vicinity of the LM on the mare surface of Palus Putredinis are the faint rays from either of the craters Aristillus and Autolycus. The spectrum of the contingency sample 15021,159 is presumably representative of the fine-grain regolith and although the spectral features were not significantly different there was a lowering of some frequencies which is usually indicative of an increasing iron content and a corresponding slight decrease in the silica and alumina content. The surface layer here may be more representative of Imbrium or other ejecta from deeper crust levels that would not have traveled as far to the Apollo 15 site compared to the Apollo 14 site. The spectrum of the Dune Crater sample 15471,68 from Station 4 shows further enhancement of the spectral feature at $\sim$450 cm^{-1} which can be associated with a slight increase in the pyroxene content; this trend is further enhanced, as seen in Fig. 7, in 15531,63 from the rim of Hadley Rille. The spectra are more similar to some of the Apollo 12 rocks such as 12002,186 or 12009,48 which have high pyroxene contents. It may be expected that these young ejecta materials are more representative of bedrock, whereas the material from the Apennine Front contains largely gardened mixtures.

The 15426,2 breccia from Station 7 at Spur Crater does not resemble the Fra Mauro breccias as it is obviously richer in pyroxene but it also does not show the same spectroscopic features as the Apollo 12 samples. There are some spectroscopic similarities between this sample and the microbreccia 12073,42 and the Cone Crater sample 14321,263. However, we have not observed any exact spectroscopic analog from any of the soil samples.

ACKNOWLEDGMENTS

We wish to thank Professor Anastassakis and Drs. Agrawal, Rastogi, and Tornberg for their collaboration in various parts of this work. This work was supported by NASA Grant NGR 22-011-069 and by a Northeastern University Grant for basic research. Partial equipment support was provided under NASA Cooperative Agreement NCAw 22-011-079.

One of us (CHP) would also like to thank Professor L. Genzel and Dr. T. P. Martin, Max-Planck Institut für Festkörperforschung, Stuttgart, Germany, for their help and encouragement while the author was a guest at the Institut and also for their illuminating discussions on surface modes.

References

Adams, J. B., and McCord, T. B. (1972). *Proc. Lunar Sci. Conf., Geochim. Cosmochim. Acta, 3rd, 19. .* Vol. 3, Suppl. 3, pp. 3021–3034.

Ade, P. A., Bastin, J. A., Marston, A. C., Pandya, S. J., and Puplett, E. (1971). *Proc. Lunar Sci. Conf., Geochim. Cosmochim. Acta, 2nd, 1971* Vol. 3, Suppl. 2, pp. 2203–2211.

Adler, I., Gerard, J., Trombka, J., Schmadebeck, R., Lowman, P., Blodgett, H., Yin, L., Lamothe, R., Gorenstein, P., Bjorkholm, P., Harris, B., and Gursky, H. (1972). *Proc. Lunar Sci. Conf., Geochim. Cosmochim. Acta, 3rd, 1972* Vol. 3, Suppl. 3, pp. 2157–2178.

Aronson, J. R., and Emslie, A. G. (1972). *Moon* **5,** 3–15.

Aronson, J. R., Emslie, A. G., Allen, R. V., and McLinden, H. G. (1967). *J. Geophys. Res.* **72,** 687–703.

Berreman, D. W. (1967). *Phys. Rev.* **163,** 855–864.

Birkebak, R. C., and Dawson, J. P. (1972). *Proc. Lunar Sci. Conf., Geochim. Cosmochim.* (C. Watkins, ed.) *Acta, 3rd, 1972* Contrib. No. 88, pp. 83–85.

Chung, D. H., Westphal, W. B., and Simmons, G. (1971). *Proc. Lunar Sci. Conf., Geochim. Cosmochim. Acta, 2nd, 1971* Vol. 3, Suppl. 2, pp. 2381–2390.

Chung, D. H., Westphal, W. B., and Olhoeft, G. R. (1972). *Proc. Lunar Sci. Conf., Geochim. Cosmochim. Acta, 3rd, 1972* Vol. 3, Suppl. 3, pp. 3161–3172.

Collett, L. S., and Katsuba, T. J. (1971). *Proc. Lunar Sci. Conf., Geochim. Cosochim. Acta, 2nd, 1971* Vol. 3, Suppl. 2, pp. 2367–2379.

Compston, W., Vernon, M. J., Berry, H., Rudowski, R., Gray, C. M., Ware, N., Chappell, B. W., and Kaye, M. (1972). *Proc. Lunar Sci. Conf., Geochim. Cosmochim. Acta, 3rd, 1972* Vol. 2, Suppl. 3, pp. 1487–1501.

Conel, J. E. (1969). *J. Geophys. Res.* **74,** 1614–1634.

Estep, P. A., Kovach, J. J., and Karr, C., Jr. (1971a). "Infrared Vibrational Spectroscopic Studies of Minerals from Apollo 11 and 12 Lunar Samples," Final Report NASA Contract No. T-88614. Nat. Aeronaut. Space Admin., Washington, D. C.

Estep, P. A., Kovach, J. J., and Karr, C., Jr. (1971b). Proc. Lunar Sci. *Conf., Geochim. Cosmochim Acta, 2nd, 1971* Vol. 3, Suppl. 2, pp. 2137–2151.

Estep, P. A., Kovach, J. J., Waldstein, P., and Karr, C., Jr. (1972). *Proc. Lunar Sci. Conf., Geochim. Cosmochim. 3rd, 1972* Vol. 3, Suppl. 3, pp. 3047–3067.

Floran, R. J., Cameron, K., Bence, A. E., and Papike, J. J. (1972). *Proc. Lunar Sci. Conf., Geochim. Cosmochim. Acta, 3rd, 1972* Vol. 1, Suppl. 3, pp. 661–667.

Fröhlich, H. (1949). "Theory of Dielectrics." Oxford Univ. Press (Clarendon), London and New York.

Genzel, L., and Martin, T. P. (1972). *Phys. Status Solidi* **51,** 91–99.

Gold, T., Bilson, E., and Yerbury, M. (1972). *Proc. Lunar Sci. Conf., Geochim. Cosmochim. Acta, 3rd, 1972* Vol. 3, Suppl. 3, pp. 3187–3193.

Hovis, W. A., Jr., and Callahan, W. R. (1966). *J. Opt. Soc. Amer.* **56,** 638–643.

Hunt, G. R., Perry, C. H., and Ferguson, J. (1964). *Phys. Rev.* **134,** A688–A691.

Jones, G. O., Martin, D. H., Mawer, P. A., and Perry, C. H. (1960). *Proc. Roy. Soc., Ser. A* **261,** 10–27.

Logan, L. M., Hunt, G. R., Balsamo, S. R., and Salisbury, J. W. (1972). *Proc. Lunar Sci. Conf., Geochim. Cosmochim. Acta, 3rd, 1972* Vol. 3, Suppl. 3, pp. 3069–3076.

Lowndes, R. P. (1970). *Phys. Rev. B* **1,** 2754–2763.

Lowndes, R. P., Rastogi, A., and Perry, C. H. (1974). To be published.

Lyon, R. J. P. (1963). *NASA Tech. Note* **NASA TN D-1871**.

Lyon, R. J. P. (1965). *Econ. Geol.* **60,** 715–736.

Mendell, W. W., and Low, F. J. (1972). *Moon* **4,** 18–27.

Mie, G. (1908). *Ann. Phys. (Leipzig)* [4] **25,** 377–445.

Morse, P. M., and Feshbach, H. (1953). "Methods of Theoretical Physics." McGraw-Hill, New York.

Murcray, F. H., Murcray, D. G., and Williams, W. J. (1970). *J. Geophys. Res.* **75,** 2662–2670.

Perry, C. H., and Lowndes, R. P. (1970). "The Study of the Reflectivity of Inorganic Materials for Remote Sensing Applications," Final Report AFCRL-70-0512. Air Force Cambridge Res. Lab., Cambridge, Massachusetts.

Perry, C. H., and Wrigley, J. D. (1967). *J. Appl. Opt.* **6,** 586–587.

Perry, C. H., and Young, E. F. (1967). *J. Appl. Phys.* **38,** 4616–4628.

Perry, C. H., Geick, R., and Young, E. F. (1966). *Appl. Opt.* **5,** 1171–1177.

Perry, C. H., Agrawal, D. K., Anastassakis, E., Lowndes, R. P., Rastogi, A., and Tornberg, N. E. (1971). *Moon* **4,** 315–336.

Perry, C. H., Agrawal, D. K., Anastassakis, E., Lowndes, R. P., and Tornberg, N. E. (1972). *Proc. Lunar Sci. Conf., Geochim. Cosmochim. Acta, 2nd, 1972* Vol. 3, Suppl. 2, pp. 3077–3096.

Robinson, T. S. (1952). *Proc. Phys. Soc., London, Ser. B* **65,** 910–911.

Ruppin, R., and Englman, R. (1970). *Rep. Progr. Phys.* **33,** 149–196.

Salisbury, J. W., Vincent, R. K., Logan, L. M., and Hunt, G. R. (1970). *J. Geophys. Res.* **75,** 2671–2682.

Warner, J. (1970). *NASA Tech. Rep.* **NASA R-353,** 75.

12

RAMAN SPECTROSCOPY OF TERRESTRIAL MINERALS

W. P. Griffith

I. Introduction

Although the Raman spectroscopy of minerals has been studied since the discovery of the technique in 1928, almost all the work has been carried out by physicists or physical chemists with a consequent emphasis on spectroscopic rather than mineralogical interest. Recently, however, a few mineralogists have started to use the technique both for mineral identification and for structural studies. It is hoped that this review will stimulate further work in this interesting and promising area; and since some readers may not have encountered Raman spectroscopy in detail before it seems appropriate to start this chapter with a brief review of the basic theory and practice of the method, and also to mention some of the newer Raman techniques which have been applied to minerals. It must be emphasized that Raman and infrared spectroscopy are complementary rather than rival techniques; any spectroscopic study of a mineral should make use of both.

It was predicted in 1923 by Smekal that monochromatic light scattered by a medium which was transparent to that light could contain radiation with frequencies which differed from the frequency of the exciting light. Such an effect was subsequently demonstrated by Raman (1928; Raman and Krishnan, 1928), using toluene as the scattering medium, the exciting radiation being sunlight focused by a lens and filtered through a blue-violet filter. The effect was named after him. Independently, and only a month later, Landsberg and Mandelstam (1928a,b) observed the phenomenon using mercury radiation on quartz, and so were the first to observe the Raman spectrum of a mineral. From 1928 until 1940 much work was done on Raman spectroscopy [in the year following the discovery 150 papers appeared on the subject; a review of these makes interesting reading (Ganestan, 1929)]. The development during the Second World War of commercial infrared machines largely supplanted the much slower and less sensitive Raman technique by infrared spectroscopy, and it was not until powerful monochromatic laser sources were introduced in the 1960's that the present renaissance and further development of Raman spectroscopy took place. Now, however, a wide range of commercial Raman instruments is available and the technique is extensively used.

II. Principles and Practice of Raman Spectroscopy

A. Basic Method

The basic technique, which is discussed in more detail in Section II,H (Fig. 1) has altered little since its inception. Monochromatic light from a suitable source is directed at the sample (powder or crystal in the case of minerals; in the general case liquids and gases may also be studied) and the light scattered at 90° or 180° is collected and analyzed by a suitable optical system. The spectrum of the scattered light thus collected is then recorded, either photographically or, more commonly these days, by photoelectric devices with pen recorders.

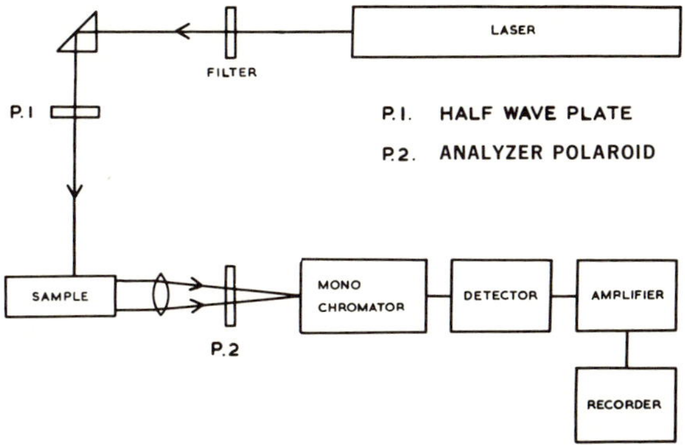

FIG. 1. Schematic diagram of Raman spectrometer with 90° scattering. P_1 and P_2 are half-wave and polarizing plates, respectively, to study the state of polarization of scattered light.

B. Theory of the Raman Effect

The scattering of monochromatic light (the "exciting line") of frequency ν by molecules or lattices to which that light is transparent (i.e., not electronically excited by it) is of two types: Rayleigh and Raman. The Raman effect involves a change of frequency whereas in Rayleigh scattering, which accounts for most of the intensity of the scattered light, there is no frequency change. The new Raman frequencies (ν') are found on either side of the exciting line; the differences ($\nu \pm \nu'$) are independent of ν and are, in the case of solids, lattice (acoustic) or vibrational modes (i.e., first-order spectra). Combination and overtones of these constitute the much weaker second-order Raman effect. These new frequencies are called Raman shifts. Those on the low frequency side of the exciting line are the Stokes lines and are generally more intense than those on the high-frequency side (anti-Stokes). The overall Raman effect is very weak, only some 10^{-6} of the incident radiation being observed as Raman scatter, and it is the Stokes lines that are normally studied because of their greater intensities. The intensity of Raman scattered light is proportional to the fourth power of the frequency of the exciting line.

C. Quantum and Classical Approaches

A simple picture of Raman and Rayleigh scattering is afforded by the quantum approach (see Woodward, 1966). This envisages the interaction of incident photons of energy $h\nu$, ν being the frequency of the exciting line, with molecules of the sample. There may be either interaction of elastic scattering without loss of energy of $h\nu$ (Rayleigh effect) or, if the molecule is in the ground state, it may abstract energy $h\nu'$ from $h\nu$ to raise itself to an excited state, ν' being a fundamental rotational, vibrational, or lattice mode. The photon thus released has been inelastically scattered, having energy $h(\nu - \nu')$ and frequency $(\nu - \nu')$, and is thus a Stokes line. Conversely, the molecule may already be in an excited state and may impart energy $h\nu'$ to the incident photon to give an anti-Stokes line of frequency $(\nu + \nu')$. Since more molecules will be in the ground state than in excited states (relative populations are governed by a Maxwell–Boltzmann distribution) we have a simple explanation of the greater intensity of Stokes lines.

The simple classical theory of Raman scattering (cf. Woodward, 1966) gives us more insight into the process. If an atom or molecule be placed in a static electric field E such as is produced by the incident light, a dipole μ will be induced caused by displacement of the nucleus within its electron cloud (such a dipole is independent of any permanent dipole which a molecule may possess). Then, to a first approximation,

$$\mu = \alpha E \tag{1}$$

where α is the polarizability of the molecule (roughly a measure of its electron density). For light of frequency ν the electric field vector E will vary according to

$$E = E_0 \cos 2\pi\nu t$$

so that

$$\mu = \alpha E_0 \cos 2\pi\nu t \tag{2}$$

If the molecule is rotating or vibrating with frequency ν', however, its polarizability will also change according to the expression

$$\alpha = \alpha_0 + \beta \cos 2\pi\nu' t \tag{3}$$

where β is the rate of change of polarizability with ν' (β may also be expressed as $(\partial\alpha/\partial Q_k)_0 Q_k$, where Q_k is the normal coordinate of the fundamental mode ν'). Then,

$$\mu = \alpha_0 E_0 \cos 2\pi\nu t + \tfrac{1}{2}\beta E_0[(\cos 2\pi(\nu - \nu')t + \cos 2\pi(\nu + \nu')t] \tag{4}$$

The first term on the right-hand side of Eq. (4) represents Rayleigh scattering, the second the Stokes–Raman scattering, and the third the anti-Stokes–Raman scattering. Although this approach, unlike quantum theory, erroneously predicts equal intensities for Stokes and anti-Stokes lines, it does give us the basic selection rule for the Raman activity of the mode ν', viz., that for it to be Raman-active there must be a change of polarizability (i.e., $\beta > 0$). This is of course quite different from the requirement for infrared activity of a mode, which is that there should be a change of dipole moment.

D. The Polarizability Tensor

If μ and E in Eq. (1) were parallel, the polarizability α would be a scalar quantity. This is not necessarily so, however; it may be that the molecule or lattice is more easily polarized in one direction than in another, so that the induced dipole and electric fields are not parallel (in such a case the molecule is anisotropic, or is undergoing a mode which destroys its isotropy). In such cases, and in fact for any general treatment of Rayleigh or Raman scattering, we must replace Eq. (1) by a set of equations

$$\begin{aligned} \mu_x &= \alpha_{xx}E_x + \alpha_{xy}E_y + \alpha_{xz}E_z \\ \mu_y &= \alpha_{yx}E_x + \alpha_{yy}E_y + \alpha_{yz}E_z \\ \mu_z &= \alpha_{zx}E_x + \alpha_{zy}E_y + \alpha_{zz}E_z \end{aligned} \tag{5}$$

So that α is now a polarizability tensor referred to coordinate axes x, y, and z fixed in the molecule; it may be expressed by a 3×3 square matrix. This matrix is symmetrical about its leading diagonal, however, since the effect of E_y on μ_x is the same as that of E_x on μ_y so that $\alpha_{yx} = \alpha_{xy}$, and similarly $\alpha_{zx} = \alpha_{xz}$ and $\alpha_{zy} = \alpha_{yz}$. Thus the polarizability tensor α has six independent components.

The components of the tensor are subject to the symmetry constraints of the point group to which the molecule belongs. The components, or linear combinations of them, are listed in the last column of group theory character tables. Thus the Raman selection rule is that a mode will be Raman-active only if at least one of the components of the polarizability belongs to the same species as that mode (e.g., for the C_{3v} point group the A_1 and E modes are Raman-active; the same modes will be infrared-active since the translation T_z transforms as A_1 and both T_x and T_y as E, the A_2 mode being neither Raman- nor infrared-active; for D_{3h}, the A_1', E′ and E″ modes are Raman-active and the A_2'' and E′ modes are active in the infrared). It is pertinent at this point to state the rule of mutual exclusion, which is that for molecules having a center of symmetry (i) the Raman- and infrared-active modes belong to different symmetry species (e.g., for C_{2h}, D_{2h}, O_h, etc.) Noncoincidence of infrared and Raman fundamentals may often indicate (though it does not prove) that a molecule belongs to a centrosymmetric point group, and this is often very useful for structural characterization.

E. State of Polarization of Raman Modes

One of the most useful features of the Raman technique is that modes which are totally symmetric, i.e., transform in the same way as the components α_{xx}, α_{yy} and α_{zz} (or combinations of these), are polarized (in the normal optical sense; it is unfortunate that polarization and polarizability have completely different connotations in Raman parlance). Such modes are totally symmetric and their symmetry species are always listed first in character tables (e.g., A_g for C_{2h}, A_1' for D_{3h}, A_1 for T_d, etc.).

If we consider incident unpolarized light traveling along the x axis (Fig. 2) it will have electric field vectors in the yz plane, E_y and E_z. Then for the light scattered at 90° along the y-axis by a sample placed at the origin, the field vectors will be in the xz plane (E_x, E_z). If the sample be isotropic, i.e., has the same polarizability in all directions, then the induced dipoles will all be in the yz plane and the scattered light will be polarized because the E_y vector cannot contribute to light traveling along y. We may define a depolarization ratio ρ so that

$$\rho = I_x/I_z \tag{6}$$

I_x and I_z being the intensities of light produced by E_x and E_z. Clearly, for an isotropic atom or molecule or for a vibrating molecule in which the isotropy is preserved during the vibration, $\rho = 0$ (i.e., for a totally symmetric mode) since in such a case the induced dipole is parallel with the incident field vector and, from Eq. (5)

$$\mu_y = \alpha_{xy}E_x + \alpha_{yz}E_z$$

$$\mu_z = \alpha_{xz}E_z + \alpha_{zz}E_z$$

with $\alpha_{xy} = \alpha_{yz} = \alpha_{xz} = 0$. For an anisotropic molecule, however, or for an isotropic molecule undergoing a mode which destroys its isotropy (e.g., a nontotally symmetric vibration) then μ and E will not be parallel, the cross-terms in Eq. (5) will not all be zero and there will be a certain amount of depolarization. In fact $\rho = \frac{6}{7}$ for incident unpolarized light or $\frac{3}{4}$ for

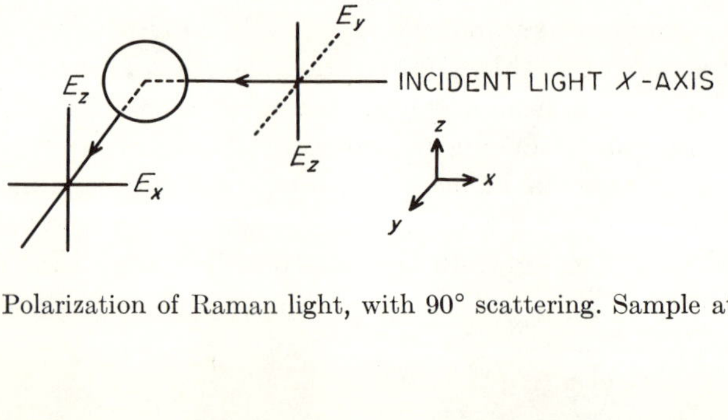

FIG. 2. Polarization of Raman light, with 90° scattering. Sample at origin.

incident polarized light. For a totally symmetric mode in an anisotropic molecule, ρ will lie between $\frac{6}{7}$ and 0 (or between $\frac{3}{4}$ and 0 if the incident light is polarized).

F. General Theory of Raman Scattering

Placzek (1934) evolved a general polarizability theory to which the quantum and classical approaches outlined above are approximations. His treatment, which lies outside the scope of this review, is particularly successful in accounting for the intensities of Raman shifts (cf. Brandmuller and Moser, 1962; Woodward, 1966).

G. Applications of Raman Spectroscopy to Minerals

The mineralogical applications have been chiefly to mineral identification and structural characterization (Forneris, 1969; Griffith, 1969a,b; Moore *et al.*, 1971). Here the approach is exactly analogous to that used for the infrared spectroscopy of minerals: Raman bands are identified and assigned, the assignment being made on the basis of site-symmetry arguments or by the more sophisticated factor group analysis [for site symmetry methods, see Halford (1946), and for the factor group analysis, see Loudon (1964) and Mitra and Gielisse (1963); an admirably clear account of these two approaches as applied to gypsum has been given by Hass and Sutherland (1956)].

The Raman spectroscopy of powdered samples gives a spectrum in which all Raman-active fundamentals should appear since all orientations of the crystal lattice are presented to the exciting radiation. This is illustrated in Fig. 3 for powdered fluorapatite; the range covered (400–1100 cm^{-1}) covers the internal modes of vibration of the phosphate group. A much more valuable technique however, is the study of oriented single crystals. The great advantage here is that the lattice is fixed with respect to the crystal, so that by changing the orientation of the crystal with respect to the exciting line various normal modes can be activated; their state of polarization together with a factor group analysis often allows identification of that mode. Detailed treatment of single crystal Raman techniques is given by Gilson and Hendra (1970); see also Beattie and Gilson (1968) and Loudon (1964).

H. Instrumentation

We have already outlined the basic technique (Section II,A); in Fig. 1 we illustrate a block outline of the main parts of a Raman spectrometer

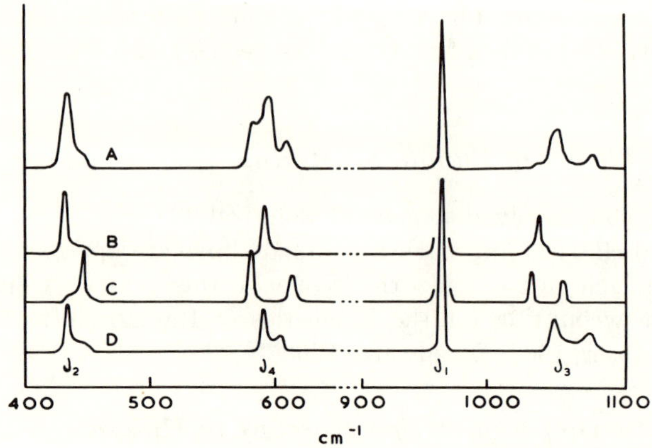

FIG. 3. Raman spectra of (A) powdered fluorapatite; (B) single crystals of fluorapatite. In the nomenclature of Kravitz *et al.* (1968), the single-crystal orientations are (B) E_{1g}; (C) E_{2g}; and (D) A_{1g}. [Reproduced by permission from *J. Chem. Soc.* (Griffith, 1970).]

with photoelectric recording. We may conveniently discuss the subject under the three sections of exciting light source; sampling; monochromator and detection. The subject of Raman instrumentation has been discussed by Hathaway (1971) in an excellent review.

1. *Excitation sources*

The gas laser has now displaced most lamp sources for conventional Raman spectroscopy. The advantages of laser sources are that the light is intense, concentrated, polarized, coherent, and monochromatic; the intensity of the line and often its frequency may be varied. For most purposes in Raman spectroscopy the laser power at the sample should be between 40 and 500 mW. The three commonest gas lasers are (abbreviations and principal wavelengths in parentheses):

Helium–neon	He/Ne; 6328 Å (red)
Argon–ion	Ar^+; 5145 Å (green), 4880 Å (blue)
Krypton	Kr; 6764 and 6471 Å (red), 5682 Å (yellow), 5308 Å (green)

Mixed gas argon–krypton lasers have been used. Higher frequency excitation is obtainable with the argon–ion laser.

The choice of exciting line is dictated by the color of the sample; the requirement that the sample be transparent to the exciting line means that for red or orange samples red light has to be used; for green or blue samples green or blue light, etc. Brown or purple samples may often be examined with red light; gray or black samples are usually very difficult to study, although the Ar^+ blue line has been used for graphite (Section IV,A,1). For colorless samples any exciting line in the visible region may be used. Since the intensity of the normal Raman effect is, however, proportional to the fourth power of the frequency, exciting lines at the blue end of the visible spectrum are better than those at the red end (the response of photomultipliers is also better for high frequencies). Sometimes, however, the sample may fluoresce (a fairly rare circumstance with minerals) and the Raman effect is swamped. Fluorescence is less likely with red- than with blue-end sources. Fluorescence and luminescence in general are described in Chapter 3.

2. *Sampling Techniques*

The technique of focusing a laser beam on to the sample means that only a small amount of material is needed (~5 mg or even less); and there is no destruction of the sample or contamination by mulling agents since dry powders may be examined. For single crystal work very small crystals only are needed, although for ease of orientation larger crystals are advantageous. There are special methods available for studying materials as melts (e.g., cryolite, see Section IV,A,3) or at low temperatures (where band resolution is often much better than at room temperatures) and there are cells for studying materials under high pressure, etc. Sampling techniques have recently been reviewed by Hester (1970).

I. Advantages and Disadvantages of the Raman Technique

It cannot be emphasized too strongly that Raman and infrared spectroscopy are complementary techniques; wherever possible both should be used together. In most cases Raman spectra of minerals may be recorded without great difficulty, although the time required to run a spectrum is generally longer than that for the infrared. It is easy to reach low frequencies (50 cm^{-1} is a normal starting point) so that lattice modes may be measured. The color of the sample is often the most difficult problem (Section II,H,1), and also the weakness of the effect in general, although most of the minerals so far studied give clear, strong spectra. The two biggest advantages of the technique are the ease of obtaining polarization data (Section II,E) and the fact that overtones and combination bands (second-order spectra) are rarely seen in the Raman. Fundamentals only are seen under normal conditions, and this contributes greatly to the simplicity and clarity of the spectra.

A more detailed introduction to the subject than it has been possible to provide here is given by Woodward (1953, 1966) and by Ware (1968). More detailed coverage is given in Brandmuller's book (see Brandmuller and Moser, 1962) and by Stoicheff (1959).

III. Special Applications of the Raman Effect

A. Conventional Raman Spectroscopy

Two particular methods have been applied to minerals using normal spectroscopic techniques: the effect of varying the temperature of the sample and the effect of external pressure. In general, as the temperature rises, Raman lines shift to lower frequencies, this shift being greater for lattice modes than for internal modes. The band-width of a Raman line is approximately proportional to $T^{1/2}$, and the line intensities increase with rising temperature for anti-Stokes lines, the reverse holding for Stokes lines. Studies of temperature effects have been made on diamond, calcite, barytes, anhydrite, topaz, and quartz (Krishnan, 1971a). The main effect of externally applied pressure on samples is to constrain the lattice so that in some cases lower site symmetries are produced; studies have been carried out on calcite and scheelite (Fong and Nicol, 1970; Nicol and Fong, 1971).

B. Recent Advances

A number of special techniques have been applied to certain minerals:

1. *Stimulated Raman Effect* (Lallemand, 1971)

If the intensity of incident light is very great, such as is produced by giant pulse lasers, such light of frequency ν can induce a gain in the sample at frequency $(\nu - \nu')$, ν' being a Raman-active vibration; under some circumstances the gain can exceed the losses and $(\nu - \nu')$ is amplified to such an extent that very powerful stimulated Raman emission is observed. In minerals the effect has been studied in diamond, calcite, and quartz.

2. *Resonance Raman Effect* (Loudon, 1965; Behringer, 1966)

This may occur when the frequency of the exciting line ν lies close to the region of an electronic or vibronic absorption of the sample. Under such conditions the intensity of some Raman lines, particularly those of the totally symmetric type and those involving large changes of polarizability, may be considerably enhanced. It is likely that the use of the mercury

2537-Å line in early studies on minerals gave anomalously high intensities for some of the lines [e.g., Ornstein and Went's early work (1935) on quartz].

3. *Electronic Raman Effect* (Krishnan, 1971a)

Although most of the Raman shifts observed arise from rotational, vibrational, or lattice transitions, it is also possible to observe transitions between two electronic states, particularly when these have low-energy levels of the same parity as the ground state. Such transitions have been observed in certain garnets doped with rare earth ions (Koningstein, 1966); a resonance electronic Raman effect has also been observed in such systems (Wadsack and Chang, 1972).

4. *Inverse Raman Effect* (Alfano and Shapiro, 1971)

This arises when the sample is irradiated with monochromatic light which, although of high intensity, was below the threshold required for stimulated scattering, together with an intense continuum of light of an appropriate frequency range. The effect has been observed in calcite (Krishnan, 1971a).

5. *Hyper Raman effect* (Long and Stanton, 1970)

As with the stimulated Raman effect this is produced by high intensity incident radiation such as is provided by pulse lasers, and it arises from the very high electric fields. The selection rules for the hyper-Raman effect are quite different from those for conventional spectroscopy and so normally inactive vibrations may be activated. The effect has been observed in quartz.

6. *Electric-field Induced Raman Scattering*

This type of scattering has been observed in diamond (Krishnan, 1971a).

7. *Brillouin-Zone Scattering* (Krishnan, 1971b)

Brillouin (1922) predicted that the scattering of monochromatic light by a sample should give rise to a very small frequency change caused by a type of Doppler effect. The observation of such an effect followed the discovery of Raman scattering and was first observed by Gross (1930a,b). The phenomenon has since been studied in a number of minerals, among them being diamond, alumina, calcite, barytes, and quartz; stimulated Brillouin scattering using a giant ruby laser has been observed in sapphire and quartz.

IV. Survey of Published Data on the Raman Spectra of Minerals

In view of the fact that only a little over one hundred minerals have been studied it is feasible to attempt to give a summary of the results obtained. The list given is as comprehensive and as up-to-date as possible; where there is more than one reference for a mineral the later or more comprehensive one has been quoted.

Arrangement of Material

The sequence of minerals is that used in other sections. For each group of minerals there is a short introductory section which aims to pick out general points and to mention studies of special significance or interest. Within each section, each mineral is listed alphabetically, and the following information provided: *Mineral name, formula. Range* of Raman measurement (cm^{-1}); *source* and *wavelength* of exciting line [abbreviations: *Hg* (mercury) arc; *He/Ne* (helium–neon) laser; Ar^+ (argon–ion) laser; *Kr* (krypton) laser; wavelengths in Å]. *Nature of sample* (xtal = crystal; pol. indicates polarization data). *Analysis of results* (site-group analysis abbreviated as SGA, factor-group analysis as FGA). IR (if reported in cited reference). (In addition, infrared spectral data are listed for 56 of these minerals in Chapter 8, Section III.) List of the ten strongest bands in Raman spectrum. Number of Raman modes observed over stated range, in parentheses. Reference.

A. Nonsilicates

1. *Native Elements and Sulfides*

Diamond is the most extensively studied member of this group. The normal Raman spectrum has a single band at 1332 cm^{-1} assigned to the symmetric C_n stretch. The theory of Raman scattering in diamond has been discussed (Maradudin and Burstein, 1967). The effect of temperature on the spectrum has been studied (Krishnan, 1971a). The second-order (stimulated) Raman spectrum has also been obtained using a giant ruby pulse laser (McQuillan *et al.*, 1970). There have also been studies on electric-field induced scattering and Brillouin-zone scattering in diamond (Krishnan 1971a,b). The Raman spectrum of graphite consists of a single line at 1575 cm^{-1}: the higher frequency for the v_n symmetric stretch is consistent with the partial double-bonding in graphite. This is a remarkable instance of the Raman spectrum of a black mineral being recorded (the powerful blue line of the argon–ion laser was used for excitation). Elemental rhombic α-sulfur has a complex spectrum; its stimulated Raman spectrum has also been recorded (Eckhardt *et al.*, 1963).

The spectrum of cinnabar has been very carefully measured using an oriented single crystal and liquid nitrogen cooling to obtain maximum resolution of bands; this is a "model" study. The first- and second-order spectra of zinc blende have been measured (Krauzmann, 1968).

Cinnabar, HgS. 40–400; He/Ne 6328. Oriented xtal., liq. N_2 cooling. Full FGA. IR (40–500). *Sp. reproduced.* 45s, 91w, 106w, 146w, 256vs, 283s, 290s, 345vs, 353s (10). Zallen *et al.* (1970).
Diamond, C_n. Range not stated; He/Ne 6328. Single xtal (type IIa). 1332vs (1). McQuillan *et al.* (1970).
Graphite, C_n. Range not stated; Ar^+ 4880. Single xtal. pol. FGA. 1575vs (1). Tuinstra and Koening (1970).
Orpiment, As_2S_3. 20–800; He/Ne 6328. Powder. Partial SGA. IR (50–500). 154s, 203s, 311s, 355vs (5). Forneris (1969).
Realgar, As_4S_4. 10–800; He/Ne 6328. Powder. Partial FGA. IR (50–400). 26s, 46s, 60s, 183vs, 193vs, 222vs, 345s, 355vs (8). Forneris (1969).
α-Sulfur, S_8. 20–500; He/Ne 6328. Single xtal, pol. Full FGA. 28m, 52s, 84m, 187s, 216s, 219vs, 247s, 438vs, 474vs (28). Ozin (1969).
Zinc Blende, ZnS. 40–800; Ar^+ 4880. Single xtal, pol., T variation. FGA. *Sp. reproduced.* 179s, 218vs, 236m, 273m, 300w, 312m, 332vs, 397m, 419w, 668s (22). Krauzmann (1968).

2. *Oxides and Hydroxides*

Quartz is considered in Section IV,B,4–6. Of the simple oxides listed the work on anatase, rutile, and cassiterite is particularly useful since the paper in which the spectra are reported (Beattie and Gilson, 1968) gives a very clear exposition of factor group analysis for this type of compound. The effect of pressure on the spectrum of rutile has been studied (Nicol and Fong, 1971). Brillouin-zone scattering in alumina has been studied (Krishnan, 1971b).

The only hydroxide which has been examined is hambergite, where the Raman spectrum in the O–H stretching region only was used to gain information on the orientation of the hydroxyl group (Cohen-Solal *et al.*, 1964).

Anatase, TiO_2. 100–700; He/Ne 6328. Single xtal. Partial FGA. 144vs, 197w, 400m, 515m, 519m, 640m (6). Beattie and Gilson (1968).
Arsenolite, As_4O_6 (synth.). 80–700; He/Ne 6328. Powder, xtal., and gas. Pol. *Sp. reproduced.* Partial SGA. 86vs, 184w, 268s, 370vs, 417w, 470w, 560s, 780w (8). Beattie *et al.* (1970).
Cassiterite, SnO_2. 40–800; He/Ne 6328. Single xtal. Full FGA. 440m, 476m, 634s, 778m (4). Beattie and Gilson (1968).
Corundum, $α\text{-}Al_2O_3$. 300–800; Ar^+ 4880. Single xtal, pol. *Sp. reproduced.* Full FGA. 378m, 418m, 432m, 451m, 578m, 645vs, 751vs (7). Porto and Krishnan (1967). Raman spectrum shown in Chapter 13, Fig. 13.

Magnesia, MgO (synth.). No range stated; Hg 4358. Powder. 358vs (1). Khambata (1956).
Rutile, TiO_2. 50–900, Ar^+ 4880. Single xtal pol. Full FGA. *Sp. reproduced.* 143vs, 447s, 612vs, 826w (4). Porto *et al.* (1967).; Beattie and Gilson (1969).
Sapphire, see corundum, above.
Senarmontite, Sb_4O_6. 40–500; He/Ne 6328. Powder. Partial SGA. 87m, 121w, 193s, 256vs, 359w, 376m, 452m, 717w (8). Beattie *et al.* (1970).

3. *Halides*

Few data are available so far. Cryolite, $Na_3[AlF_6]$, has been studied only as a melt at 1030° and under these rather extreme conditions $[AlF_4]^-$ ions are present (Solomons *et al.*, 1968).

Fluorite (*fluorspar*), CaF_2. No range stated; Ar^+ 4880. Single xtal. T-range (77–300°K). 322vs (1). Srivastan *et al.* (1971).
Rock salt, NaCl. No range stated; Hg 2536. Single xtal. Full an. 235vs (1). Raman (1961).

4. *Carbonates, Nitrates, and Borates*

a. Carbonates. There are two main types, the trigonal calcite and the less symmetrical orthorhombic aragonite types, and these provide a convenient subdivision for the main group of minerals studied. Calcite has perhaps been studied more than any other mineral. The second-order spectrum has been observed both by conventional Raman spectroscopy (Krishnamurti, 1957) and by stimulated Raman emission with a giant ruby pulse laser (Eckhardt *et al.*, 1963). The spectrum has also been studied with a sample under pressures of up to 40 kbar (Fong and Nicol, 1970); at the highest pressure ν_1 shifts from 1082 to 1108 cm^{-1}. The inverse Raman effect (Section III,B,4), Brillouin-zone scattering (Section III,B,7), and the effect of temperature (Section III,A) on the Raman spectra of calcite have also been studied. The calcite and aragonite groups provide a good demonstration of the value (and limitations) of the site group versus factor group analytical methods. The internal modes of the CO_3^{2-} group split neatly: Free carbonate has four modes under D_{3h} symmetry: ν_1 at 1063 (sym. stretch, A_1', Raman-active); ν_2 at 879 (asymmetric deformation, A_2'', IR-active; ν_3 at 1415 (asymmetric stretch, E', Raman- and IR-active) and ν_4 at 680 cm^{-1} (asymmetric deformation, E'). For nitrate ion the corresponding frequencies are at 1050, 831, 1390, and 720 cm^{-1}. Calcite has D_3 site-group symmetry and belongs to the D_{3d} factor group; the ν_2 mode remains inactive in the Raman while active in the IR, the reverse being true for ν_1. This is in fact observed for calcite and for other members of its group, and the complete spectrum in the internal mode region can be satisfactorily interpreted on the basis of D_{3d} factor group symmetry

(Couture, 1947). Aragonite has lower symmetry (C_s site group, D_{2h} factor group) and the Raman and infrared spectra of it and other members of its class are much more complex: all modes are active in the Raman and the infrared, and furthermore the triple degeneracy of the E modes is lifted so that ν_3 and ν_4 appear as triplets. Again the factor group analysis for D_{2h} describes the observed spectrum (Couture, 1947). Thus the calcite and aragonite groups are very easily distinguished on the basis of their Raman spectra; carbonates in general may be recognized by the appearance of ν_1 near 1100 cm^{-1} as a very strong Raman band, and ν_2 near 860 cm^{-1} as a very strong infrared band; appearance of 1100 cm^{-1} bands in the IR and 860 cm^{-1} bands in the Raman is good evidence for lowering of the carbonate ion symmetry.

Dolomite has a calcite-type structure, but the presence of alternating calcium and magnesium ions on the threefold axis lowers the site group symmetry to C_3 and the factor group to S_6. In vaterite, a third modification of $CaCO_3$, the material is converted to calcite under laser illumination. The minerals huntite and shortite are curious in that x-ray studies suggest that in each the carbonate ions have two different site symmetries. This is borne out by the Raman spectrum of shortite in that there are two distinct, strong symmetric stretches, but for huntite only one is seen, which may perhaps suggest that the x-ray structure is at fault or that the two environments are so similar that the two ν_1 frequencies coincide (W. P. Griffith, unpublished observations, 1973).

Calcite subgroup

Calcite, $CaCO_3$. 100–2000; Hg 2537. Single xtal. 156m, 284s, 712m, 1085vs, 1434m (5). Krishnan (1945).

Chalybite (siderite), $FeCO_3$. 600–1600; He/Ne 6328. Powder. IR same range. SGA. 736m, 1071vs, 1415w (3). Griffith (1970).

Magnesite, $MgCO_3$. 100–1600; Hg 4358. Single xtal. *Photo-sp. reproduced.* No anal., 212w, 332vs, 735w, 1096vs, 1460w (5). Krishnamurti (1956).

Nitratine, $NaNO_3$. 600–1600; He/Ne 6328. Powder. SGA. 692m, 1068vs, 1405w (3). Griffith (1970).

Otavite, $CdCO_3$. (synth.) 600–1600; He/Ne 6328. Powder. SGA. 728m, 1090vs, 1392w (3). Griffith (1970).

Rhodochrosite, $MnCO_3$. 600–1600; He/Ne 6328. Powder. SGA. 722m, 1088vs, 1420w (3). Griffith (1970).

Smithsonite, $ZnCO_3$. 600–1600; He/Ne 6328. Powder. SGA. 733m, 1032vs, 1412m (3). Griffith (1970).

Siderite, see chalybite above.

Aragonite subgroup

Aragonite, $CaCO_3$. 100–4000; Hg 2537. Single xtal, pol. FGA. *Photo-sp. reproduced.* 153vs, 191s, 206vs, 702m, 707m, 716w, 854m, 1086vs, 1415m, 1464s (10). Krishnamurti (1960).

Cerussite, $PbCO_3$. 100–4000; Hg 2537. Single xtal, Pol. *Photo-sp. reproduced.* FGA.

58vs, 73vs, 103s, 168m, 683w, 838w, 1054vs, 1363m, 1377m, 1478m. (21). Couture (1947).
Niter, KNO_3. 100–4000; Hg 2537. Single xtal., pol. *Photo-sp. reproduced.* FGA. 51s, 83vs, 124w, 713w, 1048vs, 1340w, 1354w (7). (1947).
Strontianite, $SrCO_3$. 600–1600; He/Ne 6328. Powder. SGA. 702m, 860w, 1070vs, 1400w, 1450w (5). Griffith (1970).
Witherite, $BaCO_3$. 600–1600; He/Ne 6328. Powder, SGA-FGA. 693m, 716w, 852w, 1062vs, 1424w, 1447w (6). Griffith (1970).
Miscellaneous carbonates
Dolomite, $MgCa(CO_3)_2$. 100–2000; Hg 4358. Single xtal, pol. FGA. *Photo-sp. reproduced.* 176vs, 301m, 335m, 724m, 1099vs, 1414m, 1754w (6). Couture (1947).
Huntite, $Mg_3Ca(CO_3)_4$. 40–1700; Ar^+ 4880. Powder. IR (40–1700). FGA. 85w, 247m, 254m, 706w, 726w, 741w, 881m, 1122vs, 1463w, 1531w (23). W. P. Griffith (unpublished observations, 1973).
Shortite, $Na_2Ca_2(CO_3)_3$. 40–1700; Kr 5308. Powder. IR (40–1700). FGA. 270w, 720w, 1072vs, 1094vs, 1450w, 1476m, 1485m (7). Weak spectrum. W. P. Griffith (unpublished observations, 1973).

b. Borates. There is a great need for systematic work on borate minerals; the few data so far available do not permit of any general conclusions to be drawn. The study on teepleite involved isotopic labeling with $boron^{10}$ (Klee, 1966).

Behierite, $TaBO_4$. 150–1500; He/Ne 6328. Powder. SGA. IR (200–1500). *Sp. reproduced.* 193m, 206m, 286m, 464s, 554m, 569s, 850w, 902m, 978s, 1008m (10). Ross (1972).
Borax, $Na_2B_4O_7 \cdot 10H_2O$. 40–4000; Hg 2537. Powder. 159w, 349w, 458m, 572m, 753w, 944vs (11). Krishnamurti (1955).
Cahnite, $Ca_2[B(OH)_4AsO_4]$. 150–1500; He/Ne 6328. Powder. SGA. IR (200–1500). 290, 324, 378, 395, 428, 448, 538, 759, 791, 844 (12). No intensity data. Ross (1972).
Colemanite, $Ca_2B_6O_{11} \cdot 5H_2O$. 40–4000; Hg 2537. Single xtal. *Photo-sp. reproduced.* 95vs, 181s, 318s, 538s, 565s, 670s, 745vs, 1078s (22). Krishnamurti (1955).
Danburite, $CaB_2Si_2O_8$. 80–2000; Hg 4358. Single xtal, no anal. *Sp. reproduced.* 126m, 183s, 246s, 248s, 480s, 609vs, 971s, 1107s, 1580vs (30). Narayanan (1951a).
Kernite, $Na_2B_4O_7 \cdot 4H_2O$. 40–4000; Hg 2537. Single xtal, no anal. *Photo-sp. reproduced.* 55s, 82vs, 104s, 184s, 498vs, 739vs, 932vs, 1087s, 1321s (34). Krishnamurti (1955).
Sinhalite, $AlMg(BO_4)$. 150–1500; He/Ne 6328. Powder. SGA. IR (200–1500). 376m, 488m, 554w, 864m (4). Ross (1972).
Teepleite, $Na_2[B(OH)_4]Cl$. 300–4000; He/Ne 6328 (?). Powder and aq. soln. IR (300–4000). Partial anal. 430w, 500m, 748s, 950m, 1160w, 1200w, 3536s, 3557s (9). Klee (1966).

5. *Sulfates, Tungstates, Molybdates, and Chromates*

a. Sulfates. There have been a number of studies on gypsum, the main application being the determination of orientation of the water molecules. The analysis of results by Hass and Sutherland (1956) gives an exceptionally clear and well-documented account of both site- and factor-group analysis of crystals using both infrared and Raman results. Sulfate minerals give very strong and sharp Raman spectra: the four fundamentals of the undistorted free $(SO_4)^{2-}$ ion (T_d symmetry) are given in the accompany tabulation together with those for $(WO_4)^{2-}$, $(MoO_4)^{2-}$, and $(CrO_4)^{2-}$.

	$\nu_1(A_1$; R active)	ν_2(E; R active)	$\nu_3(F_1)$	$\nu_4(F_1$; R, IR active)
$(SO_4)^{2-}$	983	450	1105	611
$(WO_4)^{2-}$	931	324	833	324
$(MoO_4)^{2-}$	897	338	841	338
$(CrO_4)^{2-}$	847	348	884	368

These four fundamentals split according to the requirements of site- and factor-group analyses. Most sulfates are typified by the appearance of the very strong asymmetric stretch ν_1 near 1000 cm^{-1}, in the Raman with the asymmetric stretch ν_3 near 1130 cm^{-1} being weak in the Raman and exceedingly strong in the infrared. The effect of temperature on the spectrum of barytes has been studied (Krishnan, 1971a) and also Brillouin scattering by the mineral (Krishnan, 1971b).

Anhydrite, $CaSO_4$. 30–1600; Hg 2537. Single xtal, pol. *Sp. reproduced.* Full FGA. 135m, 416m, 497s, 608m, 628m, 674w, 1108m, 1128s, 1160s (17). Krishnan and Kumarti (1951).

Anglesite, $PbSO_4$. No range or source details. Single xtal pol. FGA. 39s, 61s, 106m, 439vs, 451vs, 606s, 642s, 978s, 1060s, 1153s (23). Scrocco and Mathieu (1958).

Barytes, $BaSO_4$. 100–4000; Hg 4358. Single xtal., pol. *Photo-sp. reproduced.* FGA. 67w, 77m, 149m, 453s, 462s, 623w, 630w, 988vs, 1087s, 1108s (19). Couture (1947).

Celestite, $SrSO_4$. 100–4000; Hg 4358. Single xtal, pol. *Photo-sp. reproduced.* FGA. 50w, 190w, 240w, 453s, 458w, 624s, 656s, 999vs, 1094w, 1185m (14). Couture (1947).

Epsomite, $MgSO_4{\cdot}7H_2O$. No range or source stated. Single xtal., pol. Full FGA. 445w, 450s, 460m, 612s, 986vs, 1064m, 1095m, 1135s, 1148m (21). Bouhet and Lafont (1948).

Gypsum, $CaSO_4{\cdot}2H_2O$. 40–4000; He/Ne 6328. Single xtal, pol., T 77°–300°K. *Sp. reproduced.* 110w, 183s, 415s, 495s, 623m, 671s, 1006vs, 1116m, 1132s, 1140w (35). Krishnamurti and Soots (1971).

b. Tungstates, Molybdates, and Chromates. These also give very strong and distinctive Raman spectra, with strong bands near 910, 880, and 840 cm^{-1}, respectively (infrared counterparts at 840, 790, 860 cm^{-1}). The most comprehensive studies have been on members of the scheelite series (scheelite, stolzite, powellite, wulfenite). Studies have been made of the effect of pressure on scheelite and powellite (the lattice distorts to the $NiWO_4$ type under very high pressures) (Nicol and Durana, 1971). The Raman spectra of monoclinic $PbCrO_4$ (crocoite) and those of the orthorhombic form are identical, a curious circumstance (Scheuermann *et al.*, 1970).

Crocoite, $PbCrO_4$. 300–1000; He/Ne 6328. Single xtal. FGA. T 110°–300°K. *Sp. reproduced*. 325m, 337s, 347m, 358s, 377m, 400w, 823w, 839vs, 854m (9). Scheuermann *et al.* (1970).
Ferberite, $FeWO_4$. 200–1000; He/Ne 6328. Powder. FGA. 347w, 388m, 434w; 856vs (4). Very weak spectrum. Griffith (1970).
Hübnerite, $MnWO_4$. 200–1200; He/Ne 6328. Powder. FGA. 260w, 328s, 356w, 400s, 424w, 548m, 699s, 772s, 886vs (9). Griffith (1970).
Powellite, $CaMoO_4$. 50–1000; He/Ne 6328. Single xtal. FGA. 111s, 140s, 189m, 204s, 322vs, 390m, 404m, 794m, 844w, 878vs (10). Khanna *et al.* (1968).
Raspite, $PbWO_4$. 40–2000; Kr 5308. Powder. SGA. IR (40–2000). 92m, 180s, 281m, 298m, 393vs, 641m, 744w, 772m, 754w, 868vs (16). W. P. Griffith (unpublished observations, 1973).
Scheelite, $CaWO_4$. 50–1000; He/Ne 6328. Single xtal. FGA. 86s, 116w, 196m, 210s, 333vs, 397m, 403m, 795m, 838s, 911vs (11). Khanna *et al.* (1968).
Stolzite, $PbWO_4$. 50–1000; He/Ne 6328. Single xtal. FGA. *Sp. reproduced*. 52s, 76s, 86s, 187s, 322vs, 348m, 353m, 748m, 764s, 900vs (13). Khanna *et al.* (1968).
Wolframite, (Fe, Mn)WO_4. 40–2000; He/Ne 6328. Powder. FGA. 329m, 400w, 428w, 452m, 700w, 772m, 812m, 881vs (8). Griffith (1970).
Wulfenite, $PbMoO_4$. 50–1000; He/Ne 6328. Single xtal. FGA. *Sp. reproduced*. 61s, 100s, 164s, 314vs, 317m, 348m, 356m, 744w, 765s, 868vs (13). Khanna *et al.* (1968).

6. *Phosphates, Arsenates, and Vanadates*

The $[MO_4]^{3-}$ tetrahedra which most minerals in this class contain have the four fundamental frequencies for the free ions (T_d symmetry) as shown in the accompanying tabulation. As with minerals of class 5 above, these give very strong, sharp Raman spectra (see Fig. 3 for $Ca_5(PO_4)_3F$). In general the symmetric stretch ν_1 is very strong in the Raman near 960, 820, and 830 cm^{-1} for phosphates, arsenates, and vanadates, respectively, while ν_3 is seen in the infrared (and, more weakly in the Raman) centered on 1060, 800, and 740 cm^{-1}.

	$\nu_1(A_1$; R active)	ν_2(E; R active)	$\nu_3(F_1$; R, IR active)	$\nu_4(F_2)$
$(PO_4)^{3-}$	938	420	1017	567
$(AsO_4)^{3-}$	837	349	878	463
$(VO_4)^{3-}$	827	341	780	341

The group of minerals most studied are the apatites (elegant single crystal studies have been carried out on fluoroapatite and a clear discussion of its factor-group analysis given (Kravitz *et al.*, 1968), including phosphates, arsenates, and vanadates.

Adamite, $Zn_2(AsO_4)OH$. 200–2000; He/Ne 6328. Powder. FGA. 236w, 258m, 316m, 327s, 427m, 540m, 810w, 825m, 851vs, 894s (11). Griffith (1970).

Annabergite, $Ni_3(AsO_4)_2 \cdot 8H_2O$. (synth.) 200–2000; He/Ne 6328. Powder. FGA. 438m, 452w, 797m, 820w, 859vs, 880w (6). Griffith (1970).
α-Berlinite, $AlPO_4$. 80–1400; Ar^+ 4880. Single xtal. *Sp. reproduced.* FGA and Brillouin zone anal. 105vs, 216vs, 336vs, 378m, 417vs, 460vs, 568m, 729m, 1105vs, 1112vs, 1231m (21). Scott (1971).
Beryllonite, $NaBe(PO_4)$. 200–2000; He/Ne 6328. Powder. SGA. 177m, 267w, 461m, 526w, 546w, 606m, 1015vs, 1060m, 1072w, 1117w (11). Griffith (1970).
Carminite, $PbFe_2(AsO_4)_2(OH)_2$. 200–2000; He/Ne 6328. Powder. SGA. 356s, 397s, 420m, 431m, 470w, 811w, 821w, 859vs (8). Griffith (1970).
Chlorapatite, $Ca_5(PO_4)_3Cl$. 100–2000; He/Ne 6328. Powder. SGA. 183w, 270w, 434m, 449w, 578m, 588m, 961vs, 1040w, 1060m, 1080w (10). Griffith (1970).
Descloizite, (Zn, Cu)$Pb(VO_4)OH$. 200–2000; He/Ne 6328. Powder. SGA. 200w, 301m, 356m, 377m, 384w, 400w, 760w, 777s, 820w, 865vs (10). Griffith (1970).
Fluoroapatite, $Ca_5(PO_4)_3F$. 500–1200; He/Ne 6328. Single xtal, pol. FGA. *Sp. reproduced.** (For 200–500 cf. Griffith, 1970). 268w, 434m, 451w, 595w, 598m, 609m, 966vs, 1041w, 1053m, 1084w (17). Kravitz *et al.* (1968).†
Hedyphane, (Ca, Pb)$_5(AsO_4)_3Cl$. 200–2000; He/Ne 6328. Powder. SGA. 350m, 381w, 407w, 440w, 466w, 778m, 798w, 801w, 831vs, 886w (11). Griffith (1970).
Hydroxyapatite, $Ca_5(PO_4)_3OH$. 200–2000; He/Ne 6328. Powder. SGA. 270w, 436m, 456w, 589w, 595m, 612w, 966vs, 1034w, 1053w, 1083w (10). Griffith (1970).
Mimetite, $Pb_5(AsO_4)_3Cl$. 100–1000; He/Ne 6328. Powder. SGA. *Sp. reproduced.* IR (33–4000). 314m, 341m, 371w, 412w, 423w, 766m, 786m, 809vs (8). Levitt and Condrate (1970).†
Mottramite, (Cu, Zn)$Pb(VO_4)OH$. 150–2000; He/Ne 6328. Powder. SGA. 188w, 288w, 295m, 331m, 348m, 358w, 774w, 795m, 802w, 834vs (10). Griffith (1970).
Olivenite, $Cu_2(AsO_4)OH$. 200–2000; He/Ne 6328. Powder. SGA. 310w, 324w, 425w, 464w, 790w, 856vs, 880w (7). Griffith (1970).
Pyromorphite, $Pb_5(PO_4)_3Cl$. 100–1000; He/Ne 6328. Powder. *Sp. reproduced.* SGA. IR (33–4000). 394m, 413m, 541w, 552m, 576m, 920vs, 947vs, 980w, 1014w (9). Levitt and Condrate (1970).
Scorodite, $FeAsO_4 \cdot 2H_2O$. 200–2000; He/Ne 6328. Powder. SGA. 396m, 416m, 436w, 819vs, 852w, 920m, 944m (7). Griffith (1970).
Strengite, $FePO_4 \cdot 2H_2O$. 200–2000; He/Ne 6328. Powder. SGA. 322w, 350w, 570w, 618m, 665m, 983vs, 1002m, 1026m, 1095w (9). Griffith (1970).
Vanadinite, $Pb_5(VO_4)_3Cl$. 100–1000; He/Ne 6328. Powder. SGA. IR(33–4000). 287m, 321vs, 352s, 362m, 411w, 722w, 793m, 825vs (8). Levitt and Condrate (1970).†
Variscite, $AlPO_4 \cdot 2H_2O$. 150–2000; He/Ne 6328. Powder. SGA. 180w, 352m, 610w, 640w, 680w, 1033vs, 1053w, 1078m (8). Griffith (1970).
Vivianite, $Fe_3(PO_4)_2 \cdot 8H_2O$. 200–2000; He/Ne 6328. Powder. SGA. 267m, 316w, 376w, 600m, 644m, 656m, 1008w, 1018w, 1029vs (9). Griffith (1970).
Xenotime, YPO_4. 200–2000; Single xtal. FGA. *Sp. reproduced.* 292w, 330w, 481m, 578s, 997vs, 1023w, 1056s (7). Richman (1966); see also Griffith (1969b, 1970).

7. *Miscellaneous Nonsilicate Minerals*

The symmetric $(UO_2)^{2+}$ stretching frequency only has been measured in the Raman for a number of uranyl or uranite minerals and from these

*See Fig. 3.
† See also Adams, D. M. and Gardner, I. R. (1974b).

frequencies (all near 830 cm^{-1}) and the asymmetric (UO_2) infrared stretches the U—O bond distances estimated. The (synthetic) minerals studied were andersonite, beckquerelite, autunite, rutherfordine, saleite, schroedingerite, uranospinite, zappeite (Wilkins, 1971).

Bismutotantalite, $BiTaO_4$. 200–2000; He/Ne 6328. Powder. SGA. 316w, 352w, 380w, 424m, 676vs, 790s, 824m, 852s (8). Griffith (1970).
Fergusonite (Y, Yb), NbO_4. 200–2000; He/Ne 6328. Powder. SGA. IR (500–1000). 268m, 325w, 424s, 458w, 604w, 789vs, 826vs (7). Griffith (1970).
Stibiotantalite, $SbTaO_4$. 200–2000; He/Ne 6328. Powder. SGA. 277m, 320w, 330w, 600s, 638w, 802vs, 840m (7). Griffith (1970).

B. Silicate Minerals

Owing to the rather low polarizability of the Si–O bonds in most silicates, the minerals sometimes tend to give rather weak spectra: the best results are obtained with exciting lines at the high-frequency (blue end) of the spectrum. Much of the early work was carried out using the mercury 2537 Å line; it seems possible that in many cases a resonance Raman spectrum was observed under these conditions (Section III,B,2).

1. *Nesosilicates*

Most of the Raman work so far reported on silicates concerns minerals of this type, those containing isolated SiO_4 tetrahedra. The splitting of $(SiO_4)^{4-}$ frequencies is far more dependent on the nature of the cations present than is the case with carbonates, sulfates, etc., dealt with above; there is clearly considerable covalent interaction between the cations and the oxo ligands in $(SiO_4)^{4-}$ which helps to dissipate the π-electron density of the oxygen ligands (cf. Griffith, 1969a). Among this group we may single out the systematic Raman and infrared studies on garnets in which the effect of altering the composition of cations was studied (Moore *et al.*, 1971) and the very full analysis of the Raman spectrum of zircon (Dawson *et al.*, 1971). Vibrational assignments for the SiO_4 group are discussed in Chapter 13, Section II.

Almandine, $Fe_3Al_2(SiO_4)_3$. 200–1200; He/Ne 6328. Powder. SGA. 340w, 870w, 900w, 917vs, 960m. (5). Griffith (1969a); Moore *et al.* (1971). The Raman spectrum of an almandine of different composition is shown in Chapter 13, Fig. 7.
Andradite, $Ca_3Fe_2(SiO_4)_3$. 100–1200; He/Ne 6328. Powder. FGA. IR (100–1000). *Sp. reproduced*. 241m, 358m, 373m, 425vs, 501s, 581m, 659m, 816s, 875vs (20). Moore *et al.* (1971).
Chondrodite, $Mg(F, OH)_2 \cdot Mg_2SiO_4$. 200–1200; He/Ne 6328. Powder. SGA. 380w, 400s, 524w, 577s, 835vs, 860m, 868m, 910s (8). Griffith (1969a).

Clinohumite, $Mg(F, OH)_2 \cdot 4Mg_2SiO_4$. 200–1200; He/Ne 6328. Powder. SGA. 340w, 448s, 830vs, 841vs, 856w, 910w, 926w (7). Griffith (1969a).
Euclase, $BeAl(OH)SiO_4$. 200–1200; He/Ne 6328. Powder. SGA. 464w, 515w, 540w, 560m, 642m, 882vs, 912vs, 945w (8). Griffith (1969a).
Fayalite, Fe_2SiO_4. 200–1200; He/Ne 6328. Powder. SGA. Sp. very weak. 380m, 550w, 608w, 828s. (4). Griffith (1969a).
Forsterite, Mg_2SiO_4. 100–1000; Ar^+ 4880. Single xtal, pol. FGA. 317m, 374s, 438s, 585s, 591s, 825vs, 839s, 866vs, 881vs, 920vs (36). Servoin *et al.* (1972).
Grossular, $Ca_3Al_2(SiO_4)_3$. 200–1200; He/Ne 6328. Powder. SGA. 275w, 312s, 450w, 470w, 545m, 830m, 881vs, 915w (8). Griffith (1969a); see also Moore *et al.* (1971).
Monticellite, $CaMgSiO_4$. 200–1200; He/Ne 6328. Powder. SGA. 336w, 480m, 580w, 660m, 831vs (5). Griffith (1969a).
Phenacite, Be_2SiO_4. 100–1200; Hg 2537. Single xtal. *Sp. reproduced.* IR (500–1200). 161w, 223s, 383s, 446s, 526w, 776m, 879vs, 918s, 937m, 949m (25). Narayanan (1950).
Pyrope, $Mg_3Al_2(SiO_4)_3$. 150–1200; He/Ne 6328. Powder. SGA. 340w, 560w, 836w, 863w, 918vs, 982s (6). Griffith (1969a).
Spessartine, $Mn_3Al_2(SiO_4)_3$. 200–1000; He/Ne 6328. Powder. Composition-range. SGA. 360m, 456m, 470w, 560m, 880vs, 916vs. Griffith (1969a); see also Moore *et al.* (1971).
Sphene, $CaTiOSiO_4$. 100–1000; no source stated. Powder. *Sp. reproduced.* No anal. 160m, 240s, 308s, 420w, 460vs, 540vs, 605vs, 855m, 885w (9). Bobovich (1961); cf. also Griffith (1969a).
Tephroite, Mn_2SiO_4. 200–1200; He/Ne 6328. Powder. SGA. 320w, 560w, 818vs, 865s, 904w, 931w (6). Griffith (1969a).
Thorite, $ThSiO_4$. 200–1200; He/Ne 6328. Powder. SGA. 320m, 424m, 460w, 976vs, 1002w (5). Griffith (1969a).
Topaz, $Al(F, OH)_2SiO_4$. 100–4000; Hg 2537. Single xtal. Partial SGA. *Photo-sp. reproduced.* 156m, 241s, 269vs, 459m, 898m, 920s, 934vs, 985s, 3649s (32). Krishnan (1947).
Uvarovite, $Ca_3Cr_2(SiO_4)_3$. 200–1200; He/Ne 6328. Powder. SGA. 360m, 676w, 888vs, 890w (5). Griffith (1969a).
Willemite, Zn_2SiO_4. 200–1200; He/Ne 6328. Powder. SGA. 394w, 542w, 783vs, 908w, 945w, 972w (6). Griffith (1969a). The Raman spectrum of hydrothermally synthesized willemite is shown in Chapter 13, Fig. 7.
Zircon, $ZrSiO_4$. 50–1200; Ar^+ 4880. Single xtal. pol. *Sp. reproduced.* FGA and force constants. IR (40–4000). 201w, 225w, 266w, 357vs, 439vs, 885m, 974m, 989m, 1008vs (16). Dawson *et al.* (1971). Raman spectrum shown in Chapter 13, Fig. 7.

2. *Sorosilicates*

The only example studied seems to be thortveitite; some of the assignments have been disputed by Lazarev (1973).

Thortveitite, $Sc_2Si_2O_7$. (synth.) 20–2000; He/Ne 6328. Powder. SGA. force constants. 192s, 228m, 258m, 392m, 420vs, 498w, 525m, 938m, 956w, 1084w. Ross (1971).

3. *Cyclosilicates*

A number of these have been studied. The symmetric ring stretching mode is usually very strong; such bands occur in the 530–680 cm^{-1} region (the asymmetric infrared counterparts are higher, near 800 cm^{-1}) but there is no apparent correlation between frequency and ring size. The terminal SiO_2 groups have Raman-active (symmetric) vibrations at much higher frequencies (ca. 1120 cm^{-1}) than the corresponding modes in SiO_4 tetrahedra in nesosilicates (Griffith, 1969a).

Axinite, $Ca_2(Fe, Mg)Al_2(OH)(BO_3)(Si_4O_{12})$. 200–1200; He/Ne 6328. Powder. Partial SGA. 204m, 274m, 420w, 487s, 530vs, 715s, 980w, 910m, 1060w (12). Griffith (1969a).
Benitoite, $BaTi(Si_3O_9)$. 200–1200; He/Ne 6328. Powder. Partial SGA. 212m, 260w, 327w, 364m, 390w, 525w, 564m, 820w, 936s, 1116vs (10). Griffith (1969a).
Beryl, $Be_3Al_2(Si_6O_{18})$. 200–1200; He/Ne 6328. Powder. Partial SGA. 323w, 343w, 400w, 470w, 645m, 684m, 860w, 1175vs (8). Griffith (1969a).*
Catapleiite, $Na_2CaZr(Si_3O_9)\cdot 2H_2O$. 200–1200; He/Ne 6328. Powder. Partial SGA. 276w, 365w, 540w, 580m, 630m, 876w, 934s, 1000vs, 1014s (10). Griffith (1969a).
Dioptase, $Cu_6(Si_6O_{18})\cdot 6H_2O$. 200–1200; He/Ne 6328. Powder. Partial SGA. 260m, 290w, 560w, 640w, 654vs, 760w, 1050s (7). Griffith (1969a).*
Tourmaline, $Na(Mg, Fe, Al)_3Al_6(Si_6O_{18})(BO_3)_3(OH, F)_4$. 200–1200; He/Ne 6328. Powder. Partial SGA. 340w, 353s, 464vs, 569s, 682m, 929w, 1040s. (7). Griffith (1969a).
α-Wollastonite, $Ca_3(Si_3O_9)$. 200–1200; He/Ne 6328. Powder. Partial SGA. 290w, 320m, 390w, 556m, 580s, 632m, 860m, 940w, 980vs, 996w (10). Griffith (1969a).

4. *Inosilicates, Phyllosilicates, and Tectosilicates*

Only two examples have been properly studied (a spectrum has been reproduced for the single-chain inosilicate diopside, but no frequencies listed or assignments attempted (Etchepare, 1970). As has previously been mentioned, quartz was the first mineral to be subjected to Raman study, paradoxically by workers unaware of Raman's parallel initial observations of the effect (Landsberg and Mandelstam, 1928a,b). The photographic spectrum is reproduced in their original paper). Since then there have been many studies on α- and β-quartz, particularly the former (the α- and β-transformation occurs at 575°K). Assignments of the fundamentals in α-quartz have been made by Scott and Porto (1967) and correlated with those for the β-form; the latter has also been assigned by Bates and Quist (1972). Stimulated Raman scattering in α-quartz has been studied (Tannenwald, 1967), and also the resonance Raman effect (Section III,B,2). Temperature-dependent Raman scattering (Section III,A), the hyper-Raman effect (Section III,B,5) and Brillouin-zone scattering (Section III,B,7) have all been studied for α-quartz, and the second-order spectrum observed (Terhune *et al.*, 1965).

Apophyllite, $KFCa_4Si_8O_{20}\cdot 8H_2O$. 100–4000; Hg 2537. Single xtal. pol. *Sp. reproduced.* 108m, 138m, 209m, 307m, 344m, 488m, 541m, 586m, 1080s, 3520vs (19). Narayanayan (1951b).
α-Quartz, SiO_2. 80–2000; Hg 2537. Single xtal. FGA (*v. infra*). *Sp. reproduced.* 128vs,

* See also Adams, D. M. and Gardner, I. R. (1974a).

206vs, 265s, 356s, 466vs, 696m, 795m, 806m, 1082m, 1160s (15). Krishnamurti (1958); Scott and Porto (1967).

β-Quartz, SiO_2. 80–1400; Ar^+ 4880. Single xtal. pol. FGA. *Sp. reproduced.* 98s, 243w, 407m, 427m, 464vs, 686w, 786w, 1065vw, 1173vw, 1230w (10). Bates and Quist (1972).

V. Conclusions: Further Work

It will be seen that the surface of the topic has barely been touched by spectroscopists or mineralogists. Most of the work done so far relates to minerals which may be considered as having isolated groups for which internal modes may be fairly simply assigned. In many of the cases cited much more information would be desirable—fuller spectra, better complementary infrared data, more complete assignments. The use of high frequency (e.g., ultraviolet) exciting lines would be very desirable. The present reliance on helium–neon, argon–ion, or krypton lasers has lead to the paradoxical result that some of the early work using the mercury 2537 Å line is probably better than some modern work with lasers, simply because much more intense Raman spectra are obtained with high-frequency sources. The field of study of minerals which do not contain isolated units (e.g., ino-, phyllo-, and tectosilicates) is untouched as yet.

References

Adams, D. M. and Gardner, I. R. (1974a). *J. Chem. Soc.* (*Dalton*) 1502–1505.

Adams, D. M. and Gardner, I. R. (1974b). *J. Chem. Soc.* (*Dalton*) pp. 1506–1509.

Alfano, R. R., and Shapiro, S. L. (1971). *Chem. Phys. Lett.* **8,** 631–633.

Bates, J. B., and Quist, A. S. (1972). *J. Chem. Phys.* **56,** 1528–1533.

Beattie, I. R., and Gilson, T. R. (1968). *Proc. Roy. Soc., Ser. A* **307,** 407–429.

Beattie, I. R., and Gilson, T. R. (1969). *J. Chem. Soc., A* pp. 2322–2327.

Beattie, I. R., Livingston, K. M. S., Ozin, G. A., and Reynolds, D. J. (1970). *J. Chem. Soc., A* pp. 449–451.

Behringer, J. (1966). *In* "Raman Spectroscopy, Theory and Practice" (H. A. Szymanski, ed.), Vol. I, pp. 168–223. Plenum, New York.

Bobovich, Y. S. (1961). *Opt. Spectrosc.* (*USSR*) **10,** 212–213.

Bouhet, C., and Lafont, R. (1948). *Acad. Sci.* **226,** 1263–1265.

Brand, J. C., and Speakman, J. C. (1960). "Molecular Structure," pp. 126–195. Arnold, London.

Brandmuller, J., and Moser, H. (1962). "Einfuhrung in die Raman-spektroskopie," Steinkopff, Darmstadt.

Brillouin, L. (1922). *Ann. Phys.* (*Paris*) **17,** 88–122.

Cohen-Solal, G. W., Lafont R., and Bouhet, C. (1964). *Acad. Sci.* **259,** 337–340.

Couture, L. (1947). *Ann. Phys.* (*Paris*) [12] **2,** 5–94.

Dawson, P., Hargreave, M. M., and Wilkinson, G. R. (1971). *J. Phys. C* **4,** 240–256.

Eckhardt, G., Portfeld, D. P., and Geller, M. (1963). *Appl. Phys. Lett.* **3,** 137–138.

Etchepare, J. (1970). *C. R. Acad. Sci., Ser. B* **270,** 1339–1342.
Fong, M. Y., and Nicol, M. (1970). *J. Chem. Phys.* **54,** 579–585.
Forneris, R. (1969). *Amer. Mineral.* **54,** 1062–1074.
Ganestan, A. S. (1929). *Indian J. Phys,* **4,** 281–346.
Gilson, T. R., and Hendra, P. (1970). "Laser Raman Spectroscopy." Wiley, New York.
Griffith, W. P. (1969a). *J. Chem. Soc., A* pp. 1372–1377.
Griffith, W. P. (1969b). *Nature (London)* **224,** 264–266.
Griffith, W. P. (1970). *J. Chem. Soc., A* pp. 286–291.
Gross, E. (1930a). *Nature (London)* **126,** 603–604.
Gross, E. (1930b). *Z. Phys.* **63,** 685–687.
Halford, R. S. (1946). *J. Chem. Phys.* **14,** 8–15.
Hass, M., and Sutherland, G. B. (1956). *Proc. Roy. Soc., Ser. A* **236,** 427–445.
Hathaway, C. E. (1971). *In* "The Raman Effect" (A. Anderson, ed.), Vol. I, pp. 183–285. Dekker, New York.
Hester, R. E. (1970). *In* "Raman Spectroscopy, Theory and Practice" H. A. Szymanski, ed.), Vol. II, pp. 141–173. Plenum, New York.
Khambata, S. J. (1956). *Proc. Phys. Soc., London, Sect. A* **69,** 426–427.
Khanna, R. K., Brower, W. S., Guscott, B. R., and Lippincott, E. R. (1968). *J. Res. Nat. Bur. Stand., Sect. A* **72,** 81–84.
Klee, W. E. (1966). *Z. Anorg. Chem.* **343,** 58–69.
Koningstein, J. A. (1966). *J. Opt. Soc. Amer.* **56,** 1405–1406.
Krauzmann, M. (1968). *C. R. Acad. Sci., Ser. B* **266,** 1224–1226.
Kravitz, L. C., Kingsley, J. C., and Elkin, E. L. (1968). *J. Chem. Phys.* **49,** 4600–4610.
Krishnamurti, D. (1955). *Proc. Indian Acad. Sci., Sect. A* **41,** 7–11.
Krishnamurti, D. (1956). *Proc. Indian Acad. Sci., Sect. A* **43,** 210–212.
Krishnamurti, D. (1957). *Proc. Indian Acad. Sci., Sect. A* **46,** 183–202.
Krishnamurti, D. (1958). *Proc. Indian Acad. Sci., Sect. A* **47,** 276–291.
Krishnamurti, D. (1960). *Proc. Indian Acad. Sci., Sect. A* **51,** 285–295.
Krishnamurti, N., and Soots, V. (1971). *Can. J. Phys.* **49,** 885–896.
Krishnan, R. S. (1945). *Proc. Indian Acad. Sci., Sect. A* **22,** 182–193.
Krishnan, R. S. (1947). *Proc. Indian Acad. Sci., Sect. A* **26,** 460–468.
Krishnan, R. S. (1971a). *In* "The Raman Effect" (A. Anderson, ed.), Vol. I, pp. 1–67. Dekker, New York.
Krishnan, R. S. (1971b). *In* "The Raman Effect" (A. Anderson, ed.), Vol. I, pp. 343–404. Dekker, New York.
Krishnan, R. S., and Kumarti, C. S. (1951). *Proc. Indian Acad. Sci., Sect. A* **32,** 105–109.
Lallemand, P., (1971). *In* "The Raman Effect" (A. Anderson, ed.), Vol. I, pp. 287–342. Dekker, New York.
Landsberg, G., and Mandelstam, L. (1928a). *Naturwissenschaklen* **16,** 557–558.
Landsberg, G., and Mandelstam, L. (1928b). *Z. Phys.* **50,** 769–770.
Lazarev, A. N. (1973). *Spectrochim. Acta, Part A* **29,** 295–296.
Levitt, S. R., and Condrate, R. A. (1970). *Amer. Mineral.* **55,** 1562–1575.
Long, D. A., and Stanton, L. (1970). *Proc. Roy. Soc., Ser. A* **318,** 441–457.
Loudon, R. (1964). *Advan. Phys.* **13,** 423–482.
Loudon, R. (1965). *J. Phys. (Paris)* **26,** 677–683.
McQuillan, A. K., Clements, W. R. L., and Stoicheff, B. P. (1970). *Phys. Rev. A* **1,** 628–635.
Maradudin, A. A., and Burstein, E. (1967). *Phys. Rev.* **164,** 1081–1099.
Mitra, S. S., and Gielisse, P. J. (1963). *In* "Progress in Infrared Spectra" (H. A., Szymanski, ed.), p. 47–75. Plenum, New York.

Moore, R. K., White, W. B., and Long, T. V. (1971). *Amer. Mineral.* **56,** 54–71.
Narayanan, P. S. (1950). *Proc. Indian Acad. Sci. Sect. A* **31,** 354–358.
Narayanan, P. S. (1951a). *Proc. Indian Acad. Sci., Sect. A* **34,** 387–390.
Narayanan, P. S. (1951b). *Curr. Sci.* **20,** 94–95.
Nicol, M., and Durana, J. F. (1971). *J. Chem. Phys.* **54,** 1436–1440.
Nicol, M., and Fong, M. J. (1971). *J. Chem. Phys.* **54,** 3167–3170.
Ornstein, L. S., and Went, J. J. (1935). *Physica* (*Utrecht*) **2,** 391–398.
Ozin, G. A. (1969). *J. Chem. Soc., A* pp. 116–118.
Placzek, G. (1934). *In* "Handbuch der Radiologie" (E. Marx, ed.), Akad. Verlagesges., Leipzig.
Porto, S. P. S., and Krishnan, R. S. (1967). *J. Chem. Phys.* **47,** 1009–1012.
Porto, S. P. S., Fleury, P. A., and Damen, T. C. (1967). *Phys. Rev.* **154,** 522–526.
Raman, C. V. (1928). *Indian J. Phys.* **2,** 387–419.
Raman, C. V. (1961). *Proc. Indian Acad. Sci., Sect. A* **54,** 253–265.
Raman, C. V., and Krishnan, K. S. (1928). *Nature* (*London*) **121,** 501–502.
Richman, I. (1966). *J. Opt. Soc. Amer.* **56,** 1589–1590.
Ross, S. D. (1971). *Spectrochim. Acta, Part A* **27,** 1837–1843.
Ross, S. D. (1972). *Spectrochim. Acta, Part A* **28,** 1555–1561.
Scheuermann, W., Ritter, G. J., and Schutte, C. J. H. (1970). *Z. Naturforsch. A* **25,** 1856–1862.
Scott, J. F. (1971). *Phys. Rev. B* **4,** 1360–1366.
Scott, J. F., and Porto, S. P. S. (1967). *Phys. Rev.* **161,** 903–910.
Scrocco, M., and Mathieu, J. P. (1958). *Ann. Phys.* (*Paris*) [13] **3,** 463–467.
Servoin, J. L., Piriou, B., and Alain, P. (1972). *Acad. Sci., Ser. B* **274,** 135–137.
Smekal, A. (1923). *Naturwissenschaften* **11,** 873–875.
Solomons, C., Clarke, J. H. R., and Bockris, J. O'M. (1968). *J. Chem. Phys.*, **49,** 445–449.
Srivastan, R., Leuer, H. V., Chase, L. L., and Bron, W. E. (1971). *Phys. Lett. A* **36,** 333–334.
Stoicheff, B. P. (1959). *Advan. Spectrosc.* **1,** 96–109.
Tannenwald, P. E. (1967). *J. Appl. Phys.* **38,** 4788–4790.
Terhune, R. W., Maker, P. D., and Savage, C. M. (1965). *Phys. Rev. Lett.* **14,** 681–684.
Tuinstra, F., and Koening, J. L. (1970). *J. Chem. Phys.* **53,** 1126–1130.
Wadsack, R. L., and Chang, R. K. (1972). *Solid State Commun.* **10,** 45–48.
Ware, M. J. (1968). *In* "Physical Methods for Advanced Inorganic Chemistry," pp. 214–365. Wiley (Interscience), New York.
Wilkins, R. W. T. (1971). *Z. Kristallogr., Kristallgeometric, Kristallphys., Kristallchem.* **134,** 285–290.
Woodward, L. A. (1953). *Quart. Rev. Chem. Soc.* **10,** 185–207.
Woodward, L. A. (1966). *In* "Raman Spectroscopy, Theory and Practice" (H. A. Szymanski, ed.), Vol. I, pp. 1–45. Plenum, New York.
Zallen, R., Lucovsky, G., Taylor, W., Pinczuk, A., and Burstein, E. (1970). *Phys. Rev. B* **1,** 4058–4070.

13

STRUCTURAL INTERPRETATION OF LUNAR AND TERRESTRIAL MINERALS BY RAMAN SPECTROSCOPY

William B. White

I. Introduction

The vibrational spectra of silicates have been of interest since the pioneering work of Matossi and co-workers (Matossi and Krüger, 1936; Matossi and Bronder, 1938). Since then a very large volume of empirical data has accumulated. Because the strong Si–O bond places its stretching frequency comfortably in the mid-infrared, most previous work has utilized IR spectroscopy and is treated elsewhere in this volume (Chapters 5, 7, 8–9, and 11). Several very good compilations of vibrational spectra are available (Moenke, 1962, 1966; Lyon, 1962; Lazarev, 1972). Raman spectra are considerably more sparse. Except for quartz, beryl, and a few other silicates that occur as gem-quality crystals, the available specimens were just not suitable for Raman measurements until the laser-source, double-monochrometer type of spectrometer was invented in the early 1960's.

The objective of this chapter is to systematize the vibrational spectra of the rock-forming minerals, mainly silicates, in relation to their crystal structure. The emphasis will be on Raman spectra, although it will be necessary to draw on IR data to complete the analysis in many cases. Because this volume is oriented toward lunar as well as terrestrial materials, emphasis is placed on the minerals that constitute igneous rocks, mainly the anhydrous silicates.

II. Relation of Structure and Spectra

A. *Vibrational Modes of Crystals*

Historically, the vibrational spectra of crystalline solids have been approached either from a point of view of molecular vibrations (which ignores the crystal structure) or of the simple ionic solid (which ignores the complex difference in bond type that may occur within the same structure). Fortunately, several new monographs have appeared recently that provide a comprehensive treatment of crystal vibrations (Sushchinskii, 1972; Sherwood, 1972; Turrell, 1972). What follows, therefore, is only a brief summary.

1. *Homodesmic Crystals*

Homodesmic crystals are defined as those in which all chemical bonds are of equal strength. The best examples are simple ionic compounds such as the alkali halides or those with completely equivalent covalent bonds such as the diamond structure.

To get a sense of how crystal structure will influence vibrational frequencies, it is only necessary to note that the lattice energy is described by some sort of Born–Mayer potential such as

$$V = \frac{N\alpha e^2}{R}\left(1 - \frac{1}{n}\right)$$

or the similar expression used by Ohashi and Burnham (1972) to calculate site energies in crystals. Force constants are related to potential by

$$k_i = \frac{\partial^2 V}{\partial Q_i^2}$$

where Q_i are the appropriate normal coordinates. Since any Q_i is linear in interatomic distance and since the Coulomb potential is the dominant term in an ionic crystal, the force constant goes as

$$k \sim 1/R^3$$

Therefore, to a first approximation the frequency–bond length relation becomes

$$\bar{\nu}^2 \sim 1/R^3$$

Frequency obviously depends on atomic masses as well.

2. *Heterodesmic and Molecular Crystals*

If distinct molecular units can be identified in the crystal, the vibrational spectrum separates into two parts: internal modes arising from the vibrations of the molecules themselves and external or lattice modes arising from motions of the molecules and ions against each other. The assumption works best when the bonding within the molecules is very strong and the bonding between molecules is very weak.

The assumption, however, never holds perfectly, and there is always some perturbation of the internal modes: (i) the crystal field, symmetric part, distortion part; (ii) the correlation field; and (iii) the polarization field.

The crystal field arises from the influence of the nearest neighbor environment of the molecular cluster. The symmetric part of the field simply shifts the vibrational energy levels from their unperturbed values by a small amount. The distortion field, which arises if the site on which the molecule resides has lower symmetry than the molecule itself, shifts energy levels, lifts degeneracy, and may modify the molecular selection rules. It gives rise to the well-known site-group splitting of crystal vibrations. The correlation field couples equivalent molecules and may generate a multiplicity of internal modes (factor group or Davydov splitting). The long-range polarization field is a dynamic effect. It is generated by charge separation during the vibration and separates transverse and longitudinal modes.

The silicates are typical heterodesmic crystals. The basic building block is the SiO_4 tetrahedron which in its isolated state would exhibit four fundamental modes (Table I). The oxygen of the tetrahedra also coordinates the cations with moderately strong bonding, thus making the molecular model dubious even in orthosilicates. When the tetrahedra are polymerized into chains, sheets, or three-dimensional frameworks, a basic molecular building block becomes hard to define. The molecular model, although imprecise, does provide a useful label for the Si–O stretching

TABLE I

VIBRATIONAL FREQUENCIES FOR THE ISOLATED SiO_4^{4-} ION

	Mode	Symmetry	Wave number[a]
ν_1	Symmetric stretch	A_1	819
ν_2	Symmetric bend	E	340
ν_3	Asymmetric stretch	T_2	956
ν_4	Asymmetric bend	T_2	527

[a] Data from Basile *et al.* (1973).

frequencies which yield the most information about the structure. The force constants for the Si–O stretch (Table II) are typical single bond values.

B. Classification of Crystal Vibrations

1. *Factor Group Analysis*

A crystal containing r atoms in the primitive unit cell has $3r$ degrees of freedom, of which 3 appear as translations of the unit cell and thus become acoustic modes while $3r - 3$ appear as vibrations. The degrees of freedom can be classified in the irreducible representations of the factor group of the crystal's space group. This analysis takes into account both crystal field and correlation field effects but not long-range polarization field. Factor group analysis for the full Brillouin zone is discussed in depth by Maradudin

TABLE II

FORCE CONSTANTS FOR ISOLATED SiO_4^{4-} TETRAHEDRON

Force constant		UBFF[a]	OVFF[a]
K	(bond stretching)	429[b]	430[b]
H(D)	(bond bending)	26	75
F	(anion repulsion)	57	51

[a] UBFF, Urey-Bradley force field; OVFF, orbital valence force field.

[b] All values from calculations of Basile *et al.* (1973); units are newtons/meter.

and Vosko (1968) and by Warren (1968). Detailed procedures for performing the calculations for the zone center ($\kappa = O$) modes have been summarized by DeAngelis *et al.* (1972) and Fateley *et al.* (1972). Tabular results for all space groups are also available (Adams and Newton, 1970). There is a computer program that will classify normal modes and write the appropriate symmetry coordinates (Worlton and Warren, 1972).

The essentials of the technique are illustrated for the olivine structure. Olivine is orthorhombic, space group Pbnm, D_{2h}^{16} with axes for this choice of space group $a_o = 4.756$, $b_o = 10.195$, $c_o = 5.981$ Å. The atoms which remain invariant under the operations of the D_{2h} factor group are arranged in terms of their site distribution in Table III. There are two cation sites in olivine. M1 is on a center of symmetry, while M2 is on the mirror plane. The silicon atoms also lie on the 4(c) sites, so that the site group of the SiO_4 tetrahedron is C_s. The invariance conditions are used to calculate the reducible representation for the full olivine structure, and this in turn can be decomposed to distribute the degrees of freedom among the irreducible representations of the factor group (Table IV). In this analysis a distinction is made between the internal modes of the SiO_4 tetrahedron, the translatory motions of the tetrahedron against the cations, and the rotatory motions of the tetrahedron. The distortion field, described by the C_s point symmetry, lifts all degeneracy from the SiO_4 internal vibrations. Because there are four tetrahedra in the unit cell, each component of the tetrahedral modes appears four times. These components fall into different irreducible representations of the factor group and are therefore subject to different selection rules. The correlation field may give them different frequencies as well.

TABLE III

INVARIANCE CONDITIONS FOR OLIVINE, PBNM, D_{2h}^{16}

		E	$C_2(z)$	$C_2(y)$	$C_2(x)$	i	σ_{xy}	σ_{xz}	σ_{yz}
4(a) Ci		E	—	—	—	i	—	—	—
4 Mg^{2+}(1)		4	0	0	0	4	0	0	0
4(c) C_s		E	—	—	—	—	σ	—	—
4 Mg^{2+}(2) + 4 Si^{4+} + 4 O^{2-}(1) + 4 O^{2-}(2)		16	0	0	0	0	16	0	0
8(d) C_1		E	—	—	—	—	—	—	—
8 O^{2-}(3)		8	0	0	0	0	0	0	0
	ω_ρ	28	0	0	0	4	16	0	0
Full cell	χ_ρ	84	0	0	0	−12	16	0	0
Translatory	χ_ρ	36	0	0	0	−12	8	0	0
Rotatory	χ_ρ	12	0	0	0	0	−4	0	0

TABLE IV

VIBRATIONAL MODES OF OLIVINE

D_{2h}	TOTAL MODES	TRANSLATORY MODES	ACOUSTIC MODES	ROTATORY MODES	INTERNAL MODES		SELECTION RULES
A_g	11	4	0	1	6	$\nu_1 + \nu_2 + 2\nu_3 + 2\nu_4$	$\alpha_{xx}, \alpha_{yy}, \alpha_{zz}$
B_{1g}	11	4	0	1	6	$\nu_1 + \nu_2 + 2\nu_3 + 2\nu_4$	α_{xy}
B_{2g}	7	2	0	2	3	$\nu_2 + \nu_3 + \nu_4$	α_{xz}
B_{3g}	7	2	0	2	3	$\nu_2 + \nu_3 + \nu_4$	α_{yz}
A_u	10	5	0	2	3	$\nu_2 + \nu_3 + \nu_4$	Inactive
B_{1u}	10	4	1	2	3	$\nu_2 + \nu_3 + \nu_4$	IR, E || c
B_{2u}	14	6	1	1	6	$\nu_1 + \nu_2 + 2\nu_3 + 2\nu_4$	IR, E || b
B_{3u}	14	6	1	1	6	$\nu_1 + \nu_2 + 2\nu_3 + 2\nu_4$	IR, E || a

2. *Selection Rules*

The infrared-active modes of a crystal classify in the same irreducible representation of the factor group as the components of the dipole-moment vector. The orientation of the vector components gives the polarization of the IR absorption (Table IV). The Raman-active modes classify in the same irreducible representations as the polarizability tensor. In the most general (triclinic) case, α_{ij} has 6 independent components. By aligning the indicatrix axes of a single crystal with the laboratory axes (Fig. 1) and by polarizing both incident and scattered beams, all independent spectra can be obtained. In the example of olivine there are four independent spectra. All parallel polarizations (diagonal elements, α_{ii}) yield the A_g modes but the crossed-polarization (off-diagonal) terms yield B_{1g}, B_{2g}, and B_{3g} independently.

3. *The Vibrational Spectrum of Olivine*

Recent measurements of the vibrational spectra of olivine structure compounds includes systematic powder spectra (Tarte, 1962, 1963; Toropov *et al.*, 1963; Griffith, 1969; Burns and Huggins, 1972), a site-group analysis (Duke and Stephens, 1964), factor-group and normal coordinate analysis (Oehler and Günthard, 1969), and polarized IR and Raman spectra of olivine single crystals (Hohler and Funck, 1969, 1973; Hohler, 1972; Servoin and Piriou, 1973). The Raman spectrum of a synthetic forsterite single crystal is presented in Fig. 2.

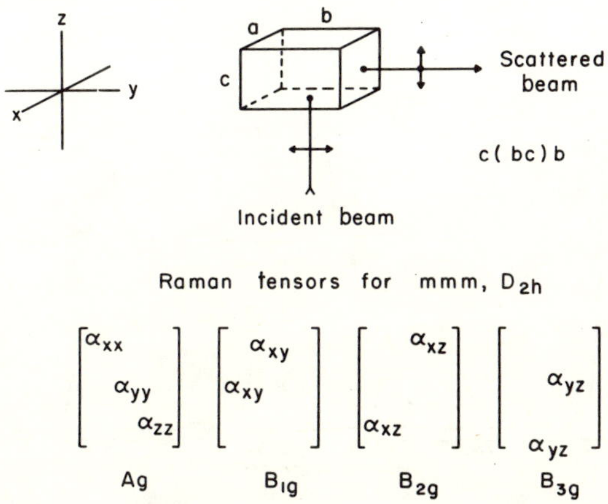

FIG. 1. Orientation of an olivine single crystal for Raman measurement. Porto symbol c(bc)b [or z(yz)y] indicates incident beam direction c, polarization directions, b(incident beam), and c (scattered beam), and scattered beam direction, b. This arrangement would couple the $\alpha_{yz} = B_{3g}$ modes. Full set of Raman tensors are from Louden (1964).

The vibrational frequencies of olivine are tabulated in Table V drawn mainly from the work of Servoin and Piriou (1973), who presented both single-crystal Raman data and Kramers–Kronig transformed polarized infrared data. The number of observed Raman bands agrees exactly with the predictions of the factor group analysis. Somewhat fewer IR bands were observed than predicted perhaps because of the low-frequency limitations of the equipment. Two additional bands of unknown polarization were observed by Oehler and Günthard (1969). Many of the vibrational modes have similar frequencies and only their polarization dependence distinguishes them.

An attempt can be made to sort out the effects of the various internal

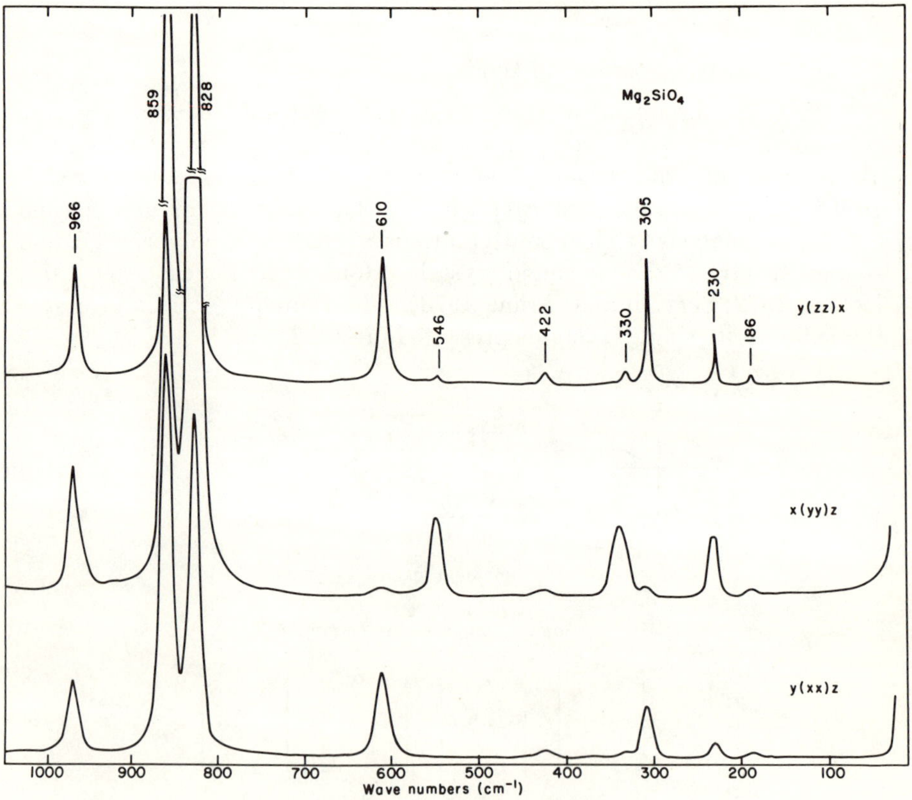

FIG. (2a).

FIG. 2. Raman spectra of oriented single crystal of synthetic forsterite. (a) Diagonal components = A_g modes. (b) Off-diagonal components = B_{1g}, B_{2g}, and B_{3g}, respectively. Crystal described by Graham and Barsch (1969).

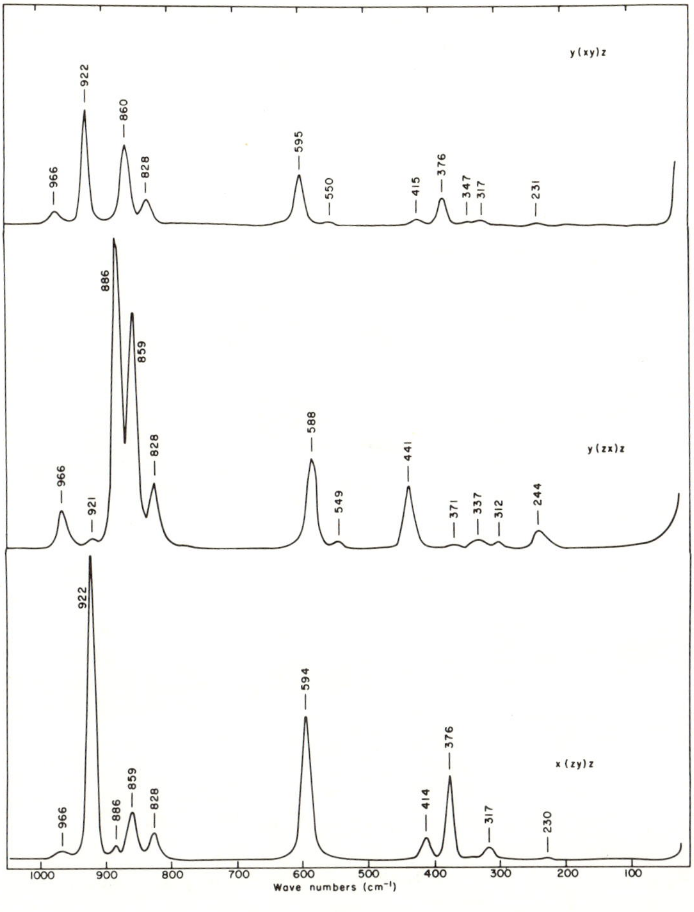

FIG. (2b).

fields on the stretching vibrations of the SiO_4 tetrahedron using the molecular model (Fig. 3). All modes in the same species can mix and for this reason only the high-frequency stretching modes are considered. The internal bending modes and the external modes would mix enough to make the analysis meaningless. On the right margin are the observed modes of single crystal Mg_2SiO_4 with their proper symmetry label, based on the data of Servoin and Piriou (1973). On the left margin are the free molecule frequencies of ν_1 and ν_3. In between an averaging has been performed to separate the effects of the different perturbation fields.

The assumption has been made that the effect of the crystal field (site-group splitting) is larger than the effect of the correlation field (factor-

TABLE V

VIBRATIONAL FREQUENCIES FOR SINGLE CRYSTAL FORSTERITE[a]

	RAMAN				IR		
MODE	A_g	B_{1g}	B_{2g}	B_{3g}	B_{1u}	B_{2u}	B_{3u}
ν_1	825	828[b]	—	—	—	838	837
ν_2	[c]				415	394	377
ν_3	966	975	—	920	—	983	978
	856	866	881	—	871	875	877
ν_4	609	632	—	591	—	527	601
	545	583	585	—	501	506	498
	—	—	—	484	473	—	—
	—	—	—	466	—	456	—
	424	434	438	—	—	—	438
	—	418	407	406	—	416	404
	—	—	365	374	—	—	—
	339	—	—	—	—	350	—
	328	317	324	314	—	—	318
	305	—	—	—	289	290	293
	—	265	—	—	274	276	274
	227	—	244	—	—	—	—
	183	197	—	—	201[d]	201[d]	201[d]
	—	164	—	—	144[d]	144[d]	144[d]

[a] All data from Servoin and Piriou (1973) except as indicated.

[b] This value from Fig. 2.

[c] Note that ν_2 is not separated from lattice modes.

[d] From powder data by Oehler and Günthard (1969).

group splitting). The tetrahedron in olivine has point symmetry C_s, but this is a small distortion on a larger trigonal distortion, point symmetry C_{3v} with the pseudo-threefold axis along *a*. First ν_3 is split into two components by the trigonal field and then the E mode is split again by the smaller C_s field. The site-group splitting thus produces two A′ and one A″ levels. The interactions of the four tetrahedra in the unit cell distribute the eight A′ and four A″ modes among the factor group representations as shown. The diagram, of course, was derived by clustering and averaging the measured lines using the symmetry combinations as a guide. There is a correlation field effect of 10–30 cm^{-1} for the A′-type modes of both ν_1 and ν_3. The position of A″ is made uncertain by the fact that one of its components, A_u, is inactive in both IR and Raman and cannot be observed, but it does appear that the effect of the correlation field on A″ is much larger than on A′.

C. *Structural Distortion*

1. *Olivine Structures*

The infrared spectrum of polycrystalline olivine contains Si–O stretching modes with the pattern illustrated in Fig. 4. The two intense bands, 1 and 3, may be correlated with unresolved components of the levels arising from the trigonal field splitting of ν_3 so that the difference between these frequencies should be a measure of the site-group splitting. The resolution of powder spectra is not sufficient to observe any factor group splitting and the number of Raman spectra of olivine structure minerals is too sparse to permit systematic trends to be observed. Band 4 is ν_1 again with the factor-group splitting unresolved.

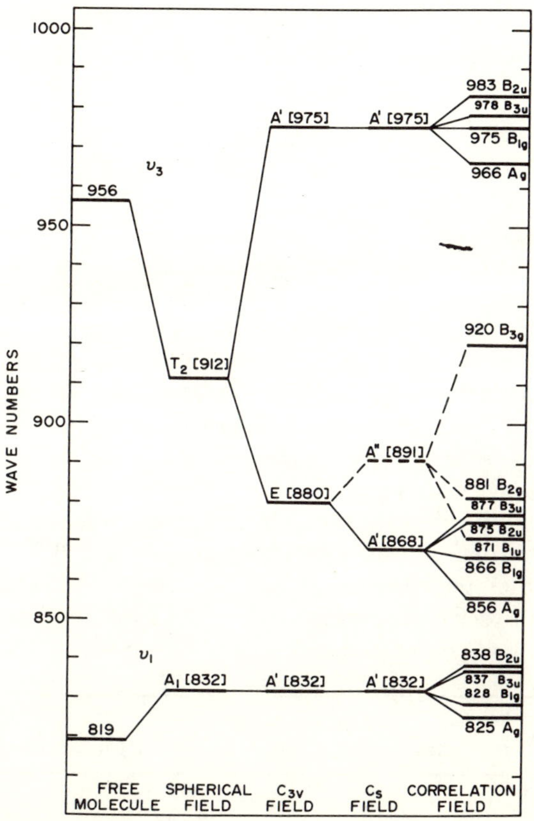

FIG. 3. Energy level diagram showing the influence of several perturbation fields on the ion levels of the SiO_4 tetrahedron in olivine. The position of the A″ level is uncertain, because one of its components is an inactive mode.

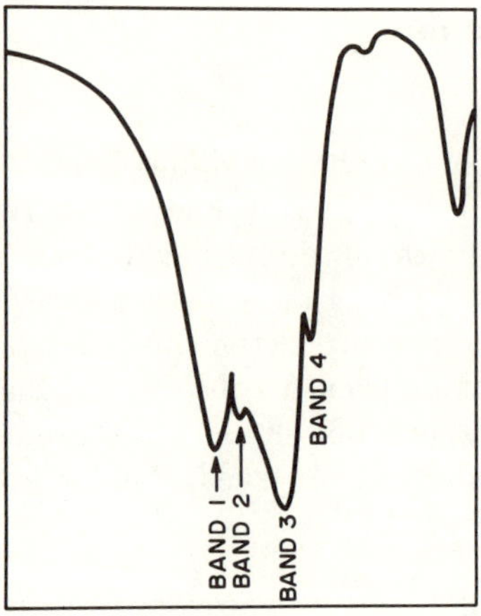

FIG. 4. Schematic IR spectrum showing general appearance of olivine Si-O bands in the 800–950 cm^{-1} region.

The frequencies vary with cation substitution (Tarte, 1963; Duke and Stephens, 1964; Burns and Huggins, 1972). Toropov *et al.* (1963) suggested that the frequency should vary as $RM^{1/2}$ where R is the cation radius and M is its mass in atomic units. Results (Fig. 5) show considerable scattering, but the curves are definitely nonlinear, showing that this function is not an exact description of the perturbation. The frequencies decrease with increasing radius and approach the free ion frequencies at large radius. This effect, observed also in the carbonate minerals (White, 1974) shows that smaller cations actually increase the force constant of the Si–O bond. This is in contrast to O–H vibrations where increased bonding of the hydrogen to adjacent oxygen weakens the primary bond and causes the well-known shift to lower frequencies.

The site group splitting appears to decrease as the cation radius increases, but the available data are not sufficiently precise (Fig. 6). The trigonal distortion of the SiO_4 tetrahedron can be measured by using the difference in edgelengths between two sets of edges. Structural refinements for two olivine minerals are tabulated below for comparison (Birle *et al.*, 1968),

MINERAL	THREE LONG EDGES–THREE SHORT EDGES	BAND 1–BAND 3
Forsterite	0.186 Å	104
Fayalite	0.162 Å	80

where the site group splittings are averages of Duke and Stephens' and Tarte's values.

2. *Other Orthosilicate Structures*

Common silicate minerals containing only isolated SiO_4 tetrahedra include those with the zircon, garnet, and phenakite structures. Many IR spectra have been obtained on these minerals (Lazarev, 1972; Lyon, 1962). Recent zircon spectra include those of Hubin and Tarte (1971), who analyze the effects of cation size on band frequency, and Dawson *et al.* (1971), who present fully determined single-crystal spectra for both IR and Raman. IR spectra of silicate garnets have been reported by Moore *et al.* (1971); Omori (1971): Suwa and Naka (1972), and Tarte and Deliens (1973). The latter show ways of determining chemical composition of the garnets from band intensities and frequency shifts. Raman spectra for some 17 garnets were presented by Moore *et al.* (1971). Tarte (1962) has examined the IR spectra of zinc silicate–germanate solid solutions with the phenakite structure and assigned the main absorption bands.

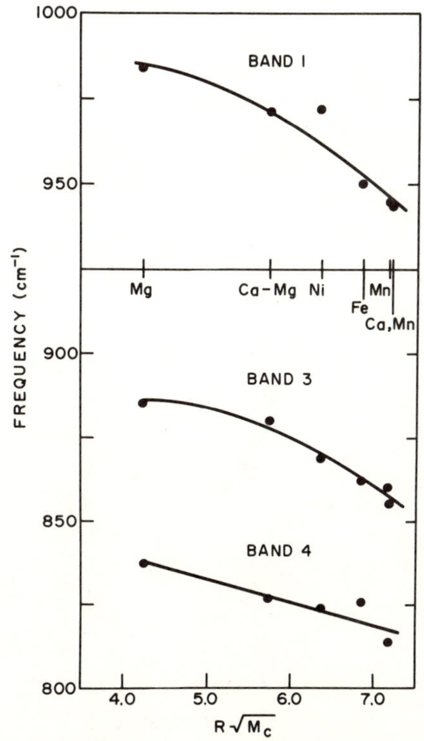

FIG. 5. IR band frequencies of olivine-structure compounds as a function of $RM^{1/2}$. Data from Duke and Stephens (1964). Tarte's (1963) data give similar plots but with a systematic shift to higher frequencies.

Raman spectra of powder samples (Fig. 7) show several striking features. All spectra are dominated by strong bands in the Si–O stretching region. Lower frequency bands in both almandine and willemite tend to be much less intense than the Si–O bands. Zircon in contrast exhibits quite intense low-frequency bands. Bands in the 500–600 cm^{-1} region of willemite which might be attributable to the stretching frequencies of the ZnO_4 tetrahedron are also very weak.

There is a very pronounced shift in the frequency of the strongest Si–O vibration which appears to reflect the next nearest neighbor environment. Each SiO_4 tetrahedron in willemite is corner shared with ZnO_4 tetrahedra to build up an overall tetrahedral framework. The SiO_4 tetrahedra in garnet share corners with an octahedron and edges with a dodecahedron (Gibbs and Smith, 1965), whereas in zircon the tetrahedra and ZrO_8 dodecahedra share edges (Robinson *et al.*, 1971). The increase in next nearest neighbor coordination appears to result in a shift of the strong Raman band to higher frequencies. This effect is interesting and puzzling in two respects: (1) The increasing coordination and interatomic distances in the next nearest neighbor metal–oxygen polyhedra should result in lower force constants and a general weakening of the metal–oxygen–silicon interaction. Rather than a shift toward the free ion value, however, the Si–O vibrations take on even higher frequencies. (2) There is a marked

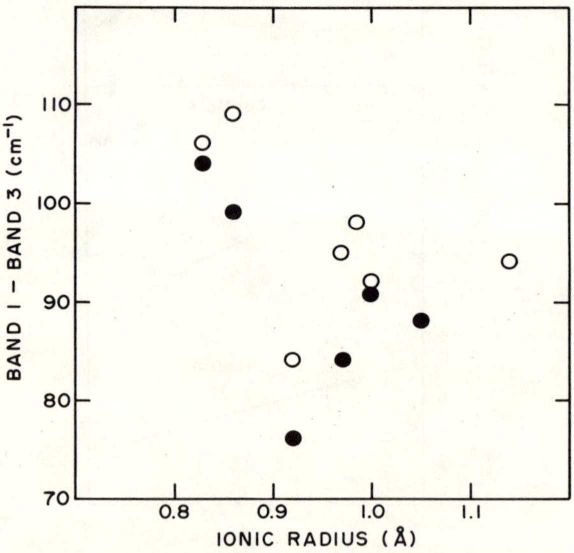

FIG. 6. Site group splitting of olivine–structure as a function of cation radius. Open circles: Tarte (1963). Solid circles: Duke and Stephens (1964).

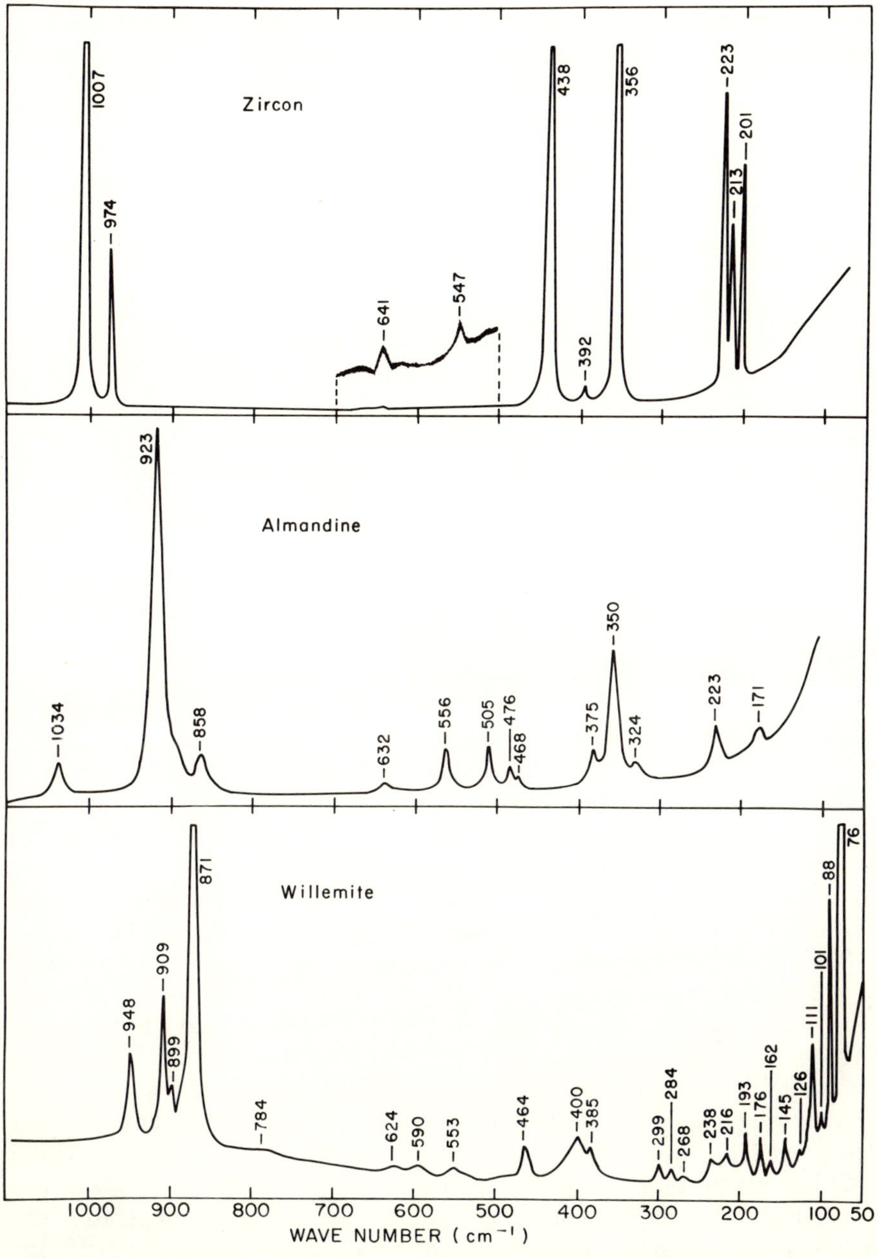

FIG. 7. Raman spectra of powder specimens of zircon, almandine garnet, and willemite. The garnet has the composition $Al_{0.68}Py_{0.23}Sp_{0.05}Gr_{0.03}An_{0.01}$. The willemite is well-crystallized synthetic Zn_2SiO_4, checked by x-ray diffraction.

contrast to the effect of silicate polymerization. In general the isolated SiO_4 tetrahedra have lower frequencies, while the polymerized tetrahedra (chains, sheets, and framework) have higher frequencies.

III. Complex Silicates

A. Role of Tetrahedral Linkages

The asymmetric stretching modes of the SiO_4 tetrahedron occur in the range of 800–950 cm^{-1} in the orthosilicates where no linking of tetrahedra occurs. Formation of polymerized silicates with corner-sharing tetrahedra separates the silicon–oxygen bands into a bridging set and a nonbridging set. Concurrently the Si–O stretching frequencies shift to higher energies reaching 1200 cm^{-1} in the framework silicates. Launer appears to have first noticed this effect in 1952.

1. *Vibrational Spectra of the Silicates*

Very good infrared data exist, mostly measured on powders, for the common silicates (Lazarev, 1972). Interpretation of the IR data is found elsewhere in this volume (Chapters 9 and 11). Raman data are considerably more sparse except for the orthosilicates already discussed and quartz. Very detailed Raman spectra on both α (Scott and Porto, 1967; Asell and Nicol, 1968; Masso *et al.*, 1970) and β (Bates and Quist, 1972) quartz, as well as a study of the phase transition (Shapiro *et al.*, 1967) and the polariton spectrum (Scott *et al.*, 1967) are available. Raman spectra of the intermediate silicates include some pyrosilicates (Tarte *et al.*, 1973), some pyroxenes (Etchepare, 1970; Bilton *et al.*, 1972), some sheet silicates (Loh, 1973), and a collection of lunar minerals (Estep *et al.*, 1972).

The spectra of the rock-forming pyroxenes (Fig. 8) are all very similar. The dominant features are a very intense band near 1000 cm^{-1} and one or two intense bands in the 650–680 cm^{-1} range. The latter have no analog in the spectra of orthosilicates or in the infrared spectra and are likely due to an etherlike symmetric stretch of the bridging oxygens in the chain. The Si–O stretching region in the Raman spectrum does not behave like IR spectrum. The IR-active Si–O stretching frequencies (Rutstein and White, 1971) occur as sets of 4–6 strong bands whose detailed arrangement is characteristic of the particular pyroxene structure. They are separated into a low-frequency group near 900 cm^{-1} that can be associated with the nonbridging oxygens and a high-frequency group at 1050–1100 cm^{-1} associated with the bridging oxygens. It seems likely that the single strong

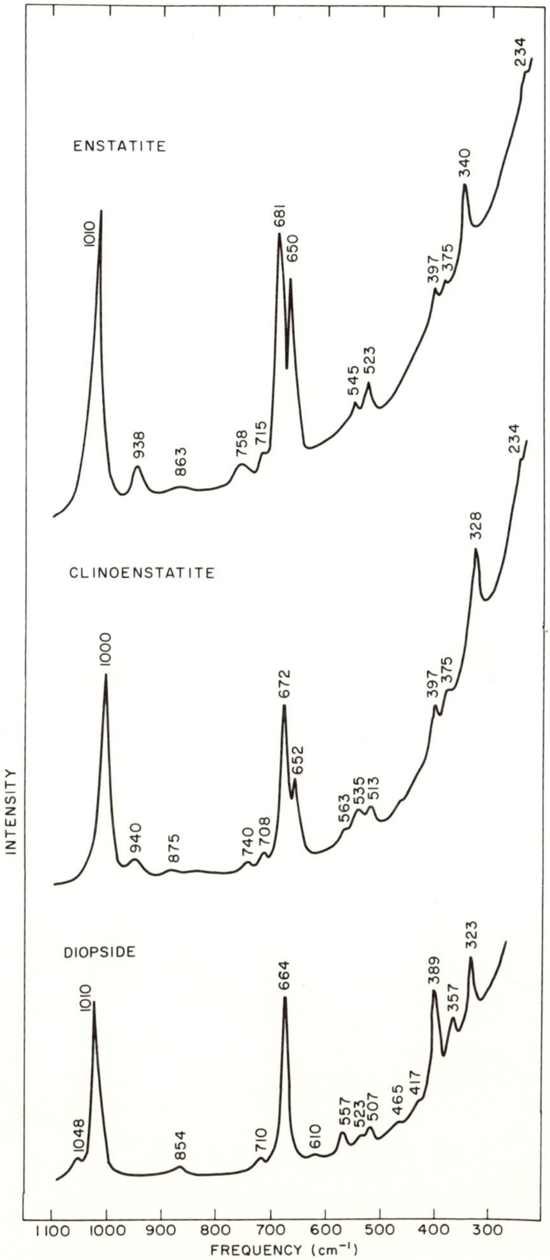

FIG. 8. Raman spectra of three pyroxene minerals.

band in the Raman spectrum is the symmetric stretch and that the bands that give the characteristic IR spectrum are subdued in the Raman spectrum. Factor group analysis of several pyroxene structures (O'Horo *et al.*, 1974) predicts 14 $A_g + 16 B_g$ Raman-active modes for diopside and 30 $A_g + 30 B_{1g} + 30 B_{2g} + 30 B_{3g}$ Raman-active modes for orthoenstatite. Single-crystal spectra reveal most of the expected 30 Raman bands of diopside and about half of the 120 expected Raman bands of orthoenstatite. Just as in the case of the orthosilicates, many modes have nearly the same frequency and are distinguished only by their polarization dependence. Powder spectra therefore reveal only the general features of the spectrum.

The Raman spectra of the amphiboles are almost completely unknown, but the powder spectrum of an actinolite (Fig. 9) is not encouraging. There are two formula units of $Ca_2(Mg,Fe)_5Si_8O_{22}(OH,F)_2$ in the C2/m monoclinic cell. Instead of the rich spectrum that might be expected from the large number of atoms in the unit cell, only a few weak bands are observed. The Si–O stretching region is almost completely subdued and only the bridging oxygen mode at 670 cm^{-1} appears with appreciable intensity. Since no extensive study of amphibole spectra has been made, no final conclusion is possible, but it appears likely that the large degree of mixed site substitution is responsible for the weak spectrum. A similar weakening of spectral bands, loss of detail, and suppression of Si–O stretching frequencies was observed in the sheet silicates (Loh, 1973).

The spectra of feldspars (Fig. 10) likewise exhibit very weak Si–O stretching frequencies but show considerably more spectral detail. Characteristic strong bands appear near 475 and 510 cm^{-1} in the alkali feldspars.

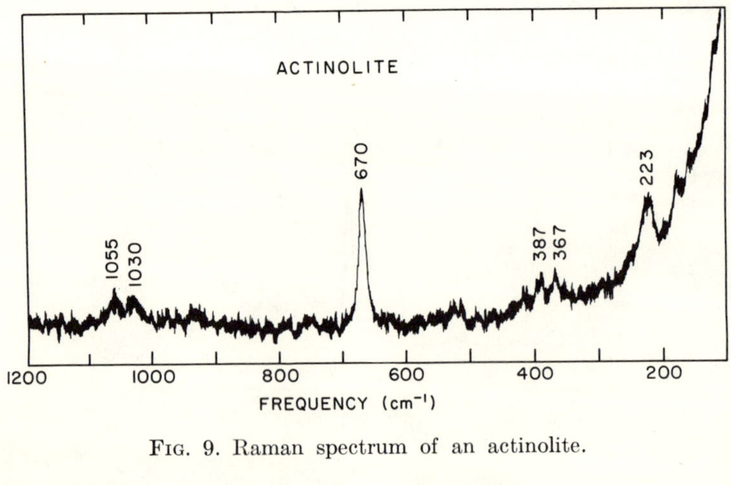

FIG. 9. Raman spectrum of an actinolite.

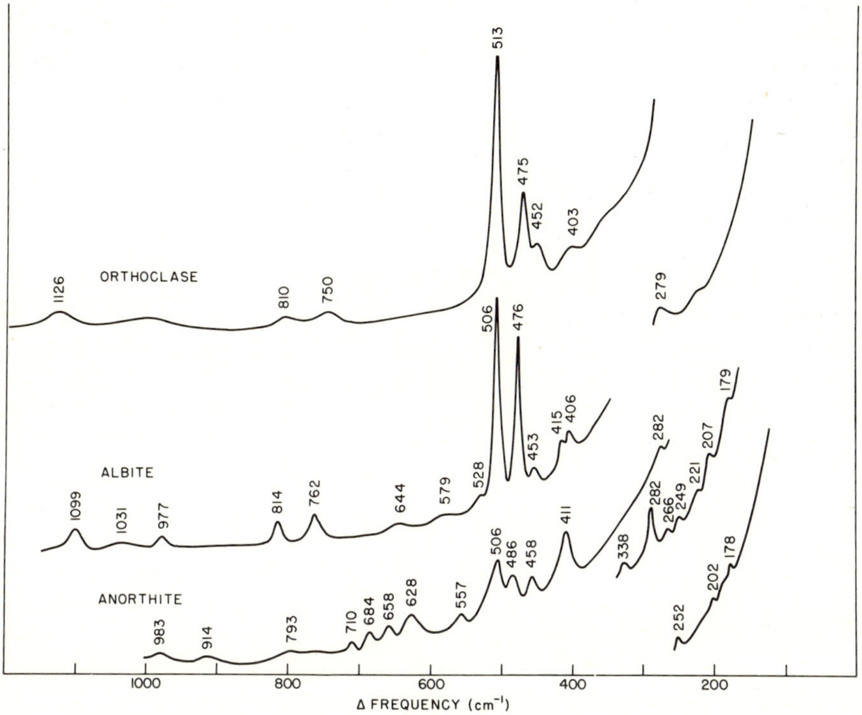

FIG. 10. Raman spectra of several common feldspars. (Data from Fabel *et al.*, 1972.)

Although no single-crystal measurements have been reported it appears likely that here also the powder spectra represent composites of overlapping bands.

2. *Silicate Polymerization*

The gradual shift of Si–O stretching bands to higher frequencies as the degree of corner-sharing of SiO_4 tetrahedra can also be observed in the Raman spectra. Each silicate structure exhibits one Raman mode distinctly more intense than the rest in powder spectra (except olivine, which has two). A plot of these against the oxygen/silicon ratio as a measure of the degree of tetrahedral polymerization (Fig. 11) shows a reasonable trend with zircon and spodumene as outlier exceptions. An alternative analysis using the average of all observable Raman bands in the Si–O stretching range produced a rather similar plot. The general conclusion is that the polymerization of tetrahedra to form Si–O–Si bridging bonds increases the

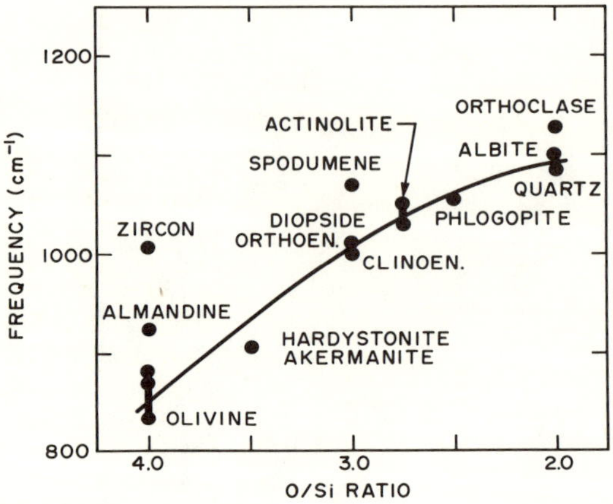

FIG. 11. Shift in strongest Raman line with degree of polymerization of silicate structures.

force constant of the Si–O bond and accounts for the observed increase in the observed vibrational frequency.

3. *Some Bond Angle Correlations*

The pyrosilicates exhibit an amazing range in Si–O–Si bond angles ranging from nearly linear bonds in the thortveitite structure to highly bent bonds in the melilite structure. Lazarev noted in 1960 that there is a relationship between the Si–O–Si stretching frequency which appears as a weak band in the 650–700 cm^{-1} region and the bond angle. This was more precisely formulated (Lazarev, 1972) as the relation

$$\Delta = \frac{\nu_{AS} - \nu_S}{\nu_{AS} + \nu_S}$$

where ν_{AS} is the asymmetric stretch associated with bridging oxygens and ν_S is the Si–O–Si vibration. This relationship, derived from Lazarev's data and augmented by Scheetz's (1972) results for the melilite minerals, seems to be quite good (Fig. 12).

4. *Ring Silicates*

IR spectra of most silicates contain a gap between the lowest frequency Si–O stretching modes near 800 cm^{-1} and the highest frequency modes due to the remainder of the structure at 600 cm^{-1} and below. Bands that do appear in this region are of especial significance. Those silicates whose structure contains closed rings of SiO_4 tetrahedra exhibit one or more intense infrared bands in this region (Matossi, 1949; Saksena *et al.*, 1963). The positions of the ring bands in some cyclosilicates are tabulated below. Bands in this region appear very weakly in the Raman spectrum.

Beryl	Si_6O_{18}	733	E ∥ c	Gervais *et al.* (1972)
		674 803	E ⊥ c	
Cordierite	Si_6O_{18}	771		Lyon (1962)
Benitoite	Si_3O_9	740		Keramidas, Scheetz and White (unpublished)

B. Range of Interactions

1. *Repeat Length in Silicate Chains*

The weak bands in the IR spectra of chain silicates that have been associated with the Si–O–Si symmetric stretch sometimes appear as multiples. Lazarev argued that the number of bands to appear would be equal to the number of tetrahedra in the repeat unit of the chain. These observations were confirmed by Ryall and Threadgold (1966), whose IR spectra indeed showed 2 bands for pyroxenes, 3 bands for wollastonite, 5 bands for rhodonite, and 7 bands for pyroxmanganite. This correlation has been ex-

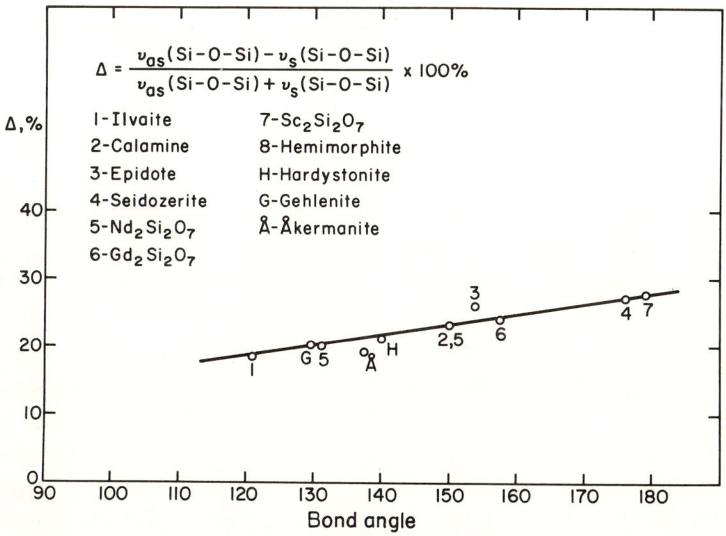

FIG. 12. Correlation of bridging oxygen band with bond angle of the linkage. (Based on data of Lazarev, 1972, and Sheetz, 1972.)

tended to the lunar mineral pyroxferroite (Estep *et al.*, 1972) and confirms its 7-tetrahedron repeat unit. Only 8 of the expected 9 bands were observed in the IR spectrum of synthetic ferrosilite-III, known from Burnham's (1966) structure analysis to have a 9-tetrahedron repeat.

If these bands are indeed due to the symmetric stretching motion of the bridging oxygen split into multiples by correlation interactions along the chain, they would be expected to have great intensity in the Raman spectrum. Such does not appear to be the case. Preliminary measurements on wollastonite and rhodonite reveal only single strong bonds at 674 and 670 cm^{-1}, respectively. There is no evidence for band multiplicity in powder spectra at room temperature.

2. *Large Unit Cells and Domain Structures*

Regarded as a structural characterization tool, it appears the IR and Raman spectroscopy are sensitive to intermediate range order. In principle the spectrum should account for all degrees of freedom in the entire unit cell. In practice this does not appear to be the case, especially in reference to complex silicate structures. In the amphiboles, micas, and feldspars in particular the spectra contain far less than the predicted number of bands. Examination of the Raman spectra of single crystals of both olivines and pyroxenes suggests the explanation. Many bands have nearly identical frequencies (accidental degeneracy) and can be separated only by their polarization dependence. Spectra of minerals with larger cells and low symmetry will be even more overlapped. The long-range forces (correlation field) within the unit cell are not sufficiently strong in most silicates to separate all motions into discrete spectral bands.

One of the most interesting discoveries of recent years has been the observation that many silicate crystals consist of domains, often on a very small scale. The main effect of the domain structure will be to scramble the polarization dependence of infrared and Raman bands, perhaps leading to apparent breakdown of selection rules. No systematic study of the effect of very small domains on spectral properties has yet been made.

IV. Structural Order

The vibrational spectrum of a crystal will be modified by small changes in crystal structure. Changes in crystal structure can be achieved in various ways: by substituting a different cation on the cation sites or by coupled substitution over several sites, by introduction of vacancies on either cation or anion sublattice, and by modifying the long-range periodicity of the structure. These structural modifications are not completely independent.

A. Substitutional Order and Disorder

1. *Cation Substitution*

The usual solid solution is formed by substitution in a random way of a second kind of cation on the same crystallographic site as the original cation. There is no change in space group and only a modest shift in lattice parameters. The situation becomes more complex if there is an ordered substitution. The possibilities shown in the following tabulation can be distinguished.

SUBSTITUTION	SYMMETRY	EFFECT ON SPECTRA
Random on equivalent or nonequivalent sites	Same cell and space group	Nearly linear frequency shifts
Ordered on nonequivalent sites	Same cell and space group	More complex frequency shifts
Ordered on equivalent sites	Derivative space group; possible multiple cell	Relaxation of selection rules; multiplicity of bands

The first case is well illustrated by the olivine series where several structural refinements (Birle *et al.*, 1968; Wenk and Raymond, 1973) show that most olivines have a disordered cation arrangement, the main exception being the lunar olivines. Infrared data (Burns and Huggins, 1972) tend to support this position. Unfortunately, Raman data for olivine sequences are too sparse to permit analysis.

Ordered orthosilicates will occur in such a series as forsterite–tephroite or forsterite–liebenbergite where considerations of size (Mn^{2+}) or strong crystal field effects (Ni^{2+}) are important. The M1 and M2 sites in the olivine structure are nonequivalent, so that the ordering of one kind of cation on M1 and a second cation on M2 will not produce any change in symmetry. The number of predicted bands and the selection rules, therefore, will remain unchanged. The shift in frequency with composition, however, is likely to be nonlinear, as has been found in the IR spectra of manganiferous olivines (Burns and Huggins, 1972). Nonlinear shifts for the forsterite–fayalite–tephroite series are shown in Chapter 9, Fig. 2. Similar arguments apply to pyroxene solid solutions where ordering between the large M2 and small M1 sites is very common (see Chapter 9, Fig. 5).

If the substituent cations enter the same equipoint in an ordered way, some of the rotational symmetry of the structure will be lost. The ordered structure will belong to a space group of lower symmetry derived from the group of the parent structure. Depending on details of the ordering scheme and whether or not a superstructure cell is required for the ordered phase (White and Keramidas, 1972), new IR and Raman bands may appear from relaxation of selection rules, from splitting of degenerate levels due to distortion, and from band multiplicity due to an enlarged unit cell.

Substitutional ordering is very nicely illustrated by a comparison between the ilmenite structure and the parent corundum structure. Ilmenite contains alternate layers of cations perpendicular to the optic axis. The space group is $R\overline{3}$ instead of $R\overline{3}c$, but the number of formula units in the unit cell remains unchanged. Factor-group analysis (Table VI) shows that the main effect of ordering is a selection rule relaxation permitting the inactive A_{2g} and A_{1u} modes to appear. Thus the Raman spectra of an ilmenite structure mineral should contain 10 bands compared with 7 in a corundum structure.

The Raman spectrum of polycrystalline corundum exhibits the expected 7 bands (Fig. 13). The spectrum of geikielite, $MgTiO_3$, in comparison, contains some 13 identifiable lines of variable intensity. The derivation of this spectrum from that of corundum is not obvious. It is apparent that the selection rule relaxation is effective and lines have high intensities. The effect of ordering, therefore, is quite different from that of distortion where, although selection rules are formally relaxed, intensities of the new modes are often very weak. Similar observations have been made on the spectra of spinels with 1:1 order on the tertrahedral sites (DeAngelis *et al.*, 1971).

TABLE VI

Normal Modes and Selection Rules for Corundum and Ilmenite Structures

CORUNDUM			ILMENITE		
D_{3d}	NUMBER OF MODES	SELECTION RULES	C_{3i}	NUMBER OF MODES	SELECTION RULES
A_{1g}	2	Raman	A_g	5	Raman
A_{2g}	3	Inactive			
E_g	5	Raman	E_g	5	Raman
A_{1u}	2	Inactive	A_u	4	IR, E $\parallel$ c
A_{2u}	2	IR, E $\parallel$ c			
E_u	4	IR, E $\perp$ c	E_u	4	IR, E $\perp$ c

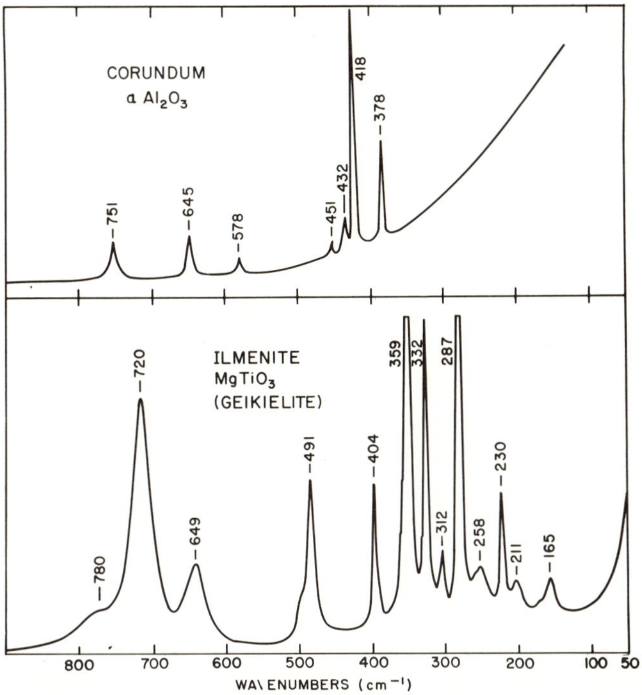

FIG. 13. Raman spectra of corundum, α-Al_2O_3, and geikielite, $MgTiO_3$, an ilmenite structure. Both spectra measured on powders.

2. *Vacancies*

Nonstoichiometric compounds contain either unoccupied cation or anion sites. Vacancies may order to produce a derivative structure in which case the effect on the spectra is very similar to other ordered compounds. If the vacancies are distributed at random in the manner of a substitutional solid solution, no spectral changes are predicted by group theory. The usual factor-group calculation, however, does not take the translational symmetry into account. The presence of the vacancy causes a loss of translational symmetry and with it a breakdown of the $k=0$ selection rule. In the translationally disordered solid it is not necessary that all first-order phonon transitions take place at the center of the Brillouin zone. Indeed, the spectrum should be a density of states continuum of all photons including those at the zone boundaries, as well as longitudinal and acoustic phonons (Whalley and Bertie, 1967; Whalley, 1972).

Small deviations from stoichiometry appear as a broadening of Raman lines. The line broadening in a series of $LiNbO_3$ and $LiTaO_3$ solid solutions can be quantitatively related to the composition of the solid (Scott and Burns, 1972) with a sensitivity of about ± 1 mole%.

Large deviations from stoichiometry do indeed produce entirely new spectra. The system ZrO_2–CaO is an example. Pure ZrO_2 is stable at low temperatures as a monoclinic form, baddeleyite, also found among the lunar minerals. At high temperatures, ZrO_2 transforms to a cubic fluorite structure which can be stabilized by the addition of CaO in solid solution.

The baddeleyite spectrum (Fig. 14) exhibits 16 sharp Raman lines of the predicted 18. The fluorite structure is expected to exhibit a single T_{2g} mode. Thorianite, ThO_2, makes a useful comparison with a single sharp band at

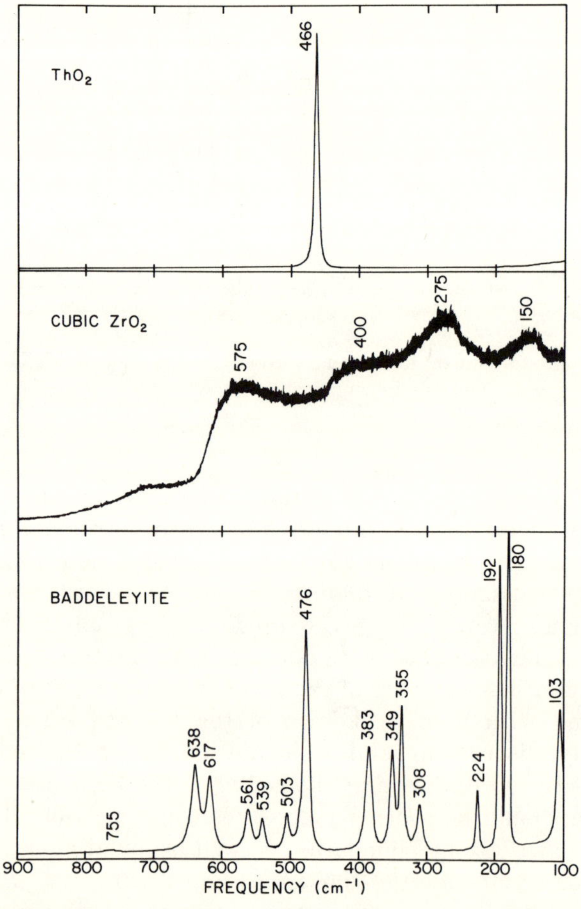

FIG. 14. Powder Raman spectra of thorianite, ThO_2, with the fluorite structure, lime-stabilized zirconia, also with the fluorite structure, and baddeleyite, ZrO_2.

465 cm^{-1}. The spectrum of $Zr_{1-x}Ca_xO_{2-x}\square_x$ is quite remarkable. It appears as a broad continuum with a few ill-defined maxima. No band occurs near 465 cm^{-1}, the expected Raman-active mode. The broad band at 275 cm^{-1}, however, is close to the IR-active T_{1u} mode which occurs at 279 cm^{-1} in ThO_2. Keramidas and White (1973) compared this spectrum with the complete set of phonons for uraninite, UO_2, determined by inelastic neutron scattering. The 575 band matches a group of longitudinal optic modes and the abrupt decrease in intensity near 625 cm^{-1} corresponds to the highest frequency at which first order phonons occur. The $k = 0$ selection rule is quite clearly broken here, and the highly defective ZrO_2 solid solution is a material with the diffraction pattern of a crystal and the Raman spectrum of a glass.

3. *Aluminum–Silicon Ordering*

Aluminum substitutes for silicon tetrahedral sites in many silicates and may do so either at random or in an ordered way. It is known from IR studies of feldspars (Laves and Hafner, 1956; Hafner and Laves, 1957; Martin, 1970) and the cordierite–indialite transformation (Langer and Schreyer, 1969) that the low-temperature equilibrated forms have sharp, well-resolved IR spectra. As the temperature of sample preparation increases, the spectra become broadened and lose detail. This effect has been associated with the degree of Al–Si ordering.

Few Raman spectra are available of silicate minerals exhibiting Al–Si order and those few are rather surprising. Spectra of members of the akermanite–gehlenite series (Scheetz, 1972) showed many changes in relative intensity with changing composition but little line broadening. Spectra of an Amelia albite taken from samples heated at various temperatures are all very similar until temperatures quite near the melting point are reached. There is little sensitivity to Al–Si ordering. The ordering in feldspars is between nonequivalent tetrahedral sites. It appears that the Al modes and Si modes are already so strongly coupled that few additional changes take place.

B. Spectra of Glass

Spectra of three rather extreme examples of glass structure are shown in Fig. 15. The bands of sodium trisilicate glass are very intense and have narrower line widths than other glasses. Two processes are responsible for weakening and broadening the Raman bands. The first is the elimination of nonbridging oxygen in the glass network. Comparison can be made with silica glass in which all oxygens bridge two silica tetrahedra. The bands

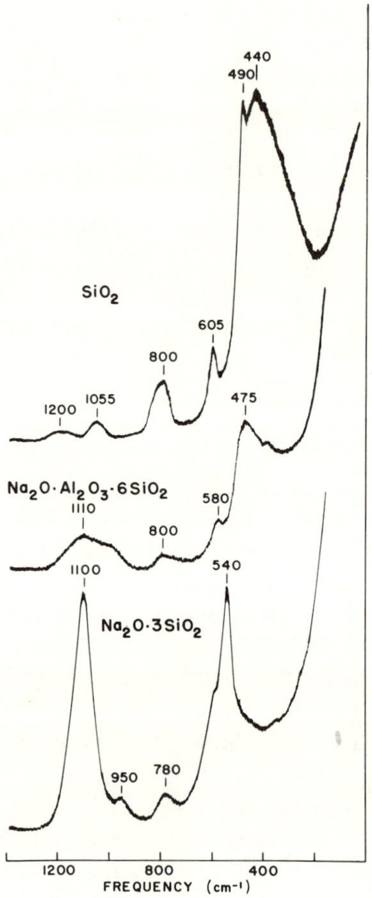

FIG. 15. Raman spectra of synthetic glasses. (Data from Sweet *et al.*, 1973.)

in silica glass are narrow, with line widths comparable to those of the sodium trisilicate glass, but the bands associated with Si–O stretching modes have greatly decreased intensities. The second factor is the introduction of more disorder into the glass network. Both sodium trisilicate and silica glass can be thought of as highly ordered glasses. This is not to imply translational symmetry but does imply an average periodicity in that every tetrahedron in the glass is much like every other tetrahedron except for a randomization of bond angles of the bridging bonds over a relatively small range. Introduction of aluminum into the network breaks down this average periodicity by further randomizing bond lengths and bond angles, and by introducing a further nonequivalence between nearby tetrahedra. The effects of aluminum in the network can be seen in the $Na_2O \cdot Al_2O_3 \cdot 6SiO_2$. The Si–O stretching region bands are broadened and smeared together. In other aluminate glasses examined, these bands are broadened until they are almost lost in the background. Likewise, much of the resolution of the strong low-frequency band is lost; it broadens and gradually merges into the widening Rayleigh tail.

Glasses occur naturally on earth in large quantities as the quick-frozen obsidians of widely varying composition in volcanic areas. Rare glasses are the fulgarites formed by fusion of desert sand by lightning strikes and the peculiar tektite glass whose origin has been a hotly debated topic. Spectra of some examples of terrestrial glasses are shown in Fig. 16.

The obsidian spectra are essentially identical with a broad, double-humped band at 1120–1140 cm^{-1} and 1020–1040 cm^{-1} associated with the Si–O stretching vibration, a well-defined band at 800 cm^{-1} and an intense broad band at 450–470 cm^{-1}. The similarity between the obsidian spectra and the synthetic albite glass spectrum is immediately apparent. The tektite spectra are similar to the obsidian spectra with the difference that the Si–O band is now skewed to lower frequencies with much of the intensity in the range of 900 cm^{-1}. The spectrum of the fulgarite is essentially identical to that of silica glass. The silica spectrum is more complex than the other compositions, with sharp bands at 605 and 490 cm^{-1}. The Si–O stretching bands of silica glass are double–humped with one component at 1200 cm^{-1}.

Glasses of great variety have been found in the lunar samples. These include black and green glassy coatings on rock fragments, irregular shards, and fragments of clear, green, yellow, orange, and brown glass, and the unique spheres and dumbbells of various colors. Fairly elaborate efforts to measure Raman spectra on lunar glasses has not met with much success (White *et al.*, 1971; Fabel *et al.*, 1972; Sweet *et al.*, 1973). Spectra are usually flat and featureless except for the Rayleigh tail and various sharp bands identified with included bits of crystalline silicate. Figure 17 shows one of the best lunar-glass spectra obtained. Two broad bands, one at 960 and one at 520, are observed. The spectrum does not resemble any of the terrestrial or synthetic glasses. The band positions are actually closer to those of the orthosilicate than of network-forming silicates. Whether the low Si–O stretching frequency is due to the presence of an exceptionally high concentration of aluminum or to the presence of nonbridging oxygen has not been established.

It is apparent that natural glasses formed by a wide variety of processes on earth exhibit spectra with well-defined features that are closely related to the more controlled glasses prepared in the laboratory. The introduction of other components such as occurs in the terrestrial glasses does not, of itself, provide extra detail or modify the spectra to any great extent. Indeed, the spectra of glasses are remarkably indifferent to details of chemical composition. How then to account for the lack of similar structures in the spectra of most lunar glasses?

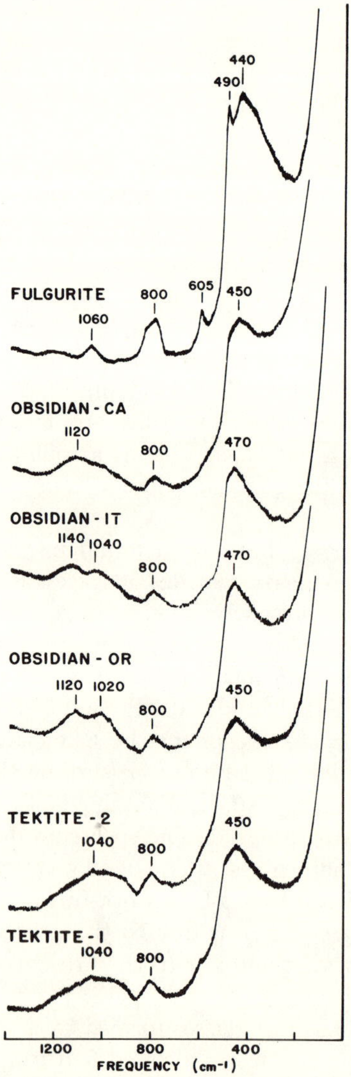

FIG. 16. Raman spectra of terrestrial glasses. (Data from Sweet *et al.*, 1973.)

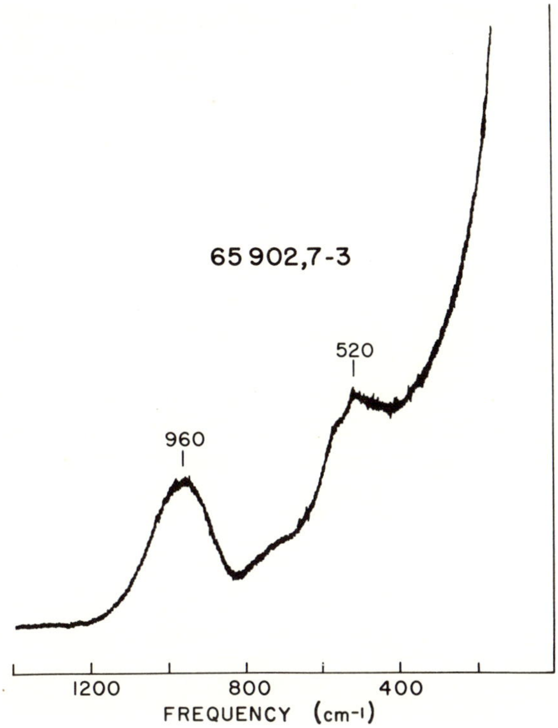

FIG. 17. Raman spectra of a lunar glass. (Data from Sweet *et al.*, 1973.)

It is argued that the lunar glasses represent a continuation of the randomization process. If the glasses are inhomogeneous on the scale of a few tens of atomic spacings, and they are known to be inhomogeneous on a scale of micrometers from electron microprobe evidence, the bands would be smeared out completely, and all that would appear in the spectra would be the extended Rayleigh tail—as is indeed observed. Specimens, such as the one whose Raman spectrum is shown in Fig. 17, are the exception and imply unusually homogeneous bits of glass.

V. Characterization of Nucleation and Crystallization

Spectra of glasses, particularly those of complex composition, exhibit only weak broad bands. In contrast, the bands of well-ordered crystals are intense and sharp. Raman spectroscopy, therefore, is a useful tool for the study of the crystallization process. Experiments with potash–silica glasses (Brawer and White, 1974) and lithia–silica glass indicate the onset of crystallization in a glassy matrix that becomes visible before evidence of crystallization can be observed from the x-ray diffraction powder pattern. There is some evidence from the Raman measurements made on lunar glasses that crystalline particles embedded in glassy fragments would produce sharp Raman spectra that could then be used as a means of identification for crystallites of submicroscopic size. The lower size limit detectable is open to some dispute, but it does appear that particles as small as 30 Å will produce a distinct Raman spectrum. The limit of resolution is controlled by the degree of crystallinity of the nuclei, by their homogeneity, and by the sharpness of the boundary between them and the surrounding medium. The Raman spectrum, it would seem, is more responsive to changes in long-range order than is the infrared. It requires a repeat unit of 5–10 unit cells to produce a sharp Raman spectrum, whereas the infrared spectrum records interactions only within the unit cell and sometimes only a subvolume of that.

ACKNOWLEDGMENTS

This Chapter is based in large part on experimental work supported by NASA Grant No. 39-009-183. The assistance of Mr. Barry E. Sheetz in the preparation of some of the spectra used to illustrate this review is greatly appreciated.

References

Adams, D. M., and Newton, D. C. (1970). *J. Chem. Soc., A* pp. 2822–2827.

Asell, J. F., and Nicol, M. (1968). *J. Chem. Phys.* **15,** 5395–5399.

Basile, L. J., Ferraro, J. R., LaBonville, P., and Wall, M. C. (1973). *Coord. Chem. Rev.* **11,** 21–69.

Bates, J. B., and Quist, A. S. (1972). *J. Chem. Phys.* **56,** 1528–1533.

Bilton, M. S., Gilson, T. R., and Webster, M. (1972). *Spectrochim. Acta, Part A* **28,** 2113–2119.

Birle, J. D., Gibbs, G. V., Moore, P. B., and Smith, J. V. (1968). *Amer. Mineral.* **53,** 807–824.

Brawer, S. A., and White, W. B. (1974). To be published.

Burnam, C. W. (1966). *Science* **154,** 513–516.

Burns, R. G., and Huggins, F. E. (1972). *Amer. Mineral.* **57,** 967–985.

Dawson, P., Hargreave, M. M., and Wilkinson, G. R. (1971). *J. Phys. C.* **4,** 240–256.

DeAngelis, B. A., Keramidas, V. G., and White, W. B. (1971). *J. Solid State Chem.* **3,** 358–363.

DeAngelis, B. A., Newnham, R. E., and White, W. B. (1972). *Amer. Mineral* **57,** 255–268.

Duke, D. A., and Stephens, J. D. (1964). *Amer. Mineral.* **49,** 1388–1406.

Estep, P. A., Kovach, J. J., Waldstein, P., and Karr, C., Jr. (1972). *Proc. Lunar Sci. Conf., Geochim. Cosmochim. Acta, 3rd, 1972* Vol. 3, pp. 3047–3067.

Etchepare, J. (1970). *In* "Amorphous Materials" (R. W. Douglas and B. Ellis, eds.), pp. 337–346. Wiley (Interscience), New York.

Fabel, G. W., White, W. B., White, E. W., and Roy, R. (1972). *Proc. Lunar Sci. Conf., Geochim. Cosmochim. Acta, 3rd, 1972* Vol. 1, pp. 939–951.

Fateley, W. G., Dollish, F. R., McDevitt, N. T., and Bentley, F. F. (1972). "Infrared and Raman Selection Rules for Molecular and Lattice Vibrations." Wiley (Interscience), New York.

Gervais, F., Piriou, B., and Cabannes, F. (1972). *Phys. Status Solid,* B **51,** 701–702.

Gibbs, G. V., and Smith J. V. (1965). *Amer. Mineral.* **50,** 2023–2039.

Graham, E. K., Jr., and Barsch, G. R. (1969). *J. Geophys. Res.* **74,** 5949–5960.

Griffith, W. P. (1969). *J. Chem. Soc., A,* pp. 1372–1377.

Hafner, S., and Laves, F. (1957). *Z. Kristallogr., Kristallgeometrie, Kristallphys., Kristallchem.* **109,** 204–225.

Hohler, V. (1972). Dissertation, Albert–Ludwigs University, Freiburg, Germany.

Hohler, V., and Funck, E. (1969). *Naturwissenschaften* **56,** 459.

Hohler, V., and Funck, E. (1973). *Z. Naturforsch. B* **28,** 125–139.

Hubin, R., and Tarte, P. (1971). *Spectrochim. Acta, Part A,* **27,** 683–690.

Keramidas, V. G., and White, W. B. (1973). *J. Phys. Chem. Solids* **34,** 1873–1878.

Langer, K., and Schreyer, W. (1969). *Amer. Mineral.* **54,** 1442–1459.

Launer, P. J. (1952). *Amer. Mineral.* **37,** 764–784.

Laves, F., and Hafner, S. (1956). *Z. Kristallogr.* **108,** 52–63.

Lazarev, A. N. (1972). "Vibrational Spectra and Structure of Silicates." Consultants Bureau, New York.

Loh, E. (1973). *J. Phys. C* **6,** 1091–1104.

Louden, R. (1964). *Advan. Phys.* **13,** 423–482.

Lyon, R. J. P. (1962). Final Report on NASA Contract No. NASr-49(04). Stanford Res. Inst., Menlo Park, California.

Maradudin, A. A., and Vosko, S. H. (1968). *Rev. Mod. Phys.* **40,** 1–37.

Martin, R. F. (1970). *Contrib. Mineral. Petrol.* **26,** 72–74.

Masso, J. D., She, C. Y., and Edwards, D. F. (1970). *Phys. Rev. B* **1,** 4179–4181.

Matossi, F. (1949). *J. Chem. Phys.* **17,** 679–685.

Matossi, F., and Bronder, B. (1938). *Z. Phys.* **111,** 1–17.

Matossi, F., and Krüger, H. (1936). *Z. Phys.* **99,** 1–23.

Moenke, H. (1962) "Mineralspektren," Vol. I. Akademie-Verlag, Berlin.

Moenke, H. (1966). "Mineralspektren," Vol. II. Akademie-Verlag, Berlin.

Moore, R. K., White, W. B., and Long, T. V. (1971). *Amer. Mineral.* **56,** 54–71.

Oehler, O., and Günthard, H. H. (1969). *J. Chem. Phys.* **51,** 4719–4728.

Ohashi, Y., and Burnham, C. W. (1972). *J. Geophys. Res.* **77,** 5761–5666.

O'Horo, M. P., Frisillo, A., and White, W. B. (1974). To be published.

Omori, K. (1971). *Amer. Mineral.* **56,** 841–849.

Robinson, K., Gibbs, G. V., and Ribbe, P. H. (1971). *Amer. Mineral.* **56,** 782–790.

Rutstein, M. S., and White, W. B. (1971). *Amer. Mineral.* **56,** 877–887.

Ryall, W. R., and Threadgold, I. M. (1966). *Amer. Mineral.* **51,** 754–761.

Saksena, B. D., Agarwal, K. C., and Jauhri, G. S. (1963). *Trans. Faraday Soc.* **59,** 276–283.

Scheetz, B. E. (1972). M. S. Thesis in Geochemistry, Pennsylvania State University, University Park, Pennsylvania.

Scott, B. A., and Burns, G. (1972). *J. Amer. Ceram. Soc.* **55,** 225–230.

Scott, J. F., and Porto, S. P. S. (1967). *Phys. Rev.* **161,** 903–910.

Scott, J. F., Cheesman, L. E., and Porto, S. P. S. (1967). *Phys. Rev.* **162,** 834–840.

Servoin, J. L., and Piriou, B. (1973). *Phys. Status Solidi* (*B*) **55,** 677–686.

Shapiro, S. M., O'Shea, D. C., and Cummins, H. Z. (1967). *Phys. Rev. Lett.* **19,** 361–364.

Sherwood, P. M. A. (1972). "Vibrational Spectroscopy of Solids." Cambridge Univ. Press, London and New York.

Sushchinskii, M. M. (1972). "Raman Spectra of Molecules and Crystals." Isr. Program Sci. Transla., Jerusalem.

Suwa, Y., and Naka, S. (1972). *J. Chem. Soc. Jap.*, 1643–1648.

Sweet, J. R., White, W. B., White, E. W., and Roy, R. (1973). *Proc. Lunar Sci. Conf., Geochim. Cosmochim. Acta, 4th, 1973* Vol. 1, pp. 389–396.

Tarte, P. (1962). *Spectrochim. Acta* **18,** 467–483.

Tarte, P. (1963). *Spectrochim. Acta.* **19,** 25–47.

Tarte, P., and Deliens, M. (1973). *Contrib. Mineral. Petrol.* **40,** 25–37.

Tarte, P., Pottier, M. J., and Proces, A. M. (1973). *Spectrochim. Acta, Part A* **29,** 1017–1027.

Toropov, N. A., Fedorov, N. F., and Shevyakov, A. M. (1963). *Russ. J. Inorg. Chem* **8,** 697–699.

Turrell, G. (1972). "Infrared and Raman Spectra of Crystals." Academic Press, New York.

Warren, J. L. (1968). *Rev. Mod. Phys.* **40,** 38–76.

Wenk, H. R., and Raymond, K. N. (1973). *Z. Kristallogr., Kristallgeometrie, Kristallphys., Kristallchem.* **137,** 86–105.

Whalley, E. (1972). *J. Chem. Soc., Faraday Trans. 2* **68,** 662–667.

Whalley, E., and Bertie, J. E. (1967). *J. Chem. Phys.* **46,** 1264–1270.

White, W. B. (1974). *In* "Infrared Spectra of Minerals." Mineral. Soc., London 227–284.

White, W. B., and Keramidas, V. G. (1972). *Nat. Bur. Stand.* (*U.S.*), *Spec. Publ.* **364,** 113–126.

White, W. B., White, E. W., Görz, H., Henisch, H. K., Fabel, G. W., Roy, R., and Weber, J. N. (1971). *Proc. Lunar Sci. Cont., Geochim. Cosmochim. Acta* 2*nd, 1971,* Vol. 3, pp. 2213–2221.

Worlton, T. G., and Warren, J. W. (1972). *Compt. Phys. Commun.* **3,** 88–117.

SUBJECT INDEX

D

E

F

M

P

T